The Chinese Roots of Linear Algebra

Roger Hart

The Chinese Roots of Linear Algebra

The Johns Hopkins University Press
Baltimore

 Published 2011
Printed in the United States of America on acid-free paper
9 8 7 6 5 4 3 2 1

The Johns Hopkins University Press
2715 North Charles Street
Baltimore, Maryland 21218-4363
www.press.jhu.edu

Library of Congress Control Number: 2010924546

ISBN 13: 978-0-8018-9755-9
ISBN: 0-8018-9755-6

A catalog record for this book is available from the British Library.

The Johns Hopkins University Press uses environmentally friendly book materials, including recycled text paper that is composed of at least 30 percent post-consumer waste, whenever possible. All of our book papers are acid-free, and our jackets and covers are printed on paper with recycled content.

With love, gratitude, and affection to my daughter, Nikki Janan Hart, and to my parents, Walter G. Hart, Jr. and Charlene Hart

Contents

Preface

My inquiry into linear algebra in China began in the context of my research on the introduction of Euclid's *Elements* into China in 1607 by the Italian Jesuit Matteo Ricci (1552–1610) and his most important Chinese collaborator, the official Xu Guangqi 徐光啟 (1562–1633). Chinese mathematics of the time, Xu had argued in his prefaces and introductions, was in a state of decline, and all that remained of it was vulgar and corrupt. Western mathematics was in every way superior, he asserted, and in the end the loss of Chinese mathematics was no more regrettable than discarding "tattered sandals." Xu's pronouncements have been so persuasive that they have, at least until recently, been accepted for the most part by historians—Chinese and Western alike—as fact.

But as I studied the Chinese mathematics of the period, I became increasingly skeptical of Xu's claims. The most interesting evidence against his claims was to be found in Chinese developments in linear algebra—arguably the most sophisticated and recognizably "modern" mathematics of this period in China. General solutions to systems of n linear equations in n unknowns had not been known in Europe at the time, and their importance was not lost on the Jesuits and their collaborators. For they copied, without attribution, one by one, linear algebra problems from the very Chinese mathematical treatises they had denounced as vulgar and corrupt. They then included the problems in a work titled *Guide to Calculation* (*Tong wen suan zhi* 同文算指, 1613), which they presented as a translation of German Jesuit Christopher Clavius's (1538–1612) *Epitome arithmeticae practicae* (1583). Chinese readers of this "translation" had no way of knowing that none of these problems on linear algebra had been known in Europe at the time.

I sought, then, to trace the specific Chinese sources from which the Jesuits and their collaborators had copied their problems. I was particularly interested in one problem, which I will call the "well problem," in which the value assigned to the depth of a well is explained as resulting from the calculation $2 \times 3 \times 4 \times 5 \times 6 + 1$. I was perplexed by this unexplained and apparently ad hoc calculation, and at first supposed that it was merely a numerological coincidence that multiplying the elements of the diagonal of a matrix together, and then adding one, would result in the value for the depth of the well. The Jesuits and their collaborators offered no explanation for this calculation, and indeed had evidently not noticed that an explanation might be necessary. The well problem had originated in the *Nine Chapters on the Mathematical Arts* (*Jiuzhang suanshu* 九章算術, c. 1st century C.E.), the earliest transmitted text on Chinese mathematics. But in examining the earliest extant version of the *Nine Chapters*, I found no mention of this perplexing calculation for finding the depth of the well. It was included, however, in a more popular text from the fifteenth century, Wu Jing's 吳敬 (n.d.) *Complete Compendium of Mathematical Arts of the Nine Chapters, with Detailed Commentary, Arranged by Category* (*Jiuzhang xiang zhu bilei suanfa da quan* 九章詳註比類算法大全, preface dated 1450, printed in 1488). However, an explanation for this calculation was not to be found in Wu Jing's *Complete Compendium* either.

It happened, at that time, that I had been accorded the great privilege of spending a year at the Institute for Advanced Study in Princeton. The Institute is a special place where one is encouraged to believe that something perplexing deserves serious inquiry. Such freedom, as I hope this book demonstrates, is essential to academic research: in any investigation, at each fork in the road, if one ignores what might be crucial evidence, one risks taking the wrong path, only, perhaps, to wind up further and further from the truth. The puzzle of the well problem proved to be one such crucial fork. For eventually, I noticed that the calculation I had stumbled upon was not merely numerological coincidence, but what we would call in modern terms the determinant of the matrix of coefficients.[1] This interested me. Not quite sure what I was looking for, or why, I began to look for general solutions to the well problem. After more than a little work, I found that the solutions to the problem could be found more easily by determinantal calculations than by the usual elimination procedures. What was more interesting still was the extraordinary symmetry of these solutions. And although determinantal solutions are impracticably difficult for matrices with 5 equations in 5 unknowns, it was only the symmetric placement of numerous zeros in the well problem that permitted a determinantal solution.

Then I noticed something that would permanently change the way I looked at these problems: not just one but five of the eighteen problems in chapter 8 of the *Nine Chapters* share a similar pattern of symmetrically placed zeros. On further investigation, I eventually found that they all have similarly symmetric solutions, at least when written in modern mathematical terminology. But there were still some problems with this explanation, and the solutions I had found seemed improbable, first because modern mathematical terminology had not been available centuries ago, and second because using modern mathematical notation seemingly rendered the solution impracticably complicated.

Because my mathematical reconstructions showed that these problems could be solved by determinantal methods, I searched through early texts and commentaries to see if I could find any evidence supporting my hypothesis. I was most fortunate to find, in an early commentary to the *Nine Chapters*, a written

[1] For a brief explanation of mathematical terminology related to linear algebra, see chapter 2 of this book.

record of a determinantal-style calculation that verified my hypothesis: the calculation $2 \times 3 \times 4 \times 5 \times 6 + 1$ was not a haphazard coincidence, but rather an explicit procedure used to solve a recognized category of problems, for which the well problem was an exemplar.

Then, in the summer of 2004, I happened upon what I consider to be one of the most important discoveries presented in this book. It is a solution to the well problem preserved in a mathematical compilation that has been ignored in the history of Chinese mathematics, simply because this compilation has seemed to be closer to popular mathematics than to the more sophisticated writings of the literati elite of the period. On my first readings, the procedure presented for solving the well problem was scarcely comprehensible: it was almost impossible to follow the procedure, and it seemed to outline a solution that has no direct modern analog. Yet it incorporated much of the perplexing terminology I had found in other treatises of the period, and provided a detailed explanation of the calculations. Because I now knew that there might be alternative solutions to the well problem—solutions not using the elimination procedure found in the original *Nine Chapters*—I spent considerable time trying to understand this procedure. At length, I was fortunate to find that this in fact appeared to be a written record of a determinantal-style solution, and indeed the earliest known extant written record of such a solution.

There remained, however, an important question about this finding. On the one hand, if this procedure, as recorded in narrative form in classical Chinese, was almost impossible to follow, it seemed to me difficult to conceive how the procedure might actually be used in practice. On the other hand, the modern mathematical notation into which I had translated this problem, so as to understand the calculation, was also quite complicated. Worse, the modern notation was not available at that earlier time—during that period in China there was no mathematical notation to denote an unknown quantity x. That is, while I was using the modern notation to help translate the problem into a form understandable for modern readers (and for me), clearly it would be anachronistic and historically false to use modern mathematics to *explain* how these calculations had been solved in practice.

But as I continued to work on this problem, something interesting occurred: I found that in order to check the complicated modern mathematical formulas I was using, I would repeatedly return to the original diagram of the problem in its two-dimensional matrix form and "cross multiply" the entries. This then provided at last an answer to my concern: when the problem is laid out on the counting board in matrix form, it is actually quite simple to follow the calculations. It is only when these calculations are translated into narrative form in classical Chinese, or into modern mathematical notation, that these solutions appear to be so unwieldy.

This finding helped lead to another important discovery, described in this book, which is now my central thesis: visualization. For as I returned to simpler problems called "excess and deficit," I noticed that here too we were instructed to place the entries on a counting board and to cross multiply. It was then that it occurred to me that these were just two different forms of cross multiplication on the counting board, and that there need not have been a sharp distinction between the two methods.

It is these results that I have presented here. I have tried to write this book in such a manner that it might reach a relatively broad audience. My primary intended audience includes professors, teachers, and students of the history of mathematics, history of science, and Chinese history, and, more broadly, those outside the academic realm who are interested in the history of science or Chinese history. To help make this book more accessible to a larger audience, the second chapter presents background information on Chinese conventions, Chinese mathematics, and modern mathematics. I have tried to explain enough about the basics of linear algebra and Chinese to make this book reasonably self-contained and in need of no prerequisites.

The approach I take is, by necessity, interdisciplinary. This book seeks to meet the highest standards of academic rigor in sinology, history, history of science, and mathematics, presenting careful research into editions, and precise translations of material in these editions, together with explanatory philological notes, critical textual analysis, technical mathematical analysis, and historical contextualization using an interdisciplinary, cultural-historical approach informed by the most recent work in history and the history of science. I should perhaps emphasize that my approach to history is ultimately that of a cultural historian. I would characterize this work as interdisciplinary research into the history of the emergence, dissemination, and legitimation of scientific knowledge in the context of "non-Western" cultures and systems of thought. Ultimately, I seek to develop better approaches for analyzing science in "non-Western" cultures.

Admittedly, parts of the book are more technical, and written for specialists. But readers need not follow every mathematical calculation or philological argument presented here: it is my hope that the lay reader may skip some of these more technical parts and nonetheless profit from the book. Hopefully, the numerous diagrams illustrating the calculations with counting rods will make the Chinese linear algebra presented here more readily understandable for a broad audience.

This entire book, then, began with my initial inquiry into a single perplexing calculation, and would not have been possible without the support provided me as a National Endowment for the Humanities Fellow and Member of the School of Historical Studies at the Institute for Advanced Study in Princeton. Further initial research was supported by a University of Texas Summer Research Assignment. Research for this book was then completed under an ACLS/SSRC/NEH International and Area Studies Fellowship from the American Council of Learned Societies along with a University of Texas Faculty Research Assignment. I am extremely grateful for this support, without which I would never have been able to explore these questions. The views expressed in this book do not necessarily represent those of the National Endowment for the Humanities or any of my other sponsors.

Preliminary findings were included in presentations at the following forums: "Matteo Ricci and After: Four Centuries of Cultural Interactions between China and the West," City University of Hong Kong, October 12–15, 2001; History of Science Colloquium, UCLA, April 15, 2002; "Disunity of Chinese Science," Franke Institute for the Humanities, University of Chicago, May 10–12, 2002; the Joint Meeting of the American Mathematical Society and the Mathematical Association of America, San Antonio, Texas, January 14, 2006; Department of Physics Colloquium, Sam Houston State University, Huntsville, Texas, October 6, 2008; History of Science Colloquium, Department of History, University of Texas at Austin, May 1, 2009; Southwest Conference on Asian Studies, October 16–17, 2009; "Go-betweens, Translations, and the Circulation of Knowledge," Wellcome Trust Centre for the History of Medicine, University College London, November 13–14, 2009; and "China in Europe, Europe in China," University of Zürich, June 14–15, 2010.

In completing this book I owe a huge debt to John N. Crossley, whose most generously offered corrections, comments, suggestions, and criticisms have considerably improved this work. I have also benefited greatly from suggestions by Karine Chemla, Geoffrey Lloyd, Jean-Claude Martzloff, and Alberto Martinez. Nathan Sivin, Donald Wagner, Christopher Cullen, and Catherine Jami offered written suggestions and criticisms. At the Johns Hopkins University Press, special thanks are due to Trevor Lipscombe, Editor-in-Chief, for his generous encouragement, enthusiasm, and support, and to Juliana McCarthy, for her expert help in producing this book. William Carver copyedited the manuscript with considerable care and erudition. Joseph Hunt and Yingwu Zhao proofread the mathematics, and Ya Zuo proofread the Chinese. Any remaining errors are my responsibility. Finally, I would like to most gratefully acknowledge the kind encouragement and constant support of Benjamin Elman, Mario Biagioli, Antony G. Hopkins, Robert Richards, Haun Saussy, and Timothy Lenoir.

The Chinese Roots of Linear Algebra

Chapter 1
Introduction

> Imagine the geometry of four-dimensional space done with a view to learning about the living conditions of spirits. Does this mean that it is not mathematics?
>
> Ludwig Wittgenstein, *Remarks on the Foundations of Mathematics*

In China, from about the first century C.E. through the seventeenth century, anonymous and most likely illiterate adepts practiced an arcane art termed *fangcheng* 方程 (often translated into English as "matrices" or "rectangular arrays"). This art entailed procedures for manipulating counting rods arranged on a counting board,[1] which enabled practitioners to produce answers to seemingly insoluble riddles. The art seems to have been closely aligned with other mathematical arts (*suanfa* 算法), including various forms of calculation, numerology, and possibly divination.

Though we know virtually nothing about these practitioners, records of their practices have been preserved. These practices were occasionally recorded by aspiring literati and incorporated in texts they compiled on the mathematical arts, which were then presented to the imperial court, along with prefaces claiming that these mathematical arts were essential to ordering the empire. In their prefaces and commentaries, these literati sometimes assigned credit to mysterious recluses, apparently to support the claimed exclusivity and authenticity of their compilation; yet at other times the literati compilers also denounced *fangcheng* practitioners for employing overly arcane techniques, apparently in an attempt to reassert their own higher status and authority. Over this period of perhaps sixteen centuries, bibliographies of imperial libraries recorded the titles of hundreds of treatises on the mathematical arts. Many of these are still extant, and many include *fangcheng* problems. The earliest of these texts were collected in imperial libraries, and later reprinted under imperial auspices. From these efforts arose a textual tradition in which literati collected the early canonical texts, offered commentaries for them, and ultimately sought to recover their original meanings.

[1] For an explanation of the counting board, see chapter 2 of this book.

Fangcheng, an art that apparently found no practical application beyond solving implausible riddles, is remarkable for several reasons: it is essentially equivalent to the solution of systems of n equations in n unknowns in modern linear algebra; the earliest recorded *fangcheng* procedure is in many ways quite similar to what we now call Gaussian elimination.

These procedures were transmitted to Japan, and there is a reasonable possibility that they were also transmitted to Europe, serving perhaps as precursors of modern matrices, Gaussian elimination, and determinants. In any case, there was very little work on linear algebra in Greece or Europe prior to Gottfried Leibniz's studies of elimination and determinants, beginning in 1678. As is well known, Leibniz was a Sinophile interested in the translations of such Chinese texts as were available to him, and in particular the *Classic of Changes* (*Yi jing* 易經).

Overview of This Book

This is the first book-length study in any language of linear algebra in imperial China; it is also the first book-length study of linear algebra as it existed before 1678, the date Leibniz began his studies. My central purpose is to reconstruct *fangcheng* practices using extant textual sources. My focus is therefore on presenting an in-depth study of the mathematics behind these early texts: extant Chinese treatises on the mathematical arts are, after all, translations into narrative form of calculations originally performed on a counting board.

The Argument

The central argument of this book is that the essential feature in the solution of *fangcheng* problems is the visualization of the problem in two dimensions as an array of numbers on a counting board and the "cross multiplication" of entries, which led to general solutions of systems of linear equations not found in Greek or early European mathematics. There are, I argue, two distinct types of "cross multiplication" found in extant Chinese mathematical treatises, which correspond to the two methods used to solve problems in modern linear algebra today: (1) *wei cheng* 維乘, the "cross multiplication" of individual entries, as found in the "excess and deficit" (*ying bu zu* 盈不足) procedure in the *Nine Chapters on the Mathematical Arts*;[2] and (2) *bian cheng* 徧乘, the "cross multiplication" by individual elements of an entire column, as found in the *fangcheng* procedure (in modern terms, "row operations," since the columns in *fangcheng* arrays correspond to rows in matrices in modern linear algebra). The latter type, the "cross multiplication" of an entire column, corresponds to Gaussian elimination,

[2] For a discussion of the *Nine Chapters on the Mathematical Arts*, see chapter 3 of this book.

as is now well known. But *fangcheng* problems were also sometimes solved, I argue, by the former type, the "cross multiplication" of individual entries. I will call calculations and solutions found in this manner "determinantal" (following the typology in Tropfke [1902–1903] 1980);[3] some of these solutions seem to follow simple diagrams, conceptually similar to what is now commonly known as the Rule of Sarrus.[4] That is, in the early history of linear algebra, determinantal solutions were not discovered suddenly in their full generality; rather, the discovery of very specialized formulas led to the generalization we now call determinants.

The Findings

This book presents several findings that I hope will contribute to a reconsideration of some of our current views on the history of Chinese mathematics. In the received historiography, the *fangcheng* procedure, which is in ways similar to Gaussian elimination, has been assumed to be rather straightforward. The consensus of historians of Chinese mathematics, summarized by one of our most eminent historians, Jean-Claude Martzloff, has been that *fangcheng* procedures saw little development beyond their earliest exposition in the *Nine Chapters*: "Chinese mathematicians never modified the *fangcheng* techniques themselves in any radical way. In particular, they never thought of the notion of determinant" (Martzloff [1987] 2006, 258). This book challenges these views in three ways:

1. The *fangcheng* procedure presented in the *Nine Chapters* is more complicated than has been recognized, differing from Gaussian elimination in that it uses a counterintuitive approach to back substitution to avoid calculations with fractions; this suggests that *fangcheng* practices were considerably more sophisticated than has previously been understood.
2. Evidence from the *Nine Chapters* suggests that by the first century C.E., determinantal-style calculations were used to supplement solutions to a certain class of problems: in the process of solving a special class of systems of n conditions in $n+1$ unknowns,[5] a value for the $(n+1)^{\text{th}}$ unknown was assigned by calculating what we would now call a determinant. The earliest extant written record of such a determinantal calculation that I have found is preserved in a commentary to the *Nine Chapters* dated 1025 C.E.
3. Evidence in the *Nine Chapters* suggests that as early as the first century C.E., determinantal-style solutions were employed for special types of *fangcheng*

[3] For an explanation of my use of the terms "determinantal," "determinantal-style," "determinantal calculations," and "determinantal solutions," see "Early Methods for Solving Systems of Linear Equations," pages 25–26.

[4] For an explanation of the Rule of Sarrus, see page 22.

[5] I will use "conditions" rather than "equations" to describe Chinese mathematical problems that we would now call linear systems of n equations in m unknowns, since early Chinese mathematics did not, strictly speaking, use equations.

problems. The earliest extant written record of such a determinantal solution that I have found is preserved in a compilation dated 1661 C.E.

Historiographic Issues

Recent Research on the History of Chinese Mathematics

Considerable research on the history of Chinese mathematics has been published during the last several decades. My debt to previous research is thus considerable: without that pioneering work, this book would certainly not have been possible. Much of this previous research has been published in Chinese. Early pioneers in the field include Li Yan 李儼 and Qian Baocong 錢寶琮 (their research is collected and republished in Li and Qian 1998). One central focus of Chinese scholarship related to my efforts in this book is research on the *Nine Chapters on the Mathematical Arts* (Bai 1983; Bai 1990; Guo 1990; Guo 1998; Li Jimin 1990; Wu 1982). Important studies on linear algebra in specific treatises have also been published (Li Yan 1930; Mei 1984; Guo 1985). In addition, important work has been published in Japanese on both traditional Chinese mathematics and traditional Japanese mathematics (Kato 1954–64; Kato 1972; Jochi 1991).

During the past several decades, a considerable amount of research has also been published in Western languages. Again, my debt to this research is considerable. There have been several important translations of Chinese mathematical treatises into Western languages (Dauben 2008; Chemla and Guo 2004; Lam and Ang 2004; Cullen 2004; Shen, Lun, and Crossley 1999; Swetz 1992; Lam 1977; Hoe 1977; Libbrecht 1973). Jean-Claude Martzloff has published an important study of Mei Wending (Martzloff 1981). Currently, Joseph Dauben is working to translate the *Ten Mathematical Classics.* Other recent publications include important surveys of the history of Chinese mathematics (Martzloff [1987] 2006; Li and Du 1987). Numerous important research articles have also been published by historians of Chinese mathematics from China, Japan, Europe, and the United States. The Bibliography (following the text) lists some of the works most important for this book. (For the most extensive bibliography in English of secondary works on Chinese mathematics, covering Western languages as well as Chinese and Japanese, see Martzloff [1987] 2006; see also Dauben and Lewis 2000 and Dauben 1985.)

The History of Linear Algebra in Imperial China

Notwithstanding the important research that has been completed in the history of Chinese mathematics, the subject of linear algebra in China has not been the focus of sustained study. Although previous studies have presented an outline of some aspects of this history, they have not explored this subject in detail. More specifically, the explanations of the *fangcheng* procedure presented in previous studies are not sufficiently detailed to show how one might go about solving all eighteen problems in "*Fangcheng*," chapter 8 of the *Nine Chapters*. In fact, one of my motivations for writing this book has been the difficulties I encountered trying to solve all the problems in chapter 8, while rigorously following the *fangcheng* procedure. I could find in the secondary historical studies no adequate general explanation of the method for back substitution, and several of the studies incorrectly use modern methods.

Previous studies of linear algebra in China have been limited to articles and book sections, either in histories of Chinese mathematics or in translations of individual Chinese mathematical treatises, most notably the *Nine Chapters*. Important studies include Chemla and Guo 2004, 599–613; Guo 1998, 400–43; Martzloff [1987] 2006, 249–58; Lam 1987b, 228–34; Lam and Ang 2004, 112–15; Lam 1994, 34–38; Lam and Shen 1989; and Bai 1983, 257–76. At present, the closest we have in Western languages to an attempt to present a history of linear algebra in imperial China is "Methods of Solving Linear Equations in Traditional China," which examines several early treatises written before 600 C.E. (Lam and Shen 1989).[6] The two best studies of linear algebra in imperial China are sections in the two recent translations of the *fangcheng* chapter of the *Nine Chapters* (Shen, Lun, and Crossley 1999; Chemla and Guo 2004). Both have made important contributions to the study of Chinese mathematics; but both also have some limitations in their presentation of Chinese linear algebra.

The translation of chapter 8 of the *Nine Chapters*, "Rectangular Arrays," in Shen, Lun, and Crossley 1999 (pages 386–438) is at the present time the best analysis in Western languages of the mathematical aspects of linear algebra in the *Nine Chapters*. It begins with a brief introduction to linear algebra, followed by a brief historical overview of linear algebra problems from early Greece and India, and from treatises from imperial China. Following the translations of the entire text of the original *Nine Chapters* and Liu Hui's commentary, it presents step-by-step solutions for the problems, along with mathematical analysis and explanations. There are, however, several limitations. First, translations of passages from the *Nine Chapters* are sometimes difficult to understand, and in places unintelligible. Second, from a historian's point of view, Shen offers too little critical contextualization of the materials he analyzes. For example, he offers little critical analysis of the many extravagant claims Liu Hui makes in his preface; Shen perhaps also fails to distinguish adequately between the original text and the commentary by Liu Hui, which is quite critical of the original *fangcheng* method. Third, perhaps the most important limitation might be that a concession was made to "the style of academic writing in China" (p. 50) rather than to the intended Western audience, as one reviewer pointed out (Swetz 2001, 675). That is, in the end, despite the considerable significance of Shen's research and the commendable

[6] Although this article is quite helpful, it is only fifteen pages long. Most of this material has been incorporated into Shen, Lun, and Crossley 1999, which I discuss next.

efforts of the authors in translating Shen's writings, their book remains in many ways discernibly a translation of a Chinese manuscript into English.

Chemla and Guo 2004, published in French, is the more recent of the two better works on the subject. The strength of their study is its bringing Guo Shuchun's considerable work on philological reconstruction of the *Nine Chapters* to a Western audience, coupled with Karine Chemla's copious notes, historical contextualization, and critical comments. This work is without question the best translation of the *Nine Chapters* into a Western language, but their study is less focused on analysis of the mathematics than are many other recent studies. Most of their analysis of the linear algebra is in their introduction, "Présentation du Chapitre 8, 'Fangcheng,' " pages 599–613, written by Guo and translated by Chemla. At fifteen pages it is quite brief; and their explanation of back substitution is incorrect. Beyond this introduction, there is little discussion of the mathematics in their translation of chapter 8, "Fangcheng," itself (pages 616–659) or in Chemla's critical endnotes (pages 861–877).

The History of Linear Algebra

Linear algebra, together with calculus, is one of the core courses in most university undergraduate mathematics curricula, yet comparatively little has been published on its history, especially the history prior to 1678. The history of calculus has been the subject of considerable research, presented in numerous articles and several books (Boyer [1949] 1959; Edwards 1979; Guicciardini 1989; for an overview of research on the history of calculus, see Guicciardini 1994; for a bibliography of the history of mathematics, see Dauben and Lewis 2000; Dauben 1985). There are in addition several important studies of the prehistory of calculus from the early Greeks up to work in early modern Europe (Jahnke 2003; for an overview, see Andersen 1994). In contrast, "matrix theory is today one of the staples of higher-level mathematics education; so it is surprising to find its history is fragmentary. ... The history of this subject has been sadly neglected," note W. Ledermann and Ivor Grattan-Guinness, editor of the comprehensive and authoritative *Companion Encyclopedia of the History and Philosophy of the Mathematical Sciences*, in their entry "Matrix Theory" (Grattan-Guinness and Ledermann 1994, 775). Of course, this is not to say that important research on developments in Europe since 1678 has not been published: important studies include the contributions made by Leibniz (Knobloch 1980), Gauss (Knobloch 1994b), Cauchy (Hawkins 1975), Weierstrass (Hawkins 1977a), and Cayley (Hawkins 1977b). There is an encyclopedic compilation of important results in the theory of determinants (Muir 1906–1923), and even a brief study of the work of Charles L. Dodgson, better known as Lewis Carroll, on determinants (Abeles 1986).

The History of Linear Algebra Prior to 1678

If the history of matrix theory as a whole is "fragmentary" and "sadly neglected," the early history of linear algebra simply has never been written: there is no book-length historical study in any language on developments in linear algebra before 1678. (For an overview of research on the history of matrices, see Grattan-Guinness and Ledermann 1994; on the history of determinants, see Knobloch 1994a; see also Dauben and Lewis 2000 and Dauben 1985). In particular, there is at present no historical study of developments before 1678 in the two main methods of linear algebra—Gaussian elimination and determinants. One reason for this is that the source materials are in Chinese.

It is now well known that what is now called "Gaussian elimination" (in modern Chinese, *Gaosi xiaoqufa* 高斯消去法) in fact appeared at least 1500 years before Gauss (1777–1855). The similarity of the approach taken in the *Nine Chapters* to Gaussian elimination, for solving what we would now call simultaneous linear systems of n equations in n unknowns, has been widely noted in works on linear algebra (including Stewart 1973, 130, and Golan 2004, 162), in studies of the history of mathematics (including Boyer and Merzbach 1991, 197; and Knobloch 1994a, 766), and in studies of the history of Chinese mathematics (including Martzloff 1994, 96; Martzloff [1987] 2006, 254; Lam 1994, 35; Shen, Lun, and Crossley 1999, 2 and 388; and Chemla and Guo 2004, 604). Gauss's calculations appear in §180 and §§183–84 of his *Theoria motus corporum coelestium in sectionibus conicis solem ambientium* (1809, reprinted in *Carl Friedrich Gauss Werke*, vol. 7, translated in Gauss [1809] 1963, 261–63 and 266–68); his *Disquisitio de elementis ellipticis palladis* (1810, reprinted in his *Werke*, vol. 6, translated in Gauss 1887); and his *Theoria combinationis observationum erroribus minimis obnoxia, Pars. I, II, cum suppl.* (1821–1826, reprinted in his *Werke*, vol. 4, translated in Gauss [1821–1826] 1995). The context of Gauss's calculations is considerably more complex mathematically than that of the *Nine Chapters*.[7]

The use of determinants to solve linear equations has previously been held to be a discovery by Leibniz (1646–1716) and the simultaneous, independent discovery by the Japanese mathematician Seki Takakazu 關孝和 (c. 1642–1708). Leibniz's work on determinants has been discovered in manuscripts dating from 1678 (Knobloch 1980, 1990). Later European works on determinants include Colin MacLaurin's *Treatise of Algebra* (1748); Gabriel Cramer's *Introduction à l'analyse des lignes courbes algébriques* (1750); and Arthur Cayley's "On a Theorem in the Geometry of Position" (1841), "On the Theory of Determinants" (1843), and "On the Theory of Linear Transformations" (1845), all three reprinted in *Collected Mathematical Papers of Arthur Cayley*.[8] Seki's work on determinants was published in 1683 in "Method for Solving Hidden Problems" (*Kaifukudai no hō*

[7] For a historical analysis of Gauss's calculations and their context, see "Gauss on Least Squares," Goldstine 1977, 212–24. For a broad overview of the history of matrices in modern linear algebra, see Grattan-Guinness and Ledermann 1994.

[8] In addition to Knobloch's works cited above, see Knobloch 1994b; for an authoritative summary of historical studies of the development of determinants, see Knobloch 1994a. A com-

解伏題之法, corrected edition; reprinted with an English translation in *Seki Takakazu zenshu* 關孝和全集—*Takakazu Seki's Collected Works Edited with Explanations*).[9]

Determinants, it has seemed, were created *ex nihilo*, a simultaneous but precipitous discovery, without significant earlier precedents or a history of development. The only calculations that have been likened to determinants are limited to solutions of linear systems of 2 conditions in 2 unknowns (in contrast, the problems analyzed in this book range from 3 conditions in 3 unknowns to 9 conditions in 9 unknowns). More specifically, it has been noted, for example, that problems in "Excess and Deficit" (*ying bu zu* 盈不足), chapter 7 of the *Nine Chapters*, are equivalent to systems of 2 equations in 2 unknowns, and that the approach to solving these problems is similar to the cross multiplication of terms in a 2×2 determinant (Shen, Lun, and Crossley 1999, 389–90). "Determinants" is the title of chapter 11 of Libbrecht 1973, in which Libbrecht notes similarities between determinants and the solution presented for an "excess and deficit" problem, again with 2 conditions in 2 unknowns, in Qin Jiushao's 秦九韶 (1202–1261) *Book of Mathematics in Nine Chapters* (*Shu shu jiu zhang* 數書九章). These results have not been placed in a larger context of the history of the development of determinants, because it has been assumed that there were no further developments in China. Thus, for example, Knobloch argues that "it was Girolamo Cardano who used the oldest determinant formulation," solving a system of 2 equations in 2 unknowns in his *Ars magna* in 1545 (Knobloch 1994a, 767).

It has been the consensus among historians—even among specialists in the history of Chinese mathematics—that Chinese techniques for solving linear equations showed little development, and that determinantal methods were never discovered. Perhaps it is because of this presumed lack of development that the history of linear algebra in imperial China has not seemed to merit deeper study. And perhaps because of the absence of a book-length historical study, even recent and comprehensive sources on the history of mathematics often devote less than one page among hundreds to the early history of linear algebra. For example, the *Companion Encyclopedia of the History and Philosophy of the Mathematical Sciences*, cited above, devotes only one paragraph to the early history of linear algebra in China, noting that the *Nine Chapters on the Mathematical Arts* presents methods identical to Gaussian elimination; and a second paragraph notes that the Japanese, "stimulated by the Chinese, applied determinant-like methods" and used "matrix-like schemes" (Grattan-Guinness and Ledermann 1994, 775). The recently revised edition of Boyer's well-known *A History of Mathematics* similarly devotes just over one page to discussing early developments in linear algebra. Under the heading "The *Nine Chapters*," it notes that "Chapter eight of the *Nine Chapters* is significant for its solution of problems of simultaneous linear equations, using both positive and negative numbers. The last problem in

---

prehensive collection of excerpts from papers on determinants (including those cited above), together with summary and analysis, can be found in Muir 1906–1923.

[9] For studies of Seki, see Mikami 1914, Kato 1972, and Horiuchi 1994a. Again, I thank Jean-Claude Martzloff for his suggestions on these points.

the chapter involves four equations in five unknowns." And under the heading "Magic Squares," it adds that "The Chinese were fond of patterns. . . . The concern for such patterns led the author of the *Nine Chapters* to solve the system of linear equations . . . " followed by one example of a system of 3 equations in 3 unknowns (Boyer and Merzbach 1991, 196–97).

This very brief treatment in these otherwise comprehensive sources is perhaps the result of the lack of a more technical and comprehensive historical work on the early history of linear algebra. This book proposes to fill that gap.

Outline of the Chapters

Chapter 2, "Preliminaries," presents background information on the Chinese language, offers a brief introduction to the methods using counting rods in Chinese mathematics, and provides some mathematical preliminaries from modern linear algebra.

Chapter 3, "The Sources: Written Records of Early Chinese Mathematics," examines the sources on which this study is based, in part to help clarify a distinction in Chinese mathematics between the textual tradition and *fangcheng* practice. Unfortunately, we know virtually nothing about the *fangcheng* practitioners themselves; we have no records of any writings by them. We do, however, have prefaces, commentaries, and sometimes biographical information about the literati who compiled treatises on the Chinese mathematical arts. Drawing upon these prefaces, commentaries, biographical materials, and other historical records, chapter 3 provides a critical historical contextualization of the sources for the study of *fangcheng*.

Chapter 4, "Excess and Deficit," focuses on "excess and deficit" problems, which are equivalent to systems of 2 equations in 2 unknowns, and seem to have been precursors to *fangcheng* problems. This chapter explores two key features later extended to *fangcheng* problems: "excess and deficit" problems were displayed in two dimensions on a counting board; and they were solved using a form of "cross multiplication," which in this case involves the "cross multiplication" (*wei cheng* 維乘) of individual entries.

Chapter 5, "*Fangcheng*, Chapter 8 of the Nine Chapters," analyzes the earliest known *fangcheng* procedure so as to elucidate three key features. First, *fangcheng* problems are displayed in two dimensions on the counting board. Second, entries are eliminated, in a manner similar to Gaussian elimination, by a form of "cross multiplication," *bian cheng* 徧乘, which involves "cross-multiplying" entire columns by individual entries. Third, the approach to back substitution in the *fangcheng* procedure differs from the intuitive approach familiar from modern linear algebra.

Chapter 6, "The *Fangcheng* Procedure in Modern Mathematical Terms," presents a detailed analysis of the *fangcheng* procedure using modern mathematical terminology. The key finding in this chapter is that the *fangcheng* procedure, as counterintuitive as it may seem, avoids calculations with fractions, which would have been quite complicated on a counting board on which the two dimensions were already used to display the entries in the form of an array.[10]

Chapter 7, "The Well Problem," examines solutions to problem 13 from chapter 8 of the *Nine Chapters*. This chapter presents the earliest known surviving written record of a determinantal calculation, which is preserved in a commentary written in 1025 C.E. to the solution of the well problem. This chapter concludes with an examination of the earliest known surviving written record of a determinantal solution (finding solutions, that is, for the unknowns), which is preserved in a text compiled in 1661 C.E.

Chapter 8, "Evidence of Early Determinantal Solutions," presents evidence that determinantal solutions were likely known at the time of the compilation of the *Nine Chapters*, about the first century C.E. Specifically, this chapter focuses on five of the eighteen problems in chapter 8 of the *Nine Chapters*, problems 3, 12, 13, 14, and 15, to argue that these problems were solved by specialized procedures that involved the type of "cross multiplication" used in "excess and deficit" problems, that is, the "cross multiplication" of individual entries.

Chapter 9, "Conclusions," presents a chronological summary of these developments, outlines areas for further research, examines methodological issues, and explores the significance of the findings.

---

[10] See chapter 2 for an explanation of calculations on the counting board.

Chapter 2
Preliminaries

This chapter presents background material necessary for understanding early Chinese linear algebra. It is divided into the following sections:

1. The first section, "Chinese Conventions," presents background information on Chinese romanization, names, book titles, and other such conventions.
2. The next section, "Chinese Mathematics," discusses Chinese mathematics, related conventions, and issues related to the translation of classical Chinese into modern English, focusing in particular on mathematical terms.
3. Finally, "Modern Mathematical Terminology" introduces some mathematical terminology and some of the basic results of linear algebra employed in this study. More specifically, we will review the following: (1) augmented matrices, which are essentially the same as the arrays of numbers found in early Chinese mathematical treatises; (2) Gaussian elimination, which is in many ways similar to the method presented in the *Nine Chapters* (though, as we will see, the method for back substitution differs); and (3) determinants, special forms of which were calculated in Chinese mathematical treatises. In addition, this section reviews the following results, which will be important for this study: (1) Gaussian elimination and determinants represent two fundamentally different approaches to solving linear equations—Gaussian elimination is a procedure, whereas determinants are calculated directly from a formula. (2) Determinants are, in general, impractical because they are too difficult to calculate. (3) Gaussian elimination is, in general, considerably more efficient than determinants for solving systems with more than 3 equations in 3 unknowns. (4) Determinants, when used, are most easily calculated by employing diagrams.

Chinese Conventions

For transcribing Chinese characters phonetically into English, I have chosen to use the Pinyin system that is currently used in the People's Republic of China. This system is now becoming the standard in Western scholarship. It should be noted that several different systems are used to transcribe Chinese characters, and this may be a source of confusion for readers not trained in Chinese studies. Older works in Western scholarship often used the Wade-Giles system, which is similar to Pinyin, but employs somewhat altered spellings. For example, Qin Jiushao (the Pinyin transliteration for the Chinese name 秦九韶) is transliterated in the Wade-Giles system as Ch'in Chiu-Shao.

Chinese names are written with the surname (employing usually one but sometimes two characters) first, followed by the personal name (employing usually two but sometimes one character). In the Pinyin transliteration system, the transliterations of the characters that comprise each name are combined, and their first letters capitalized. Chinese literati often used several different personal names, including their name given at birth (*ming* 名), their courtesy names (*zi* 字), their sobriquets (*hao* 號), and their posthumous names (*shi* 謚), among others. For example, Qin Jiushao's family name is Qin, his given name is Jiushao, and his courtesy name is Daogu 道古. Similarly, Japanese surnames come before the given name. In the Bibliography, names are always listed with the family name first, but in the case of Chinese and Japanese names I have omitted the comma used in Western names because the Chinese and Japanese names have not been inverted.

Again for the convenience of the broader intended audience, I use translations of the titles of Chinese primary sources and secondary historical research articles in the text and footnotes, instead of following the sinological tradition of using Pinyin transliterations.[1] I have used Chinese Pinyin for titles only when no reliable translation is possible. Full titles with Pinyin and characters are, of course, given in the first instance of each citation, and in the Bibliography. But in footnotes and technical appendices intended only for specialists in the field, I have followed the usual sinological conventions and have not provided translations into English.

For Chinese primary historical sources I have followed the standard bibliographic conventions, including recording the number of chapters. When these have been translated, I use the term "chapter" for *juan* 卷, "volume" for *ben* 本, and "collectanea" for *congshu* 叢書. Printing in China dates back to the early Tang Dynasty (618–907), but treatises published in the imperial period often do not include important publication facts, and publication dates are unreliable. I have used the dates recorded in prefaces where they are available. We rarely know the number of copies printed. Treatises are often preserved in collectanea compiled in later periods; the abbreviations I have used for these collections can be found at the beginning of the Bibliography. For translations of governmental offices and official titles, I have followed Charles Hucker's *A Dictionary of Official Titles in Imperial China*. I have consulted standard sources to convert to the Western calendar.

[1] This is one of the styles recommended by *The Chicago Manual of Style*, and is especially helpful in languages where, unlike Latin or French, for example, the reader of English cannot be expected to guess from cognates the meaning of a title.

In consideration of readers who are not in the field of Chinese studies, I have attempted to translate as much as possible into English. The difficulties of translation are vexing but not insurmountable. The problem of the use of translated terms here does not, I believe, fundamentally differ in nature from the problem of the use of contemporary English to discuss concepts from Latin or from seventeenth-century Italian or French. The principle I have followed is pragmatic—one simply tries to ensure maximum intelligibility and precision.

Chinese Mathematics

Mathematical problems in early China were solved using counting rods.[2] The Chinese numeral-rod system is not difficult to understand, and will be used throughout this book, so I will briefly introduce it here. The first position is for numbers 0 to 9: 0 is denoted by an empty space; 1 to 5 are denoted by the corresponding number of rods; a horizontal rod denoting 5 is placed above one to four vertical rods to denote 6 to 9, as can be seen in Table 2.1. The next position to the left is for tens: again, an empty space signifies 0; 10 to 50 are signified by placing the rods on their side; a vertical rod signifying 50 is used above the other rods for 60 to 90. The next position to the left (not shown) is for hundreds, the number of hundreds signified by using the digits 0–9; and so on.

Table 2.1: Chinese rod numerals for one through nine, and ten through ninety; zero is denoted by a blank space, as indicated here for numbers 10 through 90.

1	2	3	4	5	6	7	8	9
𝍠	𝍡	𝍢	𝍣	𝍤	𝍥	𝍦	𝍧	𝍨
10	**20**	**30**	**40**	**50**	**60**	**70**	**80**	**90**
𝍩	𝍪	𝍫	𝍬	𝍭	𝍮	𝍯	𝍰	𝍱

The form for the numbers one to nine is used for odd powers of ten, and the form for the numbers ten to ninety is used for even powers of ten. For example, the number 721 is written 𝍦𝍪𝍠, and 10815 is written 𝍠 𝍧𝍩𝍤.

[2] Standard works on the history of Chinese mathematics include Martzloff [1987] 2006 and Li and Du 1987; see also Chemla and Guo 2004, 15–20, and Shen, Lun, and Crossley 1999, 11–17. One focus of Chinese mathematics in the imperial period was algebra.

Multiplication with Counting Rods

Counting rods were used to multiply, divide, extract roots, and perform operations with fractions. The following example shows how counting rods were used to multiply, following a problem from *The Mathematical Classic of Master Sun* (*Sunzi suan jing* 孫子算經, c. 400 C.E.).[3]

STEP 1. First, we place the numbers to be multiplied on the counting board, with 81 in the "upper position" (*shang wei* 上位) and 81 in the "lower position" (*xia wei* 下位), as follows:

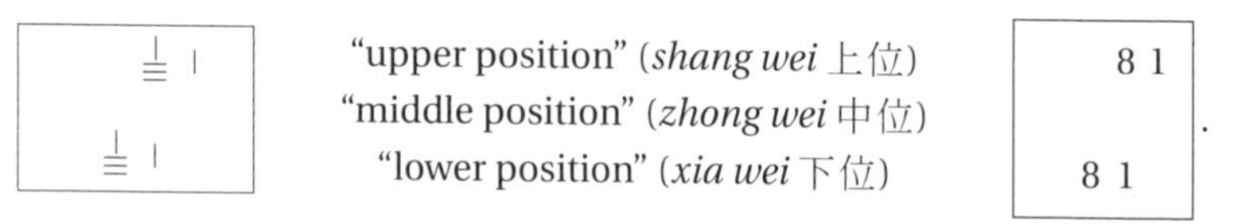

STEP 2. Next, we multiply the "upper eight" (*shang ba* 上八) by the "lower eight" (*xia ba* 下八) to yield $80 \times 80 = 6400$, which is then placed in the "middle position" (*zhong wei* 中位):

"upper position"	8 1
"middle position"	6 4 0 0
"lower position"	8 1

STEP 3. We then multiply the "upper eight" by the "lower one" (*xia yi* 下一) to yield $80 \times 1 = 80$, and add, to give the following:

"upper position"	8 1
"middle position"	6 4 8 0
"lower position"	8 1

STEP 4. We then shift the lower counting rods to the right and remove the "upper eight," as follows,

"upper position"	1
"middle position"	6 4 8 0
"lower position"	8 1

STEP 5. Next, we multiply the "lower eight" by the "upper one" and add $80 \times 1 = 80$ to the middle position:

---

[3] For a translation and analysis of *The Mathematical Classic of Master Sun*, see Lam and Ang 2004, on which the following translation and analysis are based.

"upper position"	1
"middle position"	6 5 6 0
"lower position"	8 1

STEP 6. We then multiply the "lower one" by the "upper one" and add $1 \times 1 = 1$ to the middle position:

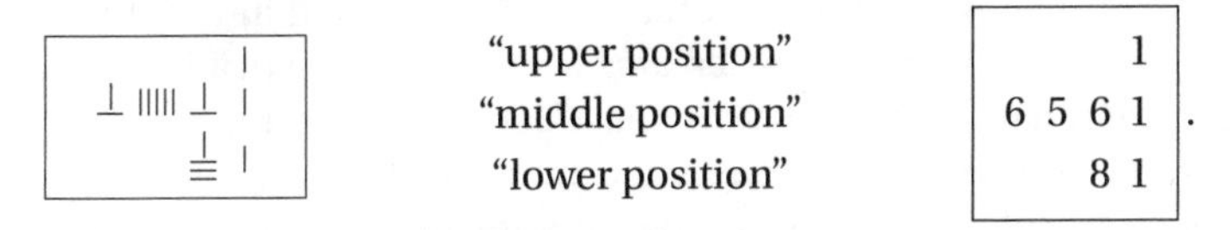

STEP 7. Finally, we remove the counting rods from the "upper position" and "lower position" to give the answer:

"upper position"	
"middle position"	6 5 6 1
"lower position"	

Chinese Mathematical Terminology

Broadly speaking, there are three main alternatives in translating Chinese mathematical terms: modern mathematical terms (e.g., "matrices" for *fangcheng* 方程); neologisms (e.g., "rectangular arrays"); or transliteration (e.g., simply using *fangcheng*). Each of these alternatives entails considerable explanatory work:

1. When modern mathematical terms are used (e.g., "matrices" for *fangcheng* 方程), considerable care must be taken to explain the differences between the Chinese term and its closest modern equivalent.
2. When neologisms are used (e.g., "rectangular arrays" for *fangcheng* 方程), an approach that is usually employed in hopes of emphasizing differences between the mathematics under study and modern mathematics, one must still explain all the properties to be attributed to the neologism, which still must be done via comparisons with modern mathematical terms (in this example, "matrices"). Too many neologisms can impose an unnecessary burden on the reader.
3. If transliteration is employed (e.g., simply using *fangcheng* for the Chinese characters 方程), again, one must still explain all the properties associated with the transliterated term, and again, do so via comparisons with modern mathematical terms. Too much transliteration places an unnecessary burden on those unfamiliar with the language, who must remember terms with which they are likely unfamiliar.

The problems encountered in translating early Chinese mathematics, I will argue, are not substantially different from those encountered in translating early Greek or "pre-modern" mathematics. In general, I will use modern terms where practicable.

The translation of Chinese mathematics is somewhat controversial, at least among sinologists. There are of course numerous and important differences between the mathematics recorded in early Chinese treatises and modern mathematics: the problems recorded in Chinese texts are given in narrative form and present a contrived context that creates a façade of practicality; the problems given are exemplars rather than the most generalized form; explicit proofs are rarely given; unknowns are represented by a position on the counting board rather than by explicit use of a variable; and fractions are handled differently, to name only a few differences. (For overviews of Chinese mathematics, see Martzloff [1987] 2006; Li and Du 1987; and Shen, Lun, and Crossley 1999.) There are also numerous and important differences between the classical Chinese language and modern English: classical Chinese lacks punctuation, often lacks explicit subjects, and is often ambiguous, again to name only a few examples. Because of these mathematical and linguistic disparities, Chinese mathematics is sometimes translated in a very literal manner, which has the advantage of highlighting these differences, but yields the unfortunate result of rendering Chinese mathematics even more difficult to understand. Again, this situation is hardly unique to the study of early Chinese mathematics: there are numerous differences between early Greek or European mathematics and modern mathematics, and differences between classical Greek or Latin languages and modern English. Recent studies of the history of science in early modern Europe often use more familiar terms in translations, for example choosing the term "science" over "natural philosophy" and adding an explanatory note.

Following the usual general principles of translation, I have attempted to render my translations of Chinese mathematics as readable as possible while remaining faithful to the Chinese original; where necessary I have added a note explaining the more literal interpretation.

Modern Mathematical Terminology

Here, I shall review some of the basics of modern linear algebra, focusing on points important for this study: augmented matrices, Gaussian elimination, determinants, and the comparatively greater efficiency of Gaussian elimination as contrasted with determinants.

Augmented Matrices

Very briefly, in modern linear algebra, we are given a system of n equations in n unknowns, $x_1, x_2, \ldots, x_n$,

$$\begin{aligned} a_{11}x_1 + a_{12}x_2 + \cdots + a_{1n}x_n &= b_1 \\ a_{21}x_1 + a_{22}x_2 + \cdots + a_{2n}x_n &= b_2 \\ \cdots\cdots\cdots\cdots\cdots\cdots\cdots\cdots \\ a_{n1}x_1 + a_{n2}x_2 + \cdots + a_{nn}x_n &= b_n , \end{aligned} \tag{2.1}$$

where the coefficients a_{ij} and b_i are constants. We can rewrite this system of equations (2.1) in matrix form as $Ax = b$, where A is the matrix of coefficients (a_{ij}), x is the column vector of unknowns (x_i), and b is the column vector of constant terms (b_i), thus

$$\begin{pmatrix} a_{11} & a_{12} & \cdots & a_{1n} \\ a_{21} & a_{22} & \cdots & a_{2n} \\ \vdots & \vdots & \ddots & \vdots \\ a_{n1} & a_{n2} & \cdots & a_{nn} \end{pmatrix} \begin{pmatrix} x_1 \\ x_2 \\ \vdots \\ x_n \end{pmatrix} = \begin{pmatrix} b_1 \\ b_2 \\ \vdots \\ b_n \end{pmatrix}. \tag{2.2}$$

To solve this system of equations,[4] we usually write the system of equations (2.1) in an even more abbreviated form, as an augmented matrix, which is formed by appending the solution column vector b to the coefficient matrix A, resulting in the $n \times (n+1)$ matrix

$$\begin{bmatrix} a_{11} & a_{12} & \cdots & a_{1n} & b_1 \\ a_{21} & a_{22} & \cdots & a_{2n} & b_2 \\ \vdots & \vdots & \ddots & \vdots & \vdots \\ a_{n1} & a_{n2} & \cdots & a_{nn} & b_n \end{bmatrix}. \tag{2.3}$$

As we will see in chapter 5, in Chinese mathematical treatises, linear algebra problems are written in a form essentially the same as that of the augmented matrix (2.3) here.

Elimination

In modern linear algebra, the most general method for solving the system (2.1) of n equations in n unknowns is elimination. There are, however, several variants of elimination; the method described below most closely corresponds to the method presented in the *Nine Chapters.*

[4] It should be noted that there is a unique solution if and only if the rows of the coefficient matrix are linearly independent, or equivalently, the determinant is not zero, $\det A \neq 0$. For a definition of the determinant, see equation (2.11); for a discussion of determinants, see page 19 of this book. For further explanation, see any standard introduction to linear algebra, for example Strang [1976] 1988.

We begin with an augmented matrix, as in equation 2.3 above. Then a series of row reductions is performed to eliminate entries below the diagonal entries. For example, in equation (2.3) above, we might first eliminate the entry a_{21}. One way to do so—which is similar to the method in the *Nine Chapters*—is to multiply each entry in the second row by a_{11}, multiply each entry in the first row by a_{21}, and subtract that result term-by-term from the second row, so that the entry in the second row and first column becomes zero:

$$\begin{bmatrix} a_{11} & a_{12} & a_{13} & \cdots & a_{1n} & b_1 \\ 0 & a_{11}a_{22} - a_{21}a_{12} & a_{11}a_{23} - a_{21}a_{13} & \cdots & a_{11}a_{2n} - a_{21}a_{1n} & a_{11}b_2 - a_{21}b_1 \\ a_{31} & a_{32} & a_{33} & \cdots & a_{3n} & b_3 \\ \vdots & \vdots & \vdots & \ddots & \vdots & \vdots \\ a_{n1} & a_{n2} & a_{n3} & \cdots & a_{nn} & b_n \end{bmatrix}.$$

For simplicity I will use primes to signify that entries have been modified, that is, writing $a'_{22} = a_{11}a_{22} - a_{21}a_{12}, \ldots, a'_{2n} = a_{11}a_{2n} - a_{21}a_{1n}$, and $b'_2 = a_{11}b_2 - a_{21}b_1$. We see that the above augmented matrix is of the same form as (2.3), with the exception that the entry a_{21} has been eliminated:

$$\begin{bmatrix} a_{11} & a_{12} & a_{13} & \cdots & a_{1n} & b_1 \\ 0 & a'_{22} & a'_{23} & \cdots & a'_{2n} & b'_2 \\ a_{31} & a_{32} & a_{33} & \cdots & a_{3n} & b_3 \\ \vdots & \vdots & \vdots & \ddots & \vdots & \vdots \\ a_{n1} & a_{n2} & a_{n3} & \cdots & a_{nn} & b_n \end{bmatrix}. \tag{2.4}$$

We can then continue these row operations repeatedly, eliminating one subdiagonal entry each time. In the approach to Gaussian elimination most similar to the method presented in the *Nine Chapters*, we continue eliminating subdiagonal entries by row operations until we arrive at an upper-triangular form. That is, all entries are zero below the diagonal represented by a_{ii} $(1 \le i \le n)$,

$$\begin{bmatrix} a_{11} & a_{12} & \cdots & a_{1,n-1} & a_{1n} & b_1 \\ 0 & a'_{22} & \cdots & a'_{2,n-1} & a'_{2n} & b'_2 \\ \vdots & \ddots & \ddots & \vdots & \vdots & \vdots \\ 0 & \cdots & 0 & a''_{n-1,n-1} & a''_{n-1,n} & b''_{n-1} \\ 0 & \cdots & 0 & 0 & a''_{nn} & b''_n \end{bmatrix}. \tag{2.5}$$

where primes a'_{2j} and double-primes a''_{ij} denote the changed value resulting from (possibly many) row operations. To solve this, one approach is to use back substitution.[5] Here I will present the approach to back substitution most com-

[5] Several approaches can be used to solve the system of linear equations represented by the augmented matrix in (2.5). For example, one of these variations is Gauss-Jordan elimination, in which row operations are continued until the augmented matrix is of reduced echelon form, that is, ones on the diagonal entries of the coefficient matrix and zeros above and below. For a more detailed discussion, see Strang [1976] 1988.

monly used in modern linear algebra. (This approach, it should be noted, differs from the approach used in the *Nine Chapters*. I am presenting this approach from modern linear algebra in order to contrast it to the approach in the *Nine Chapters*, which will be described in chapters 5 and 6.) The augmented matrix in (2.5) represents the following system of equations:

$$\begin{aligned} a_{11}x_1 + a_{12}x_2 + \cdots + a_{1n}x_n &= b_1 \\ a'_{22}x_2 + \cdots + a'_{2n}x_n &= b'_2 \\ \cdots \\ a''_{nn}x_n &= b''_n. \end{aligned}$$

The bottom row of (2.5) represents the equation

$$a''_{nn}x_n = b''_n, \tag{2.6}$$

and the solution found first is

$$x_n = \frac{b''_n}{a''_{nn}}.$$

Then this value found for x_n is substituted back into the next row from the bottom, which represents the equation

$$a''_{n-1,n-1}x_{n-1} + a''_{n-1,n}x_n = b''_{n-1}, \tag{2.7}$$

giving the solution

$$x_{n-1} = \frac{b''_{n-1} - \frac{b''_n}{a''_{nn}}a''_{n-1,n}}{a''_{n-1,n-1}}. \tag{2.8}$$

This process of back substitution continues until all the x_i $(1 \leq i \leq n)$ have been found.

Determinants

In modern linear algebra, a second approach to solving the system (2.1) of n equations in n unknowns is by means of determinants. The solution here, given by Cramer's Rule, seems elegant and simple. Assuming $\det A \neq 0$, the solution for each unknown x_j $(1 \leq j \leq n)$ is given by the formula

$$x_j = \frac{\det B_j}{\det A}, \tag{2.9}$$

where B_j is the matrix obtained by replacing the jth column of the coefficient matrix A with the solution vector $b = (b_i)$,

$$B_j = \begin{pmatrix} a_{11} & \cdots & a_{1,j-1} & b_1 & a_{1,j+1} & \cdots & a_{1n} \\ a_{21} & \cdots & a_{2,j-1} & b_2 & a_{2,j+1} & \cdots & a_{2n} \\ a_{31} & \cdots & a_{3,j-1} & b_3 & a_{3,j+1} & \cdots & a_{3n} \\ \vdots & & \vdots & \vdots & \vdots & & \vdots \\ a_{n1} & \cdots & a_{n,j-1} & b_n & a_{n,j+1} & \cdots & a_{nn} \end{pmatrix}. \tag{2.10}$$

The difficulty lies in computing the determinants given in equation (2.9). In technical terms, the general formula for the determinant of the $n \times n$ matrix $A = (a_{ij})$ is

$$\det A = \sum_{\sigma \in \mathbf{S}_n} (-1)^{\operatorname{sgn}(\sigma)} \prod_{j=1}^{n} a_{\sigma_j j} = \sum_{\sigma \in \mathbf{S}_n} (-1)^{\operatorname{sgn}(\sigma)} a_{\sigma_1 1} a_{\sigma_2 2} \cdots a_{\sigma_n n}, \tag{2.11}$$

where the sum is taken over all permutations $\sigma \in \mathbf{S}_n$.[6] In simpler terms, this means that each of the terms $a_{\sigma_1 1} a_{\sigma_2 2} \cdots a_{\sigma_n n}$ is the product of entries from the matrix $A = (a_{ij})$ in equation (2.2), chosen such that there is exactly one from each row and column; $\det A$ is the sum of all such terms (after each is multiplied by ± 1). In other words, the determinant of an $n \times n$ matrix is the sum of $n!$ terms; each term is the product of n coefficients from (a_{ij}) chosen such that precisely one element from each row and one element from each column are selected. Substituting the appropriate expansions for the determinant (2.11) into the numerator and denominator of equation (2.9), again in more technical terms, we obtain

$$x_j = \frac{\sum_{\sigma \in S_n} (-1)^{\operatorname{sgn} \sigma}\, a_{\sigma_1 1} \cdots a_{\sigma_{j-1} j-1} b_{\sigma_j j} a_{\sigma_{j+1} j+1} \cdots a_{\sigma_n n}}{\sum_{\tau \in S_n} (-1)^{\operatorname{sgn} \tau}\, a_{\tau_1 1} a_{\tau_2 2} \cdots a_{\tau_n n}}, \tag{2.12}$$

where the sums are taken over all permutations $\sigma \in \mathbf{S}_n$ and $\tau \in \mathbf{S}_n$.

The complex formula given in equation (2.12) is a generalization of specific formulas for each order n of the system of linear equations, which are, for $n = 2$ or 3, quite simple. For example, the determinants are easy to compute if the given system (2.1) has 2 equations in 2 unknowns. That is, for

$$\begin{pmatrix} a_{11} & a_{12} \\ a_{21} & a_{22} \end{pmatrix} \begin{pmatrix} x_1 \\ x_2 \end{pmatrix} = \begin{pmatrix} b_1 \\ b_2 \end{pmatrix}, \tag{2.13}$$

---

[6] Here $\mathbf{S}_n$ is the symmetric group of degree n, the group of all permutations $\sigma : \mathbb{Z}_n \to \mathbb{Z}_n$ over the set of the first n positive integers $\mathbb{Z}_n = \{1, 2, 3, \cdots, n\}$. There are $n! = 1 \cdot 2 \cdot 3 \cdots (n-1)n$ permutations in $\mathbf{S}_n$. The permutation σ is termed odd (even) if the total number of integers inverted is odd (even), that is, if the number of pairs $(i, j) \in \mathbb{Z}_n \times \mathbb{Z}_n$ such that $i < j$ and $\sigma(i) > \sigma(j)$ is odd (even). If $\operatorname{sgn}(\sigma)$ is defined as the total number of inversions, then the parity of σ, which is defined as $(-1)^{\operatorname{sgn}(\sigma)}$, is -1 when σ is odd and $+1$ when σ is even. In some treatments of linear algebra, the parity $(-1)^{\operatorname{sgn}(\sigma)}$ is written simply as $\operatorname{sgn}(\sigma)$.

the formula for calculating the determinant of the 2×2 matrix A is given by a simple and well-known formula,

$$\det A = \begin{vmatrix} a_{11} & a_{12} \\ a_{21} & a_{22} \end{vmatrix} = a_{11}a_{22} - a_{21}a_{12}.$$

For our purposes it is important to note that this result is easily visualized, and remembered as the cross multiplication of the entries along the two diagonals, that is, multiplying a_{11} by a_{22} and then subtracting from that a_{21} multiplied by a_{12},

$$\begin{matrix} a_{11} & & a_{12} \\ & \times & \\ a_{21} & & a_{22} \end{matrix}$$

In the *Nine Chapters* and later Chinese mathematical treatises, we will see, these problems are written out in a similar manner, with the instruction to "cross multiply" (*hu cheng* 互乘).

The calculation of $\det B_j$ is similar. For example, B_1 is formed by substituting the column vector $b = \binom{b_1}{b_2}$ for the first column of the coefficient matrix $A = \begin{pmatrix} a_{11} & a_{12} \\ a_{21} & a_{22} \end{pmatrix}$, and the determinant is

$$\det B_1 = \begin{vmatrix} b_1 & a_{12} \\ b_2 & a_{22} \end{vmatrix} = b_1 a_{22} - b_2 a_{12}.$$

The solution to equation (2.13) can then be written as

$$x_1 = \frac{b_1 a_{22} - b_2 a_{12}}{a_{11}a_{22} - a_{12}a_{21}}, \quad x_2 = \frac{a_{11}b_2 - a_{21}b_1}{a_{11}a_{22} - a_{12}a_{21}}. \tag{2.14}$$

It is important to note that in contrast to Gaussian elimination, a procedure that successively eliminates entries to reach the solution, solving by determinants consists in directly computing the results by substituting the coefficients into given equations, in this case equations (2.14).

The determinants are more complicated for 3 equations in 3 unknowns, but still usable. For example,

$$\det A = \begin{vmatrix} a_{11} & a_{12} & a_{13} \\ a_{21} & a_{22} & a_{23} \\ a_{31} & a_{32} & a_{33} \end{vmatrix} = \begin{matrix} a_{11}a_{22}a_{33} - a_{11}a_{23}a_{32} + a_{12}a_{23}a_{31} \\ -a_{12}a_{21}a_{33} + a_{13}a_{21}a_{32} - a_{13}a_{22}a_{31}. \end{matrix}$$

Again, the calculation of B_j is similar. For the sake of brevity, I will give only the solution for the first term,

$$x_1 = \frac{\begin{matrix} b_1 a_{22}a_{33} - b_1 a_{23}a_{32} + a_{12}a_{23}b_3 \\ - a_{12}b_2 a_{33} + a_{13}b_2 a_{32} - a_{13}a_{22}b_3 \end{matrix}}{\begin{matrix} a_{11}a_{22}a_{33} - a_{11}a_{23}a_{32} + a_{12}a_{23}a_{31} \\ - a_{12}a_{21}a_{33} + a_{13}a_{21}a_{32} - a_{13}a_{22}a_{31} \end{matrix}}. \tag{2.15}$$

In practice, the easiest way to calculate this is what is known as the Rule of Sarrus, a simple diagram that extends the determinants in the numerator and denominator, and allows us to find the solution by performing a kind of cross multiplication. More specifically, in the numerator, we repeat the entries for b_1 and a_{12} in the first row, the entry for b_2 in the second row, and the entries for b_3 and a_{32} in the third row, and then we similarly repeat the entries in the denominator, to arrive at the following:

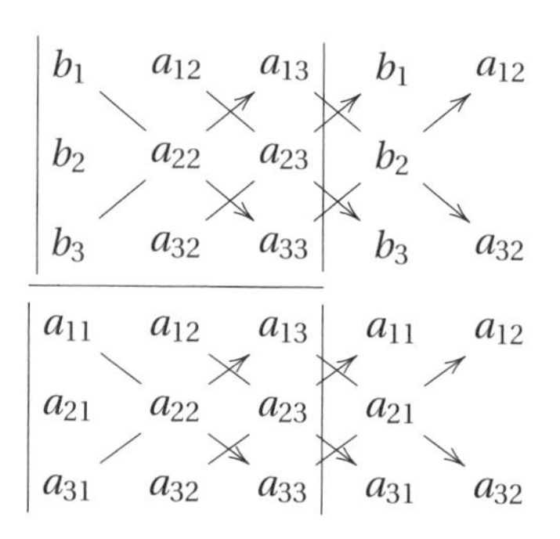

The determinant in the numerator is then found by multiplying together the entries on the main diagonal b_1, a_{22}, and a_{33}, adding to that the products of the diagonals $a_{12}a_{23}b_3$ and $a_{13}b_2a_{32}$, and then subtracting from that the products of the diagonals $b_3a_{22}a_{13}$, $a_{32}a_{23}b_1$, and $a_{33}b_2a_{12}$. The determinant in the denominator is found in a similar manner, yielding the familiar result given in equation (2.15). That is, in modern mathematics, determinants are calculated by rewriting them in a two-dimensional diagram, thus facilitating calculation.

The expansions for the variables x_2 and x_3 are similar. Because of the increased complexity here, modern texts on linear algebra sometimes omit the expansion of the determinant even for this case, where the order $n = 3$. For our purposes, it is important to note that the formulas for systems of 2 equations in 2 unknowns given in equation (2.14) and for 3 equations in 3 unknowns given above for x_1 in equation (2.15) are quite different. That is, whereas in theory, at a higher level of abstraction, both are given by Cramer's Rule in equation (2.9), in practice the formulas are different, and are remembered by cross multiplying as in the diagrams above.

For more than 3 equations in 3 unknowns, the calculation of the determinants is quite complex: it becomes too complex to be practicably usable to find solutions, and the simple mnemonic of cross multiplication fails (that is, there is no simple diagram to aid in the calculation). Again, it may be helpful to examine the expansion for the solution when $n = 4$, just to see how difficult it would be to compute, and again for the sake of brevity I have listed the solution for x_1 only, since the expansions for x_2, x_3, and x_4 are similar:

$$x_1 = \frac{\begin{aligned} &a_{12}a_{23}a_{44}b_3 - a_{12}a_{23}a_{34}b_4 + a_{12}a_{24}a_{33}b_4 - a_{12}a_{24}a_{43}b_3 \\ &+ a_{12}a_{34}a_{43}b_2 - a_{12}a_{33}a_{44}b_2 + a_{13}a_{22}a_{34}b_4 - a_{13}a_{22}a_{44}b_3 \\ &+ a_{13}a_{24}a_{42}b_3 - a_{13}a_{24}a_{32}b_4 + a_{13}a_{32}a_{44}b_2 - a_{13}a_{34}a_{42}b_2 \\ &+ a_{14}a_{22}a_{43}b_3 - a_{14}a_{22}a_{33}b_4 + a_{14}a_{23}a_{32}b_4 - a_{14}a_{23}a_{42}b_3 \\ &+ a_{14}a_{33}a_{42}b_2 - a_{14}a_{32}a_{43}b_2 + a_{22}a_{33}a_{44}b_1 - a_{22}a_{34}a_{43}b_1 \\ &+ a_{23}a_{34}a_{42}b_1 - a_{23}a_{32}a_{44}b_1 + a_{24}a_{32}a_{43}b_1 - a_{24}a_{33}a_{42}b_1 \end{aligned}}{\begin{aligned} &a_{11}a_{22}a_{33}a_{44} - a_{11}a_{22}a_{34}a_{43} + a_{11}a_{23}a_{34}a_{42} - a_{11}a_{23}a_{32}a_{44} \\ &+ a_{11}a_{24}a_{32}a_{43} - a_{11}a_{24}a_{33}a_{42} + a_{12}a_{21}a_{34}a_{43} - a_{12}a_{21}a_{33}a_{44} \\ &+ a_{12}a_{23}a_{31}a_{44} - a_{12}a_{23}a_{34}a_{41} + a_{12}a_{24}a_{33}a_{41} - a_{12}a_{24}a_{31}a_{43} \\ &+ a_{13}a_{21}a_{32}a_{44} - a_{13}a_{21}a_{34}a_{42} + a_{13}a_{22}a_{34}a_{41} - a_{13}a_{22}a_{31}a_{44} \\ &+ a_{13}a_{24}a_{31}a_{42} - a_{13}a_{24}a_{32}a_{41} + a_{14}a_{21}a_{33}a_{42} - a_{14}a_{21}a_{32}a_{43} \\ &+ a_{14}a_{22}a_{31}a_{43} - a_{14}a_{22}a_{33}a_{41} + a_{14}a_{23}a_{32}a_{41} - a_{14}a_{23}a_{31}a_{42} \end{aligned}}.$$

And for $n = 5$, the expansion for a single variable x_j fills an entire page: the denominator and numerator for each variable x_i have $5! = 120$ terms, each term being five coefficients multiplied together.

Gaussian Elimination versus Determinants

Gaussian elimination, then, is a considerably easier method by which to calculate the solution than are determinants for systems of more than 3 equations in 3 unknowns, except, as we will see, in special cases. It is well known that, in general, the number of terms in the expansion of the determinant is of the order $n! = 1 \times 2 \times 3 \ldots \times n$. For our purposes, it is helpful to have more precise estimates for the systems of linear equations found in Chinese treatises, that is, systems of order 2 to 9. For determinants, it is not difficult to show that the number of operations is given by the following:

$$n!(n^2 + 3n + 2) - 1.$$

For Gaussian elimination, the number of operations required for row reductions, as described above, is given by the following:

$$n^3 + \frac{3}{2}n^2 - \frac{5}{2}n.$$

The number of operations needed for back substitution is

$$1 + 3 + \cdots + (2n - 1) = n^2.$$

Adding the number of operations required for back substitution to the number required for row reductions, the total number of operations required for finding the solution by Gaussian elimination is

$$n^3 + \frac{5}{2}n^2 - \frac{5}{2}n.$$

Table 2.2 provides estimates of the number of operations needed to solve systems of linear equations using elimination and those needed using determinants.

Table 2.2: Comparison of the number of operations required to solve a system of n equations in n unknowns, when employing Gaussian elimination and determinants.

n	Gaussian elimination	Determinants
2	13	23
3	42	119
4	94	719
5	175	5039
6	291	40319
7	448	362879
8	652	3628799
9	909	39916799

As can be seen from Table 2.2, even for 3 equations in 3 unknowns, Gaussian elimination is considerably more efficient. For a system of 4 equations in 4 unknowns, calculating the solution using determinants requires 719 operations versus 94 for Gaussian elimination, and for the largest system found in Chinese mathematical treatises, 9 conditions in 9 unknowns, using determinants would require 39,916,799 operations, versus 909 for Gaussian elimination.

Matrices Representing Arrays of Numbers

In this book, to describe back substitution in the *fangcheng* procedure, I will also use augmented matrices in a more general manner to represent arrays of numbers that do not correspond to systems of linear equations. That is, as we will see in the following chapters, although elimination in the *fangcheng* procedure is similar to Gaussian elimination in modern linear algebra, the approach to back substitution is quite different. In the *fangcheng* procedure, the results are calculated on a counting board, and, as we will see, as the calculations proceed, the augmented matrix no longer corresponds to the original system of linear equations. This use of a matrix to store the results of calculations—which may seem peculiar since it is usually not permissible in modern mathematics—is common in computer science. One example is matrices representing LU decompositions, where the original matrix A is "overwritten." That is, in computations of the solution of an $n \times n$ matrix by Gaussian elimination, the positions of elements that have been eliminated are used to store multipliers, giving, upon the completion of the algorithm,

$$\begin{bmatrix} u_{11} & u_{12} & \cdots & u_{1n} \\ l_{21} & u_{22} & \cdots & u_{2n} \\ \vdots & \ddots & \ddots & \vdots \\ l_{n1} & \cdots & l_{n,n-1} & u_{nn} \end{bmatrix}.$$

This represents two matrices in a single matrix, namely,

$$L = \begin{bmatrix} 1 & 0 & \cdots & 0 \\ l_{21} & 1 & \ddots & \vdots \\ \vdots & \ddots & \ddots & 0 \\ l_{n1} & \cdots & l_{n,n-1} & 1 \end{bmatrix} \text{ and } U = \begin{bmatrix} u_{11} & u_{12} & \cdots & u_{1n} \\ 0 & u_{22} & \cdots & u_{2n} \\ \vdots & \ddots & \ddots & \vdots \\ 0 & \cdots & 0 & u_{nn} \end{bmatrix}.$$

(For a more detailed discussion, see Stewart 1998, 147–185).

When I use augmented matrices to represent arrays of numbers that do not correspond to systems of linear equations, I will note this in the text.

Early Methods for Solving Systems of Linear Equations

The recently revised Tropfke [1902–1903] 1980[7] presents the best overview of early methods for solving systems of linear equations in several unknowns. In that work, systems of linear equations, which I will illustrate below with the example

$$a_{11}x_1 + a_{12}x_2 = b_1$$
$$a_{21}x_1 + a_{22}x_2 = b_2,$$

are grouped into the following four major categories, based on the method of solution:

1. By "reduction," in which a system of several equations in several unknowns is transformed into 1 equation in 1 unknown.
2. By "special procedures."
3. By "elimination methods," which include the following:

[7] This is a complete revision by Kurt Vogel, Karin Reich, and Helmuth Gericke of the earlier edition by Johannes Tropfke. Again, I thank Jean-Claude Martzloff for suggesting this reference.

(i) The "substitution method"—first solving the first equation,

$$x_1 = \frac{b_1 - a_{12}x_2}{a_{11}},$$

and then substituting the result for x_1 into the second equation,

$$a_{21}x_1 + a_{22}x_2 = b_2,$$

and solving for x_2.

(ii) The "combination method"—solving both equations for x_1 and setting them equal,

$$\frac{b_1 - a_{12}x_2}{a_{11}} = \frac{b_2 - a_{22}x_2}{a_{21}}.$$

(iii) The "addition and subtraction method"—multiplying the first and second equations by a_{21} and a_{11}, respectively,

$$a_{21}a_{11}x_1 + a_{21}a_{12}x_2 = a_{21}b_1$$
$$a_{11}a_{21}x_1 + a_{11}a_{22}x_2 = a_{11}b_2,$$

subtracting the second equation from the first, and solving for x_2, yielding

$$x_2 = \frac{a_{11}b_2 - a_{21}b_1}{a_{11}a_{22} - a_{21}a_{12}}.$$

4. The "determinantal method"—characterized by "the fact that the solution for all unknown quantities is indicated at the same time in the form of a schematic rule (Cramer's Rule)" (Tropfke [1902–1903] 1980, 388–89).

In this book, to describe the early history of methods for solving systems of n conditions in m unknowns, I will follow the typology in Tropfke [1902–1903] 1980, and use the term "determinantal" broadly to refer to schematic methods similar to Cramer's Rule involving the multiplication of individual entries. I will use the term "determinantal calculation" when such a method does not yield solutions to the unknowns (see the section "The Earliest Extant Record of a Determinantal Calculation," pages 125–132); I will use the term "determinantal solution" when such a method does yield solutions to the unknowns (see the section "The Earliest Extant Record of a Determinantal Solution," on pages 132–149). To emphasize that in these Chinese sources there are no records of a general determinantal solution, I will sometimes use the term "determinantal-style."[8]

[8] I thank John Crossley for his suggestion on this point.

Chapter 3
The Sources: Written Records of Early Chinese Mathematics

In this chapter, I will present a preliminary overview of the textual records for Chinese mathematics. The chapter addresses three matters:

1. The first section, "Practices and Texts in Early Chinese Mathematics," distinguishes between the practices of *fangcheng* adepts and the texts written by the literati who recorded those practices.
2. The next, "The *Book of Computation*," presents a brief summary of the earliest extant record of Chinese mathematics, a funerary object buried in approximately 186 B.C.E.
3. The final section, and by far the longest, "The *Nine Chapters on the Mathematical Arts*," examines the earliest transmitted Chinese text devoted to mathematics, focusing on the differing editions, the commentary and subcommentary, and modern critical studies.

The chapter will focus primarily on the *Nine Chapters*, simply because there are no extant materials on the *fangcheng* adepts, and very few related to the *Book of Computation*.

Practices and Texts in Early Chinese Mathematics

We know virtually nothing about the *fangcheng* practitioners themselves: we do not even know their names; we have no records of any writings they produced; they were most likely illiterate, but if they were able to write, their writings would likely have been denounced by the literati as vulgar; and thus it is unlikely that anything they wrote would have been collected by the literati or preserved in their libraries. We do, however, have some prefaces, commentaries, and sometimes biographical information about the literati who compiled treatises on Chinese mathematical arts. We also have similar compilations of other arts, such as medicine and astronomy, which provide further relevant information.

We must first draw a distinction between the textual tradition and *fangcheng* practice. The literati who translated *fangcheng* practices into narrative form were hardly disinterested ethnographers. Although sometimes attributing credit to their unnamed sources, more often they were highly critical of the arcane *fangcheng* procedures, which they apparently failed to understand. For example, even Liu Hui 劉徽 (fl. 263 C.E.), who wrote the earliest extant commentary on the *Nine Chapters* and is often considered the most eminent mathematician of imperial China, apparently did not understand that the complexity of the *fangcheng* procedure in the *Nine Chapters* was necessary in order to avoid even more complicated calculations with fractions (see chapter 6), as is suggested by the derision he expresses toward *fangcheng* practitioners:

> Those who are clumsy in the essential principles vainly follow this original [*fangcheng*] procedure, some placing counting rods so numerous that they fill a carpet, seemingly so fond of complexity as to easily make mistakes. They seem to be unaware of the error [in their approach], and on the contrary, desire by the use of more [counting rods] to be highly esteemed. Therefore, of their calculations, all are ignorant of the establishment of understanding; instead, they are specialized to an extreme.
>
> 其拙于精理從按本術者，或用算而布氊，方好煩而喜誤，曾不知其非，反欲以多為貴。故其算也，莫不（同）〔闇〕于設通而專于一端。 (JZSS, *juan* 8, 19a; Chemla and Guo 2004, 650–51; Shen, Lun, and Crossley 1999, 426.)

The criticisms expressed above are fairly typical of the views of the literati who compiled texts on the mathematical arts.

The *Book of Computation*

The earliest extant Chinese mathematical treatise is the *Book of Computation* (*Suan shu shu* 算數書), unearthed in 1984 from Han Dynasty Tomb No. 247, at Zhangjia Mountain 張家山 in the Jingzhou district 荊州區 of the city of Jingzhou 荊州市, in Hubei Province. The tomb contained over 1,200 bamboo strips from several treatises.[1] It is believed that the tomb dates from about 186 B.C.E.; but the mathematics it contains is estimated to date as much as 300 years earlier. The *Book of Computation* has been reconstructed from these bamboo strips (Fig. 3.1): all of the strips constituting the *Book of Computation* had become immersed in silt; about half of the bamboo strips were no longer bound together by thread and had become dislodged from their original position; some of the strips have deteriorated, and portions are no longer legible. The strips were pieced together on the basis of their relative position and their neighboring strips, as well as the content and topic of the strips. There are no *fangcheng* problems in the *Book of Computation*, and the term *fangcheng* is not mentioned.[2]

[1] Other treatises unearthed include *Er nian lü ling* 二年律令, *Zou yan shu* 奏讞書, *Gai lü* 蓋廬, *Mai shu* 脈書, and *Yin shu* 引書.

[2] This paragraph summarizes material in the introduction to Peng 2001, 1–35 (in Chinese). For studies in English, see Dauben 2008 and Cullen 2004.

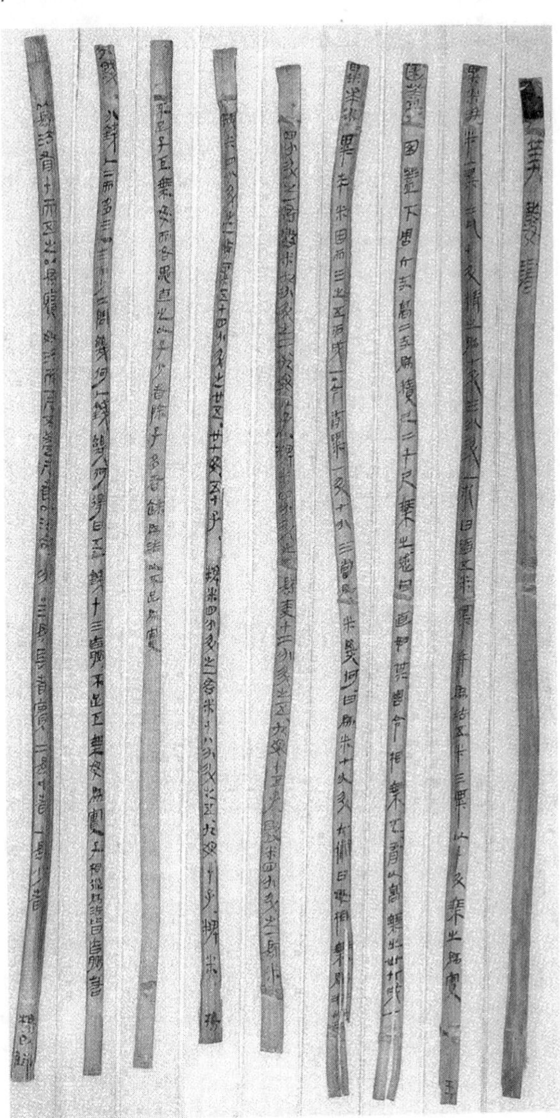

Fig. 3.1: A photograph of bamboo strips from the *Book of Computation* (*Suan shu shu* 筭數書, ca. 186 B.C.E.). The first strip pictured on the right, on which the title is written, preceded by a black square, is the backside of strip no. 6. The titles of the problems are written at the tops of the strips; continuations onto a second strip are indented from the top. For example, the problem "Dividing Coins" (discussed in the next chapter) begins on the second strip from the left (no. 133) and continues on the first strip on the left (no. 134). At the bottom of the strips the name of the compiler is occasionally noted, such as "Proofread by Yang" (*Yang yi chou* 楊已讎) on the first strip from the right. All the strips are written in the hand of a single calligrapher. The strips are approximately 30 cm long and 7 mm wide. Those shown here are in comparatively better condition than many, some of which are illegible. (Photograph reproduced from Peng 2001.)

The *Nine Chapters on the Mathematical Arts*

The *Nine Chapters on the Mathematical Arts* (*Jiuzhang suanshu* 九章算術), the earliest transmitted Chinese text devoted specifically to mathematics, is traditionally dated c. 150 B.C.E.;[3] modern scholarship places the likely date of composition during the first century C.E.[4] It now exists only in reconstructed form. Early editions of the *Nine Chapters* have not survived: in contrast to the bamboo strips on which the *Book of Calculation* was recorded, there is no surviving material record of the *Nine Chapters*, nor is any edition currently available as a reprint of an early edition. There are important differences in the texts of the differing editions, especially in the earliest surviving commentary, that by Liu Hui 劉徽 (fl. 263 C.E.). Because these differences in the text of the commentary have important consequences for the interpretation of the "method" (*shu* 術) presented to solve "excess and deficit" problems, I will briefly summarize different editions of the *Nine Chapters*.[5]

Extant Editions

The edition that scholars have sought to reconstruct is an imperially sponsored Tang Dynasty edition of the *Nine Chapters* with commentary by Liu Hui, to which Li Chunfeng 李淳風 (602-670 C.E.), together with Liang Shu 梁述 (n.d.) and Wang Zhenru 王真儒 (n.d.), added subcommentary (*zhu shi* 注釋). This edition dates from the first year of the Xianqing 顯慶 reign (656 C.E.). In fact, this one edition has been the focus of textual reconstructions and scholarly research to such an extent that it is often viewed as the definitive edition of the *Nine Chapters*.[6] All currently available editions of the *Nine Chapters* attempt to reconstruct this 656 C.E. edition. These reconstructions are based on three original sources, each with serious defects:

1. The first source for modern reconstructions of the *Nine Chapters* is the earliest surviving, but incomplete, edition of the *Nine Chapters*, the Bao 鮑 edition of 1213 C.E. During the Northern Song Dynasty, in the seventh year of the Yuanfeng 元豐 reign (1084 C.E.), the Palace Library (*Bishusheng* 秘書省)

---

[3] As noted below, in his preface, Liu Hui 劉徽 (fl. 263 C.E.) asserts that the *Nine Chapters* was compiled by Zhang Cang 張蒼 (d. 152 B.C.E.) and Geng Shouchang 耿壽昌 (fl. c. 50 B.C.E.) from materials dating from before the Qin Dynasty (221–206 B.C.E).

[4] For discussion of the dating of the *Nine Chapters*, see Chemla and Guo 2004, 43–46; Shen, Lun, and Crossley 1999, 1; Martzloff [1987] 2006, 128–31; and Cullen 1993, 16–19.

[5] Parts of this section rely on the results of Guo Shuchun 郭書春, the foremost scholar in textual research into editions of the *Nine Chapters* (see Guo 1990; Guo 1998; Chemla and Guo 2004). For a summary in English of research into the textual history of editions of the *Nine Chapters*, see Cullen 1993.

[6] Summarizing research on the *Nine Chapters*, Cullen states that "it was [Li Chunfeng's] work which fixed the version of the received text" (Cullen 1993).

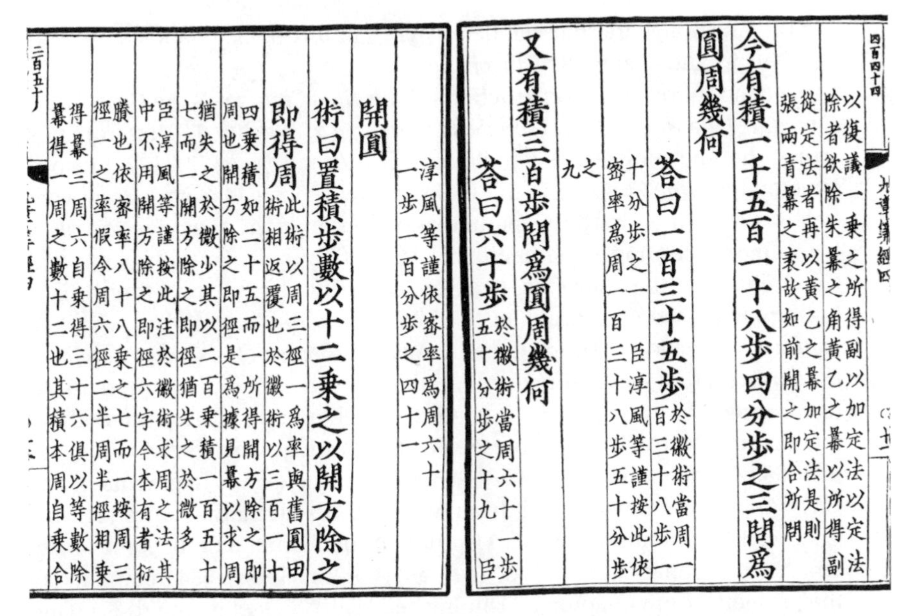

以復議一乘之所得副以加定法以定法除者欲除朱冪之角黃乙之冪以所得副
從定法者再以黃乙之冪加定法是則張兩青冪之袤故如前開之即合所問
今有積一千五百一十八步四分步之三問為
圓周幾何
荅曰一百三十五步於徽術當周一百三十八步一
十分步之一臣淳風等謹按此依密率為周一百三十八步五十分步
之九
又有積三百步問為圓周幾何
荅曰六十步於徽術當周六十一步五十分步之十九臣
淳風等謹依密率為周六十一步一百分步之四十一
開圓
術曰置積步數以十二乘之以開方除之
即得周此術以周三徑一為率與舊圓田術相返覆也於徽術以三百一十
四乘積如二十五而一所得開方除之即周也開方除之即徑是為據見冪以求周
猶失之於徽少其以二百乘積一百五十七而一開方除之即徑猶失之於徽多
臣淳風等謹按此注於徽術求周之法其中不用開方除之即徑六字今本有者衍
賸也依密率八十八乘之七而一按周三徑一之率假令周六徑二半周半徑相乘
得冪三周六自乘得三十六俱以等數除冪得一周之數十二也其積本周自乘合

Fig. 3.2: A photolithographic reproduction of the 1213 C.E. Bao 鮑 edition of the *Nine Chapters.* From right to left, a problem to be solved begins in the third column of text, in full-size characters, at the top of the page, and continues from the top of the fourth column. The answer, to the left of the problem, is also in full-size characters, but is indented from the top by a space of four characters. The answer is followed by commentary, written in half-size characters, with two columns of commentary in one column of full-size text, and indented in the same manner as the preceding text. The subcommentary of Li Chunfeng et al. is separated from the preceding commentary by a blank space, and begins "Your servant Chunfeng et al. respectfully note" 臣淳風等謹按. The eighth column states another problem, and the ninth column the answer, followed by commentary, and subcommentary by Li Chunfeng. The next page states a topic, "Factoring [the area of] a circle [to find its circumference]" (*kai yuan* 開圓), followed by the method (*shu* 術), all indented by a space of two characters from the top, followed by extensive commentary and subcommentary.

printed a collection of mathematical classics, including a reprint of this 656 C.E. edition of the *Nine Chapters* with commentaries by Liu Hui and Li Chunfeng. After the fall of the Northern Song, in the sixth year of the Jiading 嘉定 reign (1213 C.E.) of the Southern Song Dynasty, Bao Huanzhi 鮑澣之 (n.d.) reproduced this collection, including this edition of the *Nine Chapters.* By the end of the Ming Dynasty (1368-1644), only the first five chapters (*juan* 卷) of this edition of the *Nine Chapters* remained (Guo 1998, 39). This edition is currently preserved at the Shanghai Library (*Shanghai tushuguan* 上海圖書館), and was reprinted in 1981 together with five other mathematical classics (SKSJ).[7] (See Figure 3.2.) Although this is the earliest edition, and considered the most reliable, it is missing the last four chapters, including the two re-

---

[7] See the bibliography, "Collections—Chinese," for further information.

lated to linear algebra that will be analyzed in this book, chapter 7 "Excess and Deficit" and chapter 8 "*Fangcheng.*"

2. The second source for reconstructions of the *Nine Chapters* is the *Great Encyclopedia of the Yongle Reign* (*Yongle da dian* 永樂大典, 1403-07). It is from this source that the Qing Dynasty philologist Dai Zhen 戴震 (1724-1777) reconstructed the *Nine Chapters* in the 1770s. However, instead of simply copying treatises in their entirety, the mathematics section of the *Great Encyclopedia* is arranged according to mathematical content, by topic. Thus problems and commentary of the *Nine Chapters* were copied under different topic headings, interspersed with problems and commentary from other texts.[8] (See Figure 3.3 on page 34.) Unfortunately, only fragments of the *Great Encyclopedia* survive. From the section on mathematics (*suan* 筭), only chapters 16343 (the fourteenth chapter on mathematics) and chapter 16344 (the fifteenth on mathematics) are still extant, preserved at Cambridge University in England. In 1986, Zhonghua shuju 中華書局 reprinted the extant fragments of the *Great Encyclopedia*, including these two chapters on mathematics (YLDD). Chapter 16343 contains the last 11 problems of chapter 3 of the *Nine Chapters*, "Proportional Distribution" (*cui fen* 衰分); chapter 16344 contains all the problems from chapter 4, "Diminished Width" (*shao guang* 少廣) (Guo 1990, 97). Guo Shuchun argues, based on his textual research, that earlier claims that the source of the *Great Encyclopedia* was the Bao edition of the *Nine Chapters* are incorrect: the *Great Encyclopedia* was based on a different edition of the *Nine Chapters* (Guo 1990, 97).
   Although the *Great Encyclopedia* incorporated the *Nine Chapters* in its entirety, albeit as problems that were rearranged by categories and interspersed with the text from other treatises, the only remaining record of these passages is the work of Dai Zhen (1724–1777). Dai Zhen reconstructed the *Nine Chapters* in several stages, the first of which was copying the problems from the *Great Encyclopedia.*[9] Then, after copying the text, Dai made numerous emendations, and wrote copious notes explaining these changes. Unfortunately, the original and unedited copy that Dai Zhen made of the *Nine Chapters* is no longer extant; the only surviving editions of Dai Zhen's work on the *Nine*

[8] The other mathematical treatises from which the *Great Encyclopedia* contains passages, in addition to the *Nine Chapters*, include the following: *Mathematical Arts, with Details and Clarifications* (*Xiang ming suanfa* 詳明算法), attributed to An Zhizhai 安止齋 and He Pingzi 何平子 of the Yuan Dynasty; *Mathematical Classic of Xiahou Yang* (*Xiahou Yang suan jing* 夏候陽筭經); *Mathematical Classic of the Five Ministries* (*Wu cao suan jing* 五曹筭經); *Yang Hui's Collection of Strange Mathematical Arts* (*Yang Hui zhai qi suanfa* 楊輝摘奇筭法); Yan Gong's *Mathematical Arts: Penetrating to the Origin* (*Yan Gong tong yuan suanfa* 嚴恭通原筭法); Qin Jiushao's *Mathematics in Nine Chapters* (*Qin Jiushao shu shu jiuzhang* 秦九韶數書九章); *Detailed Notes for Passing Civil Examinations* (*Tou lian xi cao* 透簾細草); *Mathematical Arts of Ding Ju* (*Ding Ju suanfa* 丁巨筭法); *Sunzi's Classic of Mathematics* (*Sunzi suan jing* 孫子筭經); *Yang Hui's Detailed Explanations* (*Yang Hui xiang jie* 楊輝詳解); and *Yang Hui, Arranged by Category* (*Yang Hui zuan lei* 楊輝纂類).

[9] We cannot rule out the possibility that Dai Zhen had assistants copy or even edit the text, while later claiming credit himself.

Chapters are of the text he modified. Guo Shuchun has criticized Dai Zhen for numerous mistakes in his corrections, and worse, for making changes to the text that were not noted as Dai Zhen's alterations. Guo has attempted to reconstruct Dai Zhen's original unedited copy, apparently by working backward from Dai's editorial notes, and by comparing Dai's edition with other editions (Guo 1990, 41-45). Dai Zhen's edition has been the basis for most currently available editions.

3. The third important source of the *Nine Chapters* is Yang Hui's *Nine Chapters on the Mathematical Arts, with Detailed Explanations* (*Xiang jie Jiuzhang suanfa* 詳解九章算法), dating from the later years of the Southern Song Dynasty (1126-1279), for which Guo Shuchun offers the following historical analysis. In the Jiaqing 嘉慶 reign (1796-1820) of the Qing Dynasty, Mao Yuesheng's 毛嶽生 (1791-1841) lineage had preserved an edition with the notation "Hand-copy from the 'Studio of the Stone Inkstone'" (*Shiyan zhai chao ben* 石研齋抄本) on each page, presumably an edition from Qin Hong 秦鑀 (1722-?), who called himself "Master (*zhu ren* 主人) of the Studio of the Stone Inkstone." Guo notes that this edition includes the prefaces of Liu Hui, Rong Qi 榮棨, and later prefaces by Bao Huanzhi and Yang Hui, together with five chapters of the *Nine Chapters*: chapter 7, "Excess and Deficit"; chapter 8, "*Fangcheng*"; chapter 9, "Right-Angled Triangles" (*gou gu* 句股); chapter 5, "Construction Consultations" (*shang gong* 商功); and chapter 6, "Equitable Levies" (*jun shu* 均輸). These chapters are not arranged in their original order, and even these chapters are incomplete: "Construction Consultations" is missing the first seven problems (problems 20-27) and the "*Chu tong* Method" (*chu tong shu* 芻童術, a method for calculating the volume of a solid figure—an inverted, truncated pyramid); "Equitable Levies" is missing a problem. Guo notes that "the hand-copied version is extremely sloppy (*ji wei cu lou* 極為粗漏), with numerous errors and omissions (*wu wen duo zi shen duo* 誤文奪字甚多)," but concludes that, "despite this, because what is in the 'Studio of the Stone Inkstone' edition is exactly what is missing from the Bao edition (with the exception of the 'Construction Consultations' chapter), and adding to that the loss of the *Great Encyclopedia*, and the many errors in the copy made by Dai Zhen, the value of 'Studio of the Stone Inkstone' edition for comparison is quite significant."[10]

Liu Hui's Preface

We can now examine Liu Hui's preface as a bid for imperial patronage. The first part of the preface is as follows:

> Original preface to Liu Hui's Commentary on the *Nine Chapters on the Mathematical Arts*

[10] This paragraph on the "Studio of the Stone Inkstone" edition summarizes Guo 1990, 98.

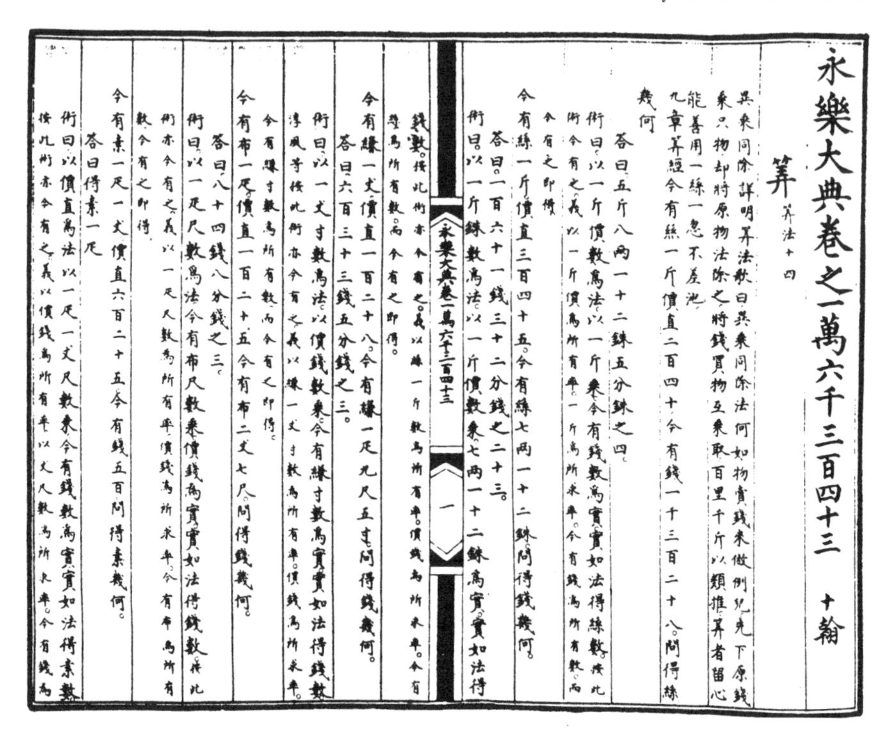

永樂大典卷之一萬六千三百四十三　十翰
算　算法十四
異乘同除詳明算法歌曰異乘同除法何如物貨錢來做例先下原錢乘只物却將原物法除之將錢買物互乘取百里千斤以類推算者留心
能善用一絲一忽不差池
九章算經今有絲一斤價直二百四十今有錢一千三百二十八問得絲幾何
答曰五斤八兩一十二銖五分銖之四
術曰以一斤價數為法以一斤乘今有錢數為實實如法得絲數按此術今有之義以一斤價為所有率一斤為所求率今有錢為所有數而今有之即得
今有絲一斤價直三百四十五今有絲七兩一十二銖問得錢幾何
答曰一百六十一錢三十二分錢之二十三
術曰以一斤銖數為法以一斤價數乘七兩一十二銖為實實如法得錢數按此術亦今有之義以絲一斤銖數為所有率價錢為所求率今有絲為所有數而今有之即得
永樂大典卷一萬六千三百四十三　一
今有縑一丈價直一百二十八今有縑一疋九尺五寸問得錢幾何
答曰六百三十三錢五分錢之三
術曰以一丈寸數為法以價錢數乘今有縑寸數為實實如法得錢數淳風等按此術亦今有之義以縑一丈寸數為所有率價錢為所求率今有縑寸數為所有數而今有之即得
今有布一疋價直一百二十五今有布二丈七尺問得錢幾何
答曰八十四錢八分錢之三
術曰以一疋尺數為法今有布尺數乘價錢為實實如法得錢數按此術亦今有之義以一疋尺數為所有率價錢為所求率今有布為所有數今有之即得
今有素一疋一丈價直六百二十五今有錢五百問得素幾何
答曰得素一疋
術曰以價直為法以一疋一丈尺數乘今有錢數為實實如法得素數按此術亦今有之義以價錢為所有率以丈尺數為所求率今有錢為

Fig. 3.3: A photolithographic reproduction of the first page of chapter 16343, the fourteenth chapter on mathematics, one of the few surviving fragments of the *Great Encyclopedia of the Yongle Reign* (*Yongle da dian* 永樂大典, 1403-07). Mathematics in the *Great Encyclopedia* is organized by content, so problems from different texts are copied together. The chapter begins with a verse on multiplication and division taken from the *Mathematical Arts, with Details and Clarifications* (*Xiang ming suanfa* 詳明算法), dating from the Yuan Dynasty (1279-1368). Following the verse, the first problem is preceded by the notation "*Classic of Nine Chapters on Mathematics*." The answer to the problem begins on a new column, and is indented two characters from the top. The "method" (*shu* 術) begins a new column, and is indented by a single character. Commentary and subcommentary are in smaller characters, for example in the sixth (double) column from the left. Further notes on problems, for example from *Yang Hui's [Nine Chapters on the Mathematical Arts with] Detailed Explanations* (*Yang Hui xiang jie* 楊輝詳解, 1261), are sometimes copied following the problems, beginning in a new column, following the title of the source, and not indented; occasionally, these further notes are copied immediately following the commentary, without a new column.

劉徽九章算術注原序

In the past, when Pao Xi first drew the eight trigrams, in order to fully comprehend the spirit brightness of virtue, in order to categorize the state of the myriad beings, [Pao Xi] created the Art of the Nine Nines, in order to complete the transformations of the six lines [hexagrams].

昔在庖犧氏始畫八卦，以通神明之德，以類萬物之情，作九九之數，以合六爻之變。

Then the Yellow Emperor miraculously transformed it [the Art of the Nine Nines], developing and extending it, to establish calendrical regularities, regulate the pitch pipes, and investigate the origins of the Dao. Following this, the subtle essence of the *qi* of the two *yi* and the four *xiang* was obtainable and became efficacious.

暨于黃帝神而化之，引而伸之，於是建歷紀，協律呂，用稽道原，然後兩儀四象精微之氣可得而效焉。

It is recorded that Li Shou created numbers, but the details remain unknown. Comment: only when the Duke of Zhou established the rituals did the mathematical arts come into being; arts related to the nine numbers are just the *Nine Chapters.*

記稱隸首作數其詳未之聞也。按：周公制禮而有九數，九數之流，則九章是矣。

Previously, the brutal [First Emperor of the] Qin Dynasty burned the books, and classical scholarship was scattered and destroyed. After this, in the [Former] Han Dynasty, the Marquis of Beiping Zhang Cang [c. 250–152 B.C.E.] and the Chamberlain for the National Treasury and Palace Aide to the Censor-in-Chief Geng Shouchang [fl. 1st century B.C.E.] both achieved fame throughout their generation for their excellence in calculation. [Zhang] Cang, and others, using the surviving remnants of the old texts, each edited and supplemented [the Nine Chapters]. Comparing its table of contents, it differs in places from the ancient, and what is discussed contains considerable modern language.

往者暴秦焚書，經術散壞。自時厥後，漢北平侯張蒼、大司農中丞耿壽昌，皆以善算命世。蒼等因舊文之遺殘，各稱刪補。故校其目則與古或異，而所論者多近語也。

Liu Hui's preface promotes the mathematical arts as the semidivine invention of sage kings, fundamental to understanding cosmogony, and essential to ordering the empire. There is little discussion of any technical aspects of Chinese mathematics. There is no mention of "excess and deficit" or *fangcheng* problems. Instead, the first paragraph provides for the Art of the Nine Nines illustrious origins in their creation by divine sage kings of antiquity.[11] As is conventional in the prefaces of compilations submitted to the imperial court for patronage, important inventions that presumably date back to Chinese prehistory are attributed to divine sage rulers. (It should be noted that there is no evidence beyond the imaginings of patronage-seeking literati for these attributions.) These attributions function to simultaneously legitimate both the importance of the discovery and the sageness of the purported discoverer, thereby also legitimating the importance of the

[11] The subject of the first paragraph is in the Art of the Nine Nines, that is, the precursor to the mathematical arts presented by Liu (this is somewhat obscured by the grammar and sentence structure of the classical Chinese). The Nine Nines, in the literal sense as found in later mathematical treatises, is simply the multiplication tables from nine to one (the order in the Chinese texts is reversed from our usual order). The Art of the Nine Nines refers more broadly to all the mathematical arts.

compilation for the ordering of the empire. For example, in this case, the eight trigrams, used for divination by rulers and the elite, are attributed to Pao Xi; the purported efficacy of the eight trigrams in divination is attributed to them by asserting those functions as the reasons for the sage ruler's discovery. Through the eight trigrams, then, the sage fully comprehends the spirit brightness of virtue, that is, in Confucian terms, the original brightness and purity of man's Confucian nature; through the eight trigrams, the ordered state of the myriad beings is brought into perceptible order by elucidating its categories—the order of the myriad beings exemplifies, reinforces, and is codified in the order of the Han Dynasty.

The Imperially Sponsored Tang Dynasty Edition of **656 C.E.**

From a mathematical point of view, there is no reason to assume a priori that the imperially sponsored Tang edition of 656 C.E., which all modern editions have attempted to reconstruct, is necessarily superior. Instead, it achieved preeminence among literati-officials by imperial sponsorship, largely irrespective of its mathematics. This edition was compiled under the leadership of Li Chunfeng, and to better understand the historical context, we need to examine the biographical information recorded in the historical records that concern Li.

Li Chunfeng in Historical Context

The biography of Li Chunfeng suggests that he was primarily a diviner for the imperial court rather than a mathematician. Further examination of this biography, included in the "serial biographies" (*lie zhuan* 列傳) of the *Old Official History of the Tang* (*Jiu Tang shu* 舊唐書), will help illuminate the circumstances of the compilation of this text.[12] Li Chunfeng, this brief biography states, was the son of an official who resigned from office to become a Daoist adept (*dao shi* 道士). Li Chunfeng was appointed to a series of astrological posts: after prevailing in a dispute over astronomical methods, he was appointed to the post of Attendant Gentleman (*shi lang* 仕郎) in the Astrological Service (*taishi ju* 太史局), and later promoted to Gentleman for Rendering Service (*chengwu lang* 承務郎), Erudite of the Court Imperial Sacrifices (*taichang boshi* 太常博士), and Aide to the Grand Astrologer (*tai shi cheng* 太史丞).

Much of Li Chunfeng's biography focuses on his purported interpretation of a portent cryptically prophesying that Wu Zhao 武曌 (624–705 C.E., better known as Empress Wu Zetian 武則天), would one day wrest power from the Tang Dynasty and rule. (Her reign lasted from 684 to 705.) The biography of Li Chunfeng in the *Old Official History of the Tang* records the following account:

---

[12] A similar but abbreviated version is included in the *New Official History of the Tang* (*Xin Tang shu* 新唐書).

During the period of the Taizong [literally, "Great Ancestor"] Emperor, there was an occult record which stated: "After three generations of the Tang Dynasty, a woman master, a warrior-ruler, will replace [the Tang] and rule all under heaven."[13] [The emperor] once secretly summoned Li Chunfeng to ask about this, and Li stated: "Your servant has calculated and according to the portent, the omen is already complete. This person has already been born, and is within the palace of your highness, and in less than 30 years will have all under heaven, and will kill the sons and grandsons of the Tang, exterminating them completely." The king stated: "If I kill all suspected of resembling her, how would that be?" [Li] Chunfeng answered: "What heaven has ordained necessarily cannot be exorcised by sacrifices or circumvented. If this ruler-to-be still is not killed, there is much to fear that the injustices will include the innocent. Furthermore, according to the portent, now already complete, again she is inside the palace, already among those adored by your highness. Thirty years from now, [she] should be feeble and old, and being old, will be humane and benevolent; although in the end the surname of the ruling family will change, perhaps the loss will not be great. If she is killed, [she] must be reborn, young, powerful, severe, and venomous, with vengeance for having been killed. If this happens, killing the sons and grandsons of your highness, then there will certainly be nothing left!" The Emperor Taizong appreciated his statements and stopped at that.

太宗之世有祕記云：「唐三世之後，則女主武王代有天下。」太宗嘗密召淳風以訪其事，淳風曰：「臣據象推算，其兆已成。然其人已生，在陛下宮內，從今不踰三十年，當有天下，誅殺唐氏子孫殲盡。」帝曰：「疑似者盡殺之，如何？」淳風曰：「天之所命，必無禳避之理。王者不死，多恐枉及無辜。且據上象，今已成，復在宮內，已是陛下眷屬。更三十年，又當衰老，老則仁慈，雖受終易姓，其於陛下子孫，或不甚損。今若殺之，即當復生，少壯嚴毒，殺之立讎。若如此，即殺戮陛下子孫，必無遺類。」太宗善其言而止。

The only writings of Li Chunfeng's that have been preserved in commonly available collectanea, beyond the annotations of mathematical treatises to which he contributed, are on divination.[14] Further fragments of his writings, again on divination, are preserved in various texts, such as the work *Han Learning on "Changes"* (*Yi Han xue* 易漢學), which includes passages of Li's commentary on the *Classic of Changes* (*Yijing* 易經).

---

[13] "Warrior" (*wu* 武) here is the same character as the family name of the Empress Wu, and thus presumably a prophetic pun.

[14] The *General Catalogue of Chinese Collectanea* (*Zhongguo congshu zonglu* 中國叢書綜錄), 578, lists four works by Li, all on various forms of divination: *Zhou yi xuan yi* 周易玄義; *Yi si zhan* 乙巳占; *Yu li tong zheng jing* 玉曆通政經; and *Zhi gui lun* 質龜論. It also lists two Daoist works that he annotated: *Jin suo liu zhu yin* 金鎖流珠引; and *Tai shang chi wen dong sheng san lu*太上赤文洞神三籙.

Compilation of the Ten Mathematical Classics

The biography of Li Chunfeng in the *Old History of the Tang* offers the following account of the compilation of the *Ten Mathematical Classics*:

> In the first year of the Xianqing reign (656 C.E.) [of the Tang Dynasty], [Li Chunfeng] was rewarded for contributions as State Historiographer by being appointed District Baron of Changle. Previously, Wang Sibian, the Expectant Supervisor of the Directorate of Astrology, submitted a memorial stating that the *Ten Mathematical Classics* were in disrepair. Li Chunfeng, with Liang Shu, Erudite of the Mathematics School of the Directorate of Education, and Wang Zhenru, Instructor at the National University, accepted an imperial edict to add [sub]commentary to the *Five Ministries, Sunzi,* [and the other] *Ten Classics of Mathematics.* When the books were complete, the Gaozong Emperor ordered the National Universities to use them.
>
> 顯慶元年，復以修國史功封昌樂縣男。先是，太史監候王思辯表稱五曹、孫子十部算經理多踳駮。淳風復與國子監算學博士梁述、太學助教王真儒等受詔注五曹、孫子十部算經。書成，高宗令國學行用。

Given the edict, and Li Chunfeng's background, it would be most surprising if this imperially sponsored text, to be used at the National Universities, did not omit methods or content considered too difficult or inappropriate for its purpose. Indeed, the above account does not even mention the *Nine Chapters*, which is the most sophisticated and difficult text mathematically. Instead, it mentions the *Five Ministries* and *Sunzi*, two texts with more elementary mathematics, which were considered more relevant to scholar-officials.

Subcommentary by Li Chunfeng et al. to the Nine Chapters

The subcommentary, written by Li Chunfeng et al. (which is always preceded by the notation, "[Your] servants [Li] Chunfeng et al. [respectfully] note" 臣淳風等〔謹〕按), most frequently addresses the earlier and more elementary chapters, chapters that in the formulation of their problems at least purport to address issues related to governance. As Table 3.1 shows, Li offers little or no subcommentary for the most difficult final chapters.

Table 3.1: Number of subcommentaries for each chapter in the *Nine Chapters*. Subcommentaries are preceded by "Note" or "Your servants Chunfeng et al. [respectfully] note." There are no subcommentaries explicitly attributable to Li Chunfeng for chapter 8, "*Fangcheng*."

Chapter:	1	2	3	4	5	6	7	8	9
"Note" (*an* 按):	34	29	15	20	34	37	12	5	7
"Your servants Chunfeng et al. [respectfully] note" 臣淳風等〔謹〕按:	24	26	5	14	16	9	0	0	3

There is not a single subcommentary attributed to Li Chunfeng et al. for the two chapters related to linear algebra, chapter 7 “Excess and Deficit” and chapter 8 “Fangcheng”; there are only three short and elementary passages of subcommentary for chapter 9 “Right Triangles (*Gou gu* 句股).”

This orientation to scholar-officials not interested in the technicalities of mathematics is perhaps best demonstrated by the subcommentary by Li Chunfeng et al. on perhaps the most difficult problem in the first chapter, estimates of π, in “Field Measurement” (*fang tian* 方田), where a series of rules is given for finding the area of a circle. The most difficult computations, and a place where Liu Hui improves considerably on the original text of the *Nine Chapters*, is on the calculation of the area of the circle, given the radius. Liu Hui’s commentary calculates the area of an inscribed hexagon as a lower limit less than the area of a circle. Liu then calculates the area of inscribed regular polygons with 12, 24, 48, and 96 sides, noting that these give increasingly accurate approximations of the area. In addition, Liu adds to each of the sides of the inscribed polygons a rectangle so as to cover the entire area of the circle, which is then greater than the area of the circle. He arrives at the following estimate of the area of a circle with a radius of 10 units: $314\frac{64}{625} \leq$ the area of the circle $\leq 314\frac{169}{625}$. (For comparison in decimal notation, the actual value is 314.1593, versus Liu’s lower estimate of $314\frac{64}{625} = 314.1024$, and his upper estimate of $314\frac{169}{625} = 314.2704$.) The subcommentary of Li Chunfeng et al. offers no mathematical analysis, nor any criticism of Liu Hui’s complex calculation. Instead, the subcommentary concludes,

> [Taking the] diameter as 1 and circumference as 3, in principle is not the finest in precision, the reason being that this method follows the simplified essentials, taking the larger guidelines, stating it in summary. [But] Liu Hui especially took this as imprecise, and then found its ratio in an alternative manner. But the number [π] to multiply the circumference and diameter by [to transform one into the other] is hard to make fit. Although [Liu] Hui put forward this one method, in the end [he] was not able to reach the end of its minutiae. Zu Chongzhi, because of the lack of precision, then again calculated the number. Now [we] have collected all the various results, examined their correctness, and that of [Zu] Chongzhi is the most accurate. Therefore we have displayed this below the method of [Liu] Hui, hoping scholars will understand the reason we made the arrangement.
>
> 徑一周三，理非精密。蓋術從簡要，舉大綱略而言之。劉徽特以為疏，遂改張其率。但周徑相乘數難契合。徽雖出斯一法，終不能究其纖毫也。祖沖之以其不精，就中更推其數。今者修撰攈摭諸家，攷其是非，沖之為密。故顯之于徽術之下，冀學者知所裁焉。

But throughout the subcommentary, Li Chunfeng et al. use as the “precise rate” $\frac{22}{7} \approx 3.142857$, a worse estimate than that calculated by Liu, instead of Liu’s estimate. Li Ji’s *Nine Chapters on the Mathematical Arts, Pronunciations, and Mean-*

ings (*Jiuzhang suanshu yin yi* 九章算術音義) offers the following description of the "precise ratio":[15]

> Precise ratio (*mi lü*): *mi* is pronounced with the initial of *mei* and the final of *bi*. Precise ratio: take 7 times the circumference, [divide by] 22, and convert an improper fraction into a mixed fraction, giving the radius. Take 22 times the radius, [divide] by 7, and convert an improper fraction into a mixed number, giving the circumference. This ratio originates from Zu Chongzhi, a [minor official].[16] [Zu] Chongzhi took 10000000 as one *zhang* [a unit of measure approximately equal to one meter], and found the upper limit of the circumference of the circle to be 31415927; the lower limit is 31415926.[17] The precise ratio is 355/113 [≈ 3.14159292]; the approximate ratio is 22/7 [≈ 3.1428571428]. This then is the ratio at its highest precision.
>
> 密率：美畢切。密率：以七乘周，二十二而一，即徑。以二十二乘徑，七而一，即周。此率本於宋南徐州從事史祖沖之。沖之以圓徑一億為一丈，圓周盈數三丈一尺四寸一分五釐九毫二秒七忽。朒數三丈一尺四寸一分五釐九毫二秒六忽。正數在盈朒二限之間。密率圓徑一百一十三圓周三百五十五。約率圓徑七周二十二。此乃率之最密也。

To summarize, the subcommentary by Li Chunfeng et al. offers little mathematical analysis, mistakes Zu Chongzhi's rough rate for his precise rate, and throughout chooses this rough rate over Liu Hui's more accurate calculations. Therefore, Guo Shuchun is far too generous when he asserts that Li Chunfeng's *Nine Chapters on the Mathematical Arts, with Commentary and Subcommentary* is "the collective creation of first-rate mathematicians of the period," but accurate in stating that "its level is far, far below (*yuan yuan di yu* 远远低于) that of the commentary of Liu Hui from about 400 years before" (Guo 1998, 39).

The prestige conferred on the 656 C.E. edition derives not from its mathematical content, but rather from its state sponsorship, its imperial approval, and the resulting high regard of later scholar-officials. The importance of attempts at the textual reconstruction of this edition notwithstanding, for the purposes of understanding Chinese mathematics this edition is simply one among many mathematical texts, and need not be assigned any special priority. In particular, there is no reason to suppose that the Tang Dynasty-sponsored editing and subcommentary by Li Chunfeng is necessarily mathematically superior to the Qing Dynasty reconstruction and commentary by Dai Zhen. And as the extant bamboo strips of the *Book of Computation* remind us, we should expect all editions, including the earliest, to contain errors and omissions that inaccurately convey the mathematical content. Instead of attempting to return to an imagined origin through

---

[15] The *History of the Sui Dynasty* (*Sui shu* 隋書), in the "Record of Music and Calendrics" (*Yue li zhi* 樂歷志), records that Zu Chongzhi had calculated the value to be between 31415927 [which is mistakenly transcribed as 21415927] and 31415926, and gave as the "precise rate" $\frac{355}{113}$ (≈ 3.1415929), and the "rough rate" $\frac{22}{7}$.

[16] The exact nature of his official post is not clear; it appears he was a scribe.

[17] More literally, both of these numbers are written with units of measure for each decimal place: the first number is written 3 *zhang* 1 *chi* 4 *cun* 1 *fen* 5 *li* 9 *hao* 2 *miao* 7 *hu*.

the commentaries, it will prove more profitable to view the main text, the commentaries, and the subcommentaries as expressions of possibly different mathematical approaches.

Other Editions of the Nine Chapters

As noted above, modern reconstructions of the *Nine Chapters* have focused on recovering one among the many editions of this treatise—the imperially sponsored Tang Dynasty edition with subcommentary by Li Chunfeng et al. of 656 C.E. As shown in Appendix A, "Mathematical Treatises Listed in Chinese Bibliographies" (see pages 230–254), records up to the early Qing Dynasty record a considerable number of other editions, including the following:

- Xu Yue 徐岳 (fl. 220 C.E.), *Nine Chapters on the Mathematical Arts* (*Jiuzhang suanshu, jiu juan* 九章算術九卷), no longer extant.
- Zhen Luan 甄鸞 (fl. 560 C.E.), *Nine Chapters, the Mathematical Classic* (*Jiuzhang suan jing, jiu juan* 九章算經九卷), no longer extant.
- Song Quanzhi 宋泉之 *Nine Classics, Methods and Subcommentaries* (*Jiu jing shu shu, jiu juan*九經術疏九卷, from the Tang Dynasty 618-906 C.E.), no longer extant.
- Jia Xian 賈憲 (fl. 1050 C.E.), *Jia Xian's Nine Chapters* (*Jia Xian jiuzhang* 賈憲九章), no longer extant (though excerpts are preserved in YHJZ); the *Yellow Emperor's Nine Chapters* (*Huang di jiuzhang* 黃帝九章), no longer extant, may be the same edition.
- Xu Rong 許榮, *Nine Chapters on the Mathematical Arts, with Detailed Commentary* (*Jiuzhang suanshu zhu jiu juan* 九章算法詳註九卷, ca. 1478), no longer extant.
- Yu Jin 余進 (n.d.), *Nine Chapters of Detailed and Comprehensive Mathematical Arts* (*Jiuzhang xiang tong suanfa*九章詳通筭法), no longer extant.
- Liu Shilong 劉仕隆, *Nine Chapters, Mathematical Arts of Penetrating Clarity* (*Jiuzhang tong ming suanfa* 九章通明筭法), ca. 1424), no longer extant.
- Wu Jing 吳敬, *Complete Compendium of Mathematical Arts of the Nine Chapters, with Detailed Commentary, Arranged by Category* (*Jiuzhang xiang zhu bilei suanfa da quan* 九章詳註比類算法大全, ca. 1450), extant.

Editions Consulted

Among the sources reviewed, probably the most important for the reconstructions of the *Nine Chapters* is the set of notes by the eminent Qing Dynasty philologist Dai Zhen,[18] completed in 1774, when he recovered the original text of the

[18] For a biography of Dai, see Hummel 1943.

Nine Chapters and Liu's commentary from extracts preserved in the *Great Encyclopedia of the Yongle Reign* (*Yongle da dian* 永樂大典, 1403–07).[19] Dai's reconstruction of the *Nine Chapters* and Liu's commentary, together with his notes, were published in a series of three successive editions: the 1774 *Hall of Martial Eminence Collectanea, Movable Type Edition* (*Wuying dian juzhen ban congshu* 武英殿聚珍版叢書);[20] the 1777 Kong Jihan 孔繼涵 edition;[21] and the 1784 edition included in the widely available *Complete Collection of the Four Treasuries* (*Siku quan shu* 四庫全書, hereafter SKQS). Each of these editions contains errors, defects, and misprints.

Modern editions have often been based on Dai Zhen's *Complete Collection of the Four Treasuries* (*Siku quan shu* 四庫全書) edition, especially for the chapters "Excess and Deficit" and "*Fangcheng*." There have been many attempts to reconstruct the *Nine Chapters*. Two of the most recent, and most important, are Guo Shuchun's 1990 *Nine Chapters on the Mathematical Arts, Critical Edition* (*Jiuzhang suanshu huijiaoben* 《九章算術》匯校本) and 1998 *Nine Chapters on the Mathematical Arts* (*Jiuzhang suanshu* 九章算术).[22] These two editions differ significantly, however, in their reconstruction of the commentaries on the "Excess and Deficit" chapter, which concerns us in this book: Guo 1990 is closer to the "Studio of the Stone Inkstone" edition of Yang Hui's *Nine Chapters on the Mathematical Arts, with Detailed Explanations*; Guo 1998 is closer to the Dai Zhen edition. In addition, as noted above, Guo and Karine Chemla have published a Chinese and French critical edition, Chemla and Guo 2004.

I have chosen to base my translations on the earliest extant source, the *Hall of Martial Eminence Collectanea* edition. Although I have consulted all of Guo's critical editions, and although I am indebted to the considerable research that he and

---

[19] For more on Dai's reconstruction, see Chemla and Guo 2004, 74–79.

[20] Reprinted in ZKJD (the sources corresponding to these abbreviations can be found in the Bibliography).

[21] Reprinted by Shanghai guji chubanshe 上海古籍出版社 in 1990.

[22] Guo Shuchun summarizes his views on editions of the *Nine Chapters* as follows: "Seeing that the chaos (*hunluan* 混乱) created by Dai Zhen et al. over editions of the *Nine Chapters on the Mathematical Arts* had not received fundamental correction in the space of over 210 years, Guo Shuchun thought that any edition after Dai Zhen was not suitable to serve as the basis for a critical edition. For the first five chapters (*juan*), [Guo] took the Bao Huanzhi 鮑澣之 engraved edition of the Southern Song Dynasty. For the later four chapters and the introduction by Liu Hui, [Guo] compared the Quzhen 聚珍 edition and the *Complete Collection of the Four Treasuries* (*Siku quan shu* 四庫全書) edition to recover Dai Zhen's copy [made from the *Great Encyclopedia* without any editorial changes], and according to the textual criticism notes of Li Huang 李潢, recovered the original text of the *Great Encyclopedia* as the basic text, and again using the *Great Encyclopedia* edition and the Yang Hui edition for comparision, incorporating the numerous correct emendations of Dai Zhen, Li Huang, Qian Baocong et al., correcting numerous erroneous emendations by Dai Zhen et al., removing Dai Zhen's [spurious and unnoted] additions and changes, recovering over 400 instances of erroneous corrections of originally correct text by Dai Zhen et al., recorrecting many places where previous editors made inappropriate emendations, and made emendations in places that previous editors had overlooked, and at the same time, preserving editorial material from over 20 different editions, to become [Guo 1990]." Guo then offers further criticisms of the edition by Li Jimin. Guo 1998, 43–45).

Karine Chemla have undertaken, I have chosen to use the earliest extant source, the *Hall of Martial Eminence Collectanea* edition reprinted in ZKJD, rather than the modern critical editions, for the following reasons: (1) Dai Zhen's notes will be a focus of my analysis; (2) continued uncertainty about some passages has necessitated Guo's changing some of the emendations of his earlier critical editions; (3) there are on occasion places where I do not agree with Guo's emendations; and (4) it seems to me more convenient to adopt (or reject) Guo's emendations to the earliest version of the text, rather than to offer emendations to Guo's own emendations. In addition, where appropriate, I have consulted other editions of the *Nine Chapters*, including YHJZ; CSJC; SJSS; SKSJ; YLDD; and JGG.

Chapter 4
Excess and Deficit

"Excess and deficit" problems, which seem to have been a precursor to *fangcheng*, are equivalent to systems of 2 conditions in 2 unknowns. They are found in the earliest surviving record of Chinese mathematics, the *Book of Computation* (c. 186 B.C.E.), and thus seem to predate *fangcheng* problems, which are more general problems of *n* conditions in *n* unknowns, and first appear in the *Nine Chapters* (c. 1st century C.E.). Many *fangcheng* problems are formulated using the terms "excess" and "deficit," which further suggests a linkage. This chapter focuses on the "excess and deficit" procedure to explore two key features later extended to *fangcheng* problems: (1) "excess and deficit" problems were displayed in two dimensions on a counting board; and (2) they were solved using "cross multiplication." The chapter is divided into two sections:

1. The first, "Excess and Deficit Problems in the *Book of Computation*," examines several problems from that work to show that these problems were arranged on a counting board and were solved using "cross multiplication."
2. The second, " 'Excess and Deficit,' Chapter 7 of the *Nine Chapters*," examines exemplars from the *Nine Chapters*.

Excess and Deficit Problems in the *Book of Computation*

The earliest examples of "excess and deficit" problems appear in the *Book of Computation*. These problems are equivalent to a system of 2 conditions in 2 unknowns. There are three sets of "excess and deficit" problems in the *Book of Computation*: "Dividing Coins" (*fen qian* 分錢), two problems under the title "Coins for Millet" (*mi chu qian* 米出錢), and "Rectangular Fields" (*fang tian* 方田).

"Dividing Coins"

The first problem we will examine from the *Book of Computation,* "Dividing Coins," is a typical "excess and deficit" problem, similar to those we will encounter in the *Nine Chapters.*[1]

Translation of "Dividing Coins"

As noted in Peng 2001, the text of this problem is corrupt: though the statement of the problem is clear enough, the method offered for its solution is not just obscure but incorrect—apparently, portions have been omitted and excrescent phrases have been added.[2] The following translation presents the problem and solution with a minimum of interpretation and without corrections. I will also use the conventional translation of terms— "numerator" for *zi* 子, "denominator" for *mu* 母, "dividend" for *shi* 實, and "divisor" for *fa* 法. But it should be noted that there are no fractions in this problem, and hence no "numerators" or "denominators"; the "divisor" is not divided into the "dividend." So these terms are being used in ways that differ from the conventional meanings. Following this translation I will offer corrections and reconstruct the solution to the problem.

In the following translation (Fig. 4.1), the Chinese text has been placed in columns representing the bamboo strips on which it was written.[3]

"Dividing Coins" in Modern Mathematical Terminology

The method provided in the *Book of Computation* for solving this problem is erroneous, or at least incomplete, even though the stated solution is correct. Fortunately, the stated method does provide us enough information to reconstruct the likely solution to the problem.

To reconstruct the method used to solve this problem, we will first rewrite it in modern mathematical terms as follows. Denote the first unknown (the number of persons) by x_1; denote the second unknown (the number of coins) by x_2. The first condition can then be written as follows: if each person receives a_1, the result is a surplus of e, so we have the equation $a_1x_1 + e = x_2$, where $e > 0$. Second, if each person receives a_2, the result is a shortage of d, and this condition can be written as $a_2x_1 - d = x_2$, where $d > 0$. That is, if we write this in modern terms as a system

[1] See " 'Excess and Deficit,' Chapter 7 of the *Nine Chapters,*" which begins on page 57, below.

[2] This should remind us, when we examine the controversies surrounding the *Nine Chapters,* that the original text we are attempting to reconstruct itself undoubtedly contained errors.

[3] Note that for technical reasons, these strips are ordered from left to right, and not from right to left, as they might be ordered in a Chinese text.

分錢

分錢人二而多三人三而少二問幾何人錢幾何得曰五人錢十三贏不足互乘母為實子相從為法皆贏若

不足子互乘母而各異直之以子少者除子多者餘為法以不足為實

Dividing Coins:
In dividing coins, if each person receives 2, there is a surplus of 3; if each person receives 3, there is a shortage of 2. It is asked how many persons and coins are there? The answer: 5 persons and 13 coins. Excess and deficit: cross multiply (*hu cheng* 互乘) the "denominators" (*mu* 母) to become the "dividend" (*shi* 實); the "numerators" (*zi* 子) are added (*xiang cong* 相從) to become the "divisor" (*fa* 法). Both excess or[4] [both] deficit: the "numerators" are cross multiplied by the "denominators" and each is set aside. The lesser of the "numerators" is subtracted from the greater of the "numerators," and the remainder is the "divisor." Take the deficit as the "dividend."

[4] Reading *ruo* 若 as *huo* 或.

Fig. 4.1: Translation of "Dividing Coins."

of 2 equations in 2 unknowns, x_2 and x_1, we have

$$a_1 x_1 - x_2 = -e \tag{4.1}$$

$$a_2 x_1 - x_2 = d, \tag{4.2}$$

where a_1, a_2, d, and e are all assumed to be positive integers, or occasionally, positive integral fractions. Written in matrix form, we have

$$\begin{pmatrix} a_1 & -1 \\ a_2 & -1 \end{pmatrix} \begin{pmatrix} x_1 \\ x_2 \end{pmatrix} = \begin{pmatrix} -e \\ d \end{pmatrix}. \tag{4.3}$$

We can easily find the solution to (4.3), giving

$$x_1 = \frac{e+d}{a_2 - a_1}, \tag{4.4}$$

$$x_2 = \frac{ea_2 + da_1}{a_2 - a_1}. \tag{4.5}$$

It should be noted, however, that in the problem "Dividing Coins," we were given the constants $a_1 = 2$ and $a_2 = 3$. If we note that in this case $a_2 - a_1 = 1$, and assume that this problem is an exemplar for such cases, solutions (4.4) and (4.5) above become somewhat simpler,

$$x_1 = e + d, \tag{4.6}$$

$$x_2 = ea_2 + da_1. \tag{4.7}$$

More concretely, if we substitute $a_1 = 2$, $a_2 = 3$, $e = 3$, and $d = 2$ into equations (4.1) and (4.2), we obtain the system of 2 equations in 2 unknowns

$$2x_1 - x_2 = -3$$

$$3x_1 - x_2 = 2.$$

Substituting the excess $e = 3$ and the deficit $d = 2$ into the solutions given by equations (4.6) and (4.7) above, we have

$$x_1 = 3 + 2 = 5$$

$$x_2 = 3 \times 3 + 2 \times 2 = 13,$$

which is the solution given in the *Book of Computation*.

"Dividing Coins" also includes instructions for the case where there are two excesses, or two deficits. For the case with two excesses, we have, in modern notation,

$$\begin{pmatrix} a_1 & -1 \\ a_2 & -1 \end{pmatrix} \begin{pmatrix} x_1 \\ x_2 \end{pmatrix} = \begin{pmatrix} -e_1 \\ -e_2 \end{pmatrix}, \tag{4.8}$$

where we will assume, without loss of generality, that $a_2 > a_1$, and therefore that

$e_1 > e_2$. This then has the solutions

$$x_1 = \frac{e_1 - e_2}{a_2 - a_1}, \tag{4.9}$$

$$x_2 = \frac{a_2 e_1 - a_1 e_2}{a_2 - a_1}. \tag{4.10}$$

Again, assuming that $a_2 - a_1 = 1$, we have the simple formulas

$$x_1 = e_1 - e_2,$$
$$x_2 = a_2 e_1 - a_1 e_2.$$

For two deficits, we have, in modern notation,

$$\begin{pmatrix} a_1 & -1 \\ a_2 & -1 \end{pmatrix} \begin{pmatrix} x_1 \\ x_2 \end{pmatrix} = \begin{pmatrix} d_1 \\ d_2 \end{pmatrix}, \tag{4.11}$$

where we will again assume, without loss of generality, that $a_2 > a_1$, and therefore that $d_2 > d_1$. This then has the solutions

$$x_1 = \frac{d_2 - d_1}{a_2 - a_1}, \tag{4.12}$$

$$x_2 = \frac{a_1 d_2 - a_2 d_1}{a_2 - a_1}. \tag{4.13}$$

and once again assuming $a_2 - a_1 = 1$, this reduces to

$$x_1 = d_2 - d_1,$$
$$x_2 = a_1 d_2 - a_2 d_1.$$

"Dividing Coins" on a Counting Board

Having rewritten and solved the problem using modern notation, we are now in a much better position to reconstruct the method used to solve the problem. First, we should note that the conventional terms we used in the translation above are misleading: neither *zi* 子, which is conventionally translated as "numerator," nor *mu* 母, which is conventionally translated as "denominator," were components of a fraction at any stage in this calculation. Instead, *mu* 母 denoted the "contributions" a_1 and a_2; the *zi* 子 denoted the excess(es) or deficit(s), that is, e and d, or e_1 and e_2, or d_1 and d_2. However, this does accord with the general sense of *mu* 母, which is "mother" or "origin" (*genyuan* 根源), and *zi* 子, which is "baby" (*ying'er* 嬰兒): the former, a_1 and a_2, are in some sense primary or independent; the latter, excess(es) and deficit(s), are dependent, the result of the calculation once the "contributions" have been chosen. These terms also appear in later treatises. I will translate *zi* 子 as "minor term," and the *mu* 母 as "major term."[5]

Similarly, in division, the term *shi* 實 denotes the dividend, and *fa* 法 denotes the divisor; in operations on fractions, the term *shi* is used to denote the numerator, and the term *fa* is used to denote the denominator. For example, the commonly used mathematical phrase *shi ru fa de [er] yi* 實如法得〔而〕一 means "convert an improper fraction to a mixed fraction." But in "Dividing Coins," and other excess and deficit problems, the common translation of these terms as "dividend" and "divisor" or "numerator" and "denominator" is again misleading, since they are not treated there as a fraction, nor as elements of a division: as noted above, the former, *shi*, is never at any stage of the calculation divided by the latter, *fa*.

Instead, these terms, possibly by analogy to fractions, seem to denote the place where the calculations are made;[6] presumably this was done on a counting board. Thus, in the context of excess and deficit problems, I will translate *shi* as "upper quantity," and *fa* as " lower quantity," understanding that these are the places on the counting board where the unknowns are calculated. Indeed, the *Book of Computation* employs several terms that denote position, which seems to offer further confirmation that the problem was laid out on a counting board: along with the terms noted above commonly used for fractions (*zi* 子 and *mu* 母) and division (*shi* 實 and *fa* 法), this includes the terms "cross multiplied" (*hu cheng* 互乘) and "set it aside" (*yi zhi zhi* 異直之). Unfortunately, the *Book of Computation* does not record the positions on the counting board in any detail. But if we assume that numerators were placed above denominators, and dividends were placed above divisors (see Martzloff [1987] 2006, 218), the positioning of the quantities might look something like the following:

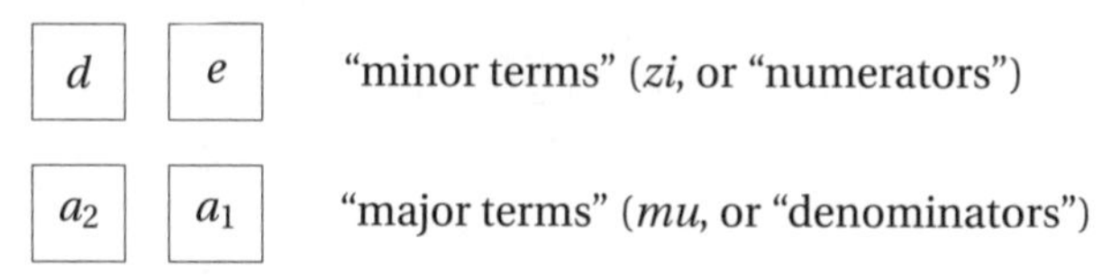

where the columns are placed from right to left (that is, following the Chinese ordering of right-to-left, the quantities corresponding to the first equation are placed on the right, those corresponding to the second equation are placed on the left). Here the "minor terms" (*zi*) have been placed above the "major terms" (*mu*), just as numerators are placed above denominators. "Dividing Coins" gives the instruction "cross multiply (*hu cheng* 互乘) the 'denominators' (*mu* 母) to become the 'dividend' (*shi* 實)" (lines 5 and 6 on page 46), omitting mention of what

[5] Peng 2001 suggests that the terminology used included *bu zu zi* 不足子 ("deficit minor term"), *bu zu mu* 不足母 ("deficit major term"), *ying zi* 贏子 ("excess minor term"), and *ying mu* 贏母 ("excess major term"). However plausible this may be, there is no direct usage of these precise terms in the passage.

[6] The term *shi* is used in a variety of ways in later texts, including the following: "(1) Literally, 'the full.' The dividend (the number of full divisors in the original number). (2) The constant term in an equation" (Libbrecht 1973, 487).

is cross multiplied, namely the "minor terms" or "numerators." On the counting board, the calculation might be visualized as follows:

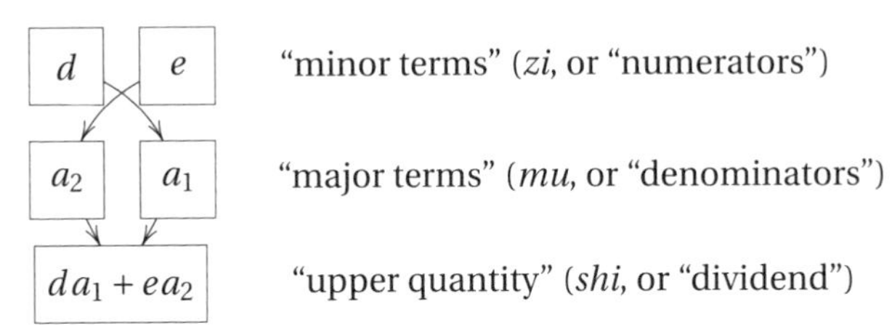

The following instruction, "the 'numerators' (*zi* 子) are added (*xiang cong* 相從) to become the 'divisor' (*fa* 法)" (lines 7 to 9 on page 46), might then look as follows:

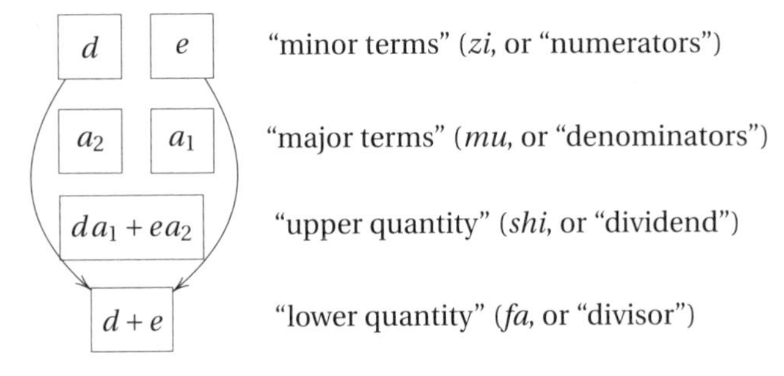

where the "lower quantity" has been placed below the "upper quantity." In more concrete terms, we have

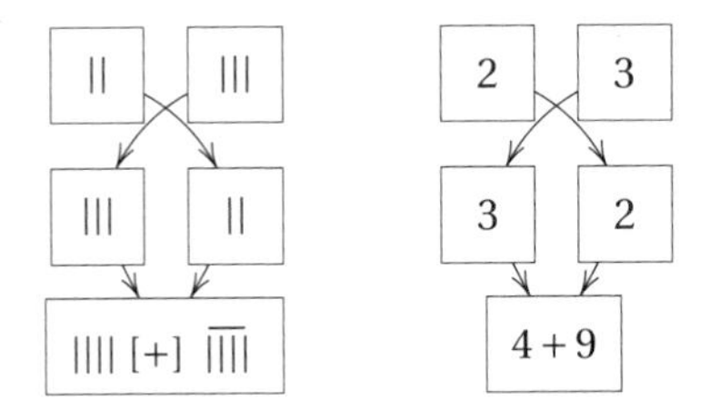

where the diagram on the left uses counting rods (with the modern symbol for addition in brackets). The calculation of the second unknown is then given in the following diagram,

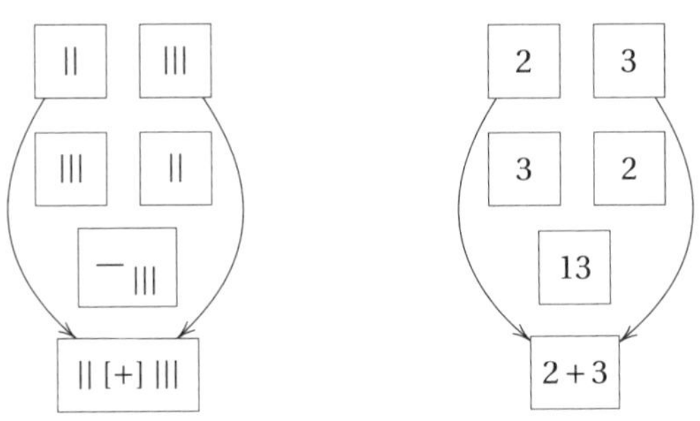

Reconstructing the Solution to "Dividing Coins"

As noted above, the solution given in the *Book of Computation* for this problem is incomplete. Drawing upon the above analysis, and using alternative terms in the translation, we might reconstruct the practice described in "Dividing Coins" as follows:[7]

> "Excess and deficit" [method]: cross multiply the "major terms" [by the "minor terms," and add the results together] to become the "upper quantity" [that is, the value of the second unknown, $x_2 = da_1 + ea_2$, here $2 \times 2 + 3 \times 3 = 13$]; and the "minor terms" are added to become the "lower quantity" [that is, the value of the first unknown, $x_1 = e + d$, here $3 + 2 = 5$]. "Both excess" or "[both] deficit" [method]:[8] cross multiply the "major terms" [with the "minor terms"], keep both results[9] [then subtract the lesser from the greater, and take this as the "upper quantity," that is, the value of the second unknown]. Subtract the lesser of the "minor terms" from the greater "minor term," and take the remainder as the "lower quantity"[10] [that is, the value of the first unknown].
>
> 贏不足〔術〕：〔子〕互乘母為實〔即錢數〕；子相從為法〔即人數〕。皆贏若〔皆〕不足〔術〕：子互乘母而各異直之，〔以少者除多者，餘為實〕；[11] 以子少者除子多者，餘為法。[12]

"Dividing Coins": Conclusions

The corrected solution, expressed in modern terms, is as follows. If we write b_1 and b_2 for the "minor terms" (that is, the "excess" and "deficit," or two "excesses," or two "deficits"), we can rewrite the system of 2 equations in 2 unknowns in equations (4.1) and (4.2) as

$$a_1 x_1 - x_2 = b_1 \tag{4.14}$$

$$a_2 x_1 - x_2 = b_2, \tag{4.15}$$

[7] In the following translation, I have consulted Peng's (2001) suggested corrections, and added my own.

[8] Specification of these as separate methods is necessitated by the lack of negative numbers: because the result of addition and subtraction must be positive, different methods are offered, depending on the signs of the values.

[9] This instruction is not entirely clear. It is apparently included because it is necessary first to compare the two results to determine which is larger, and thus which will be subtracted from which.

[10] Again, note here that the instruction "Subtract the lesser of the 'minor terms' from the greater 'minor term,'" must specify "greater" and "lesser," since there are no negative numbers.

[11] Following the interpolation suggested in Peng 2001.

[12] Removing the five characters 以不足為實, following the correction suggested in Peng 2001.

where we allow b_1 and b_2 to be positive or negative integers, or integral fractions. If, as in "Dividing Coins," $a_2 - a_1 = 1$, the solution is then simply of the form

$$x_1 = b_2 - b_1 \quad (4.16)$$

$$x_2 = a_1 b_2 - a_2 b_1. \quad (4.17)$$

The calculation given in the *Book of Computation*, then, is important, because it is here that we find the earliest instruction to "cross multiply" the terms.

"Rice for Coins"

Another example of excess-deficit problems in the *Book of Computation* is the topic "Rice for Coins" (*mi chu qian* 米出錢), under which two separate problems are presented. Here we will examine the first problem only. As with the previous problems, there are several mistakes in the text; but the mistakes in this problem are of a much simpler sort: the terms for the two unknowns, *bai* 粺 (fine rice) and *li* 糲 (coarse rice), are at several points interchanged (all of these are corrected in Peng 2001). In my translation (Fig. 4.2), I have indicated corrections and other interpolations by brackets. Except for these transcription errors, the text is complete and the methods correct.

"Rice for Coins" in Modern Mathematical Terms

To better understand this problem, we will again first write it in modern terms, and then solve it directly (in this case, however, the direct solution will differ considerably from the previous method presented in the *Book of Computation*). Let x_1 represent the number of *dou* of fine rice (that is, the volume of fine rice measured in *dou*), and let x_2 represent the number of *dou* of coarse rice. In the problem, we are given the following prices: fine rice is three coins for two *dou*, or $\frac{3}{2}$ coins per *dou*; coarse rice is two coins for three *dou*, or $\frac{2}{3}$ coins per *dou*. We are then told that the combined volume of fine rice plus coarse rice is 10 *dou*, and that the total cost of the fine rice plus the coarse rice is 13 coins. This can be written as the following 2 equations in 2 unknowns,

$$x_1 + x_2 = 10 \quad (4.18)$$

$$\frac{3}{2}x_1 + \frac{2}{3}x_2 = 13, \quad (4.19)$$

or, in terms of matrices,

$$\begin{pmatrix} 1 & 1 \\ \frac{3}{2} & \frac{2}{3} \end{pmatrix} \begin{pmatrix} x_1 \\ x_2 \end{pmatrix} = \begin{pmatrix} 10 \\ 13 \end{pmatrix}.$$

"Rice for Coins."
[Fine] rice costs 3 coins[13] for 2 *dou*;[14] coarse rice costs 2 coins for 3 *dou*. Now given 10 *dou* of coarse and fine rice [combined] that is sold for 13 coins, how much coarse rice and fine rice is there?
It is stated: $7\frac{3}{5}$ *dou* of fine rice, $2\frac{2}{5}$ *dou* of coarse rice.
The method states: If all [10 *dou*] of the rice is [fine], the excess is 2 coins; if all the rice is [coarse], the deficit is $6\frac{1}{3}$ coins.[15] Combine the excess [2] and deficit [$\frac{19}{3}$] to become the denominator [$\frac{25}{3}$], take the excess [2] multiplied by ten *dou* [20] as [the numerator, and divide, giving] [coarse] rice [$\frac{20}{25/3} = \frac{12}{5}$]. Take the deficit [$\frac{19}{3}$] multiplied by ten *dou* [$\frac{190}{3}$] as [the numerator, and divide, giving] [fine] rice [$\frac{190/3}{25/3} = \frac{38}{5}$]. Convert the [improper fractions into mixed] fractions [$7\frac{3}{5}, 2\frac{2}{5}$].

米出錢　糲[16]米二斗三錢糲米三斗二錢今有糲粺十斗賣得十三錢問糲粺各幾何曰粺七斗五分三

糲二斗五分二术曰令偕糲[17]也錢贏二令偕粺[18]也錢不足六少半同贏不足以為法以贏乘十斗為粺[19]以不

足乘十斗為糲[20]皆如法一斗

Fig. 4.2: Translation of "Rice for Coins"

[13] "Coins" (*qian* 錢) is a monetary unit, apparently first used for gold pieces molded in the shape of an ancient farming tool similar to a spade, then later used for coins.

[14] A unit of volume, equal to ten *sheng* 升. Currently, a *dou* is equal to one decaliter.

[15] Literally, six and a "lesser half" (*shao ban* 少半), a mathematical term commonly used for $\frac{1}{3}$.

[16] Substituted throughout for a rare variant form of this character. Here, *li* 糲 (coarse rice) is mistaken, and should be *bai* 粺 (fine rice).

[17] *Bai* 粺 (fine rice) should be *li* 糲 (coarse rice).

[18] *Li* 糲 (coarse rice) should be *bai* 粺 (fine rice).

[19] *Bai* 粺 (fine rice) should be *li* 糲 (coarse rice).

[20] *Li* 糲 (coarse rice) should be *bai* 粺 (fine rice).

If we consider the general system of 2 conditions in 2 unknowns for which this problem is an exemplar, then written in modern mathematical terms we have

$$x_1 + x_2 = b_1 \tag{4.20}$$

$$a_1 x_1 + a_2 x_2 = b_2. \tag{4.21}$$

In matrix form, this can be written as

$$\begin{pmatrix} 1 & 1 \\ a_1 & a_2 \end{pmatrix} \begin{pmatrix} x_1 \\ x_2 \end{pmatrix} = \begin{pmatrix} b_1 \\ b_2 \end{pmatrix}. \tag{4.22}$$

The solution, which is easily found, is

$$x_1 = \frac{a_2 b_1 - b_2}{a_2 - a_1}, \tag{4.23}$$

$$x_2 = \frac{b_2 - a_1 b_1}{a_2 - a_1}. \tag{4.24}$$

It would not have been difficult to describe these computations, which involve "cross multiplication." For example, the *Book of Computation* could have presented the following instructions: multiply the amount of combined rice, 10 *dou*, by the cost of one *dou* of coarse rice, $\frac{2}{3}$, and subtract this from the cost of the combined rice, 13, to obtain the dividend, $13 - 10 \times \frac{2}{3} = \frac{19}{3}$; then subtract the cost of one *dou* of fine rice, $\frac{3}{2}$, from the cost of one *dou* of coarse rice, $\frac{2}{3}$, and take that as the divisor, $\frac{5}{6}$; divide the divisor into the dividend, giving $\frac{19/3}{5/6} = \frac{38}{5}$ as the amount of fine rice in *dou*. But as noted above, this is not how this problem is solved: the original problem is first used to create an "excess and deficit" problem, which is subsequently solved.

The following is a reconstruction, in modern terms, of the solution to "Rice for Coins." Although the calculations here are elementary, because the solution is found by methods perhaps unfamiliar to the modern reader, this deserves some explanation. The solution consists of two main steps: (1) transforming the problem into an "excess and deficit" problem, and (2) then solving the resulting "excess and deficit" problem. More specifically, the two steps are as follows:

STEP 1. The first step is to transform the given problem into an "excess-deficit" problem as follows:

1. First, we assume (falsely, as it will turn out) that all the rice is fine rice. Since the price of fine rice is $\frac{3}{2}$ coins per *dou*, for 10 *dou* of fine rice we would pay $\frac{30}{2} = 15$ coins, which is $15 - 13 = 2$ coins more than was paid, that is, an excess of 2 coins.
2. Second, we (again falsely) assume that all the rice is coarse rice. Since the price of coarse rice is $\frac{2}{3}$ coins per *dou*, for 10 *dou* of coarse rice we would pay $\frac{20}{3}$ coins, which is $13 - \frac{20}{3} = \frac{19}{3}$ coins less than was paid, or a deficit of $\frac{19}{3}$ coins.

In order to better understand this method, we can reconstruct this calculation in abstract terms: we will first set x_2 equal to zero and calculate the "excess," then we will set x_1 equal to zero and calculate the "deficit."

1. First, we set the fine rice to the total amount of rice, $x_1 = b_1$, which is equivalent to setting the amount of coarse rice to zero, $x_2 = 0$, so that equations (4.20) and (4.21) become the following equation,

$$a_1 b_1 - e = b_2,$$

where e is the "excess" resulting from the assignment $x_1 = b_1$ and $x_2 = 0$ (we take $e > 0$ and subtract it from $a_1 b_1$, since it is to be an "excess"). Rewriting this equation, we can solve for the "excess" e,

$$e = a_1 b_1 - b_2, \tag{4.25}$$

as was found in the problem "Rice for Coins."

2. Second, we set the coarse rice to the total amount of rice, $x_2 = b_1$, which is equivalent to setting the fine rice to zero, $x_1 = 0$, so that equations (4.20) and (4.21) become the following equation,

$$a_2 b_1 + d = b_2, \tag{4.26}$$

where $d > 0$ is the "deficit" resulting from the assignment of $x_2 = b_1$ and $x_1 = 0$. If, in a manner similar to the above, we solve for the "deficit" d, we obtain

$$d = b_2 - a_2 b_1, \tag{4.27}$$

again, as was found in the problem "Rice for Coins."

STEP 2. The second step is to solve the "excess and deficit" problem. The *Book of Computation* presents the following formula to solve "Paying Coins for Rice":

$$x_1 = \frac{10 \times \frac{19}{3}}{\frac{19}{3} + 2} = \frac{38}{5}$$
$$x_2 = \frac{10 \times 2}{\frac{19}{3} + 2} = \frac{12}{5}.$$

In modern notation, this represents the following,

$$x_1 = \frac{b_1 d}{d + e} \tag{4.28}$$
$$x_2 = \frac{b_1 e}{d + e}. \tag{4.29}$$

Unfortunately, since the *Book of Computation* offers no explanation as to why this problem is first transformed into an excess-deficit problem, nor does it offer any explanation as to why the formula provided might work, the best we can do is to verify that the calculation is correct. This can be verified simply by taking equations (4.20) and (4.21) in their matrix form, equation (4.22), and multiplying both sides on the left by a constant matrix,

$$\begin{pmatrix} 1 & 0 \\ b_2 & -b_1 \end{pmatrix} \begin{pmatrix} 1 & 1 \\ a_1 & a_2 \end{pmatrix} \begin{pmatrix} x_1 \\ x_2 \end{pmatrix} = \begin{pmatrix} 1 & 0 \\ b_2 & -b_1 \end{pmatrix} \begin{pmatrix} b_1 \\ b_2 \end{pmatrix},$$

which yields the result

$$\begin{pmatrix} 1 & 1 \\ b_2 - a_1 b_1 & b_2 - a_2 b_1 \end{pmatrix} \begin{pmatrix} x_1 \\ x_2 \end{pmatrix} = \begin{pmatrix} b_1 \\ 0 \end{pmatrix}. \tag{4.30}$$

Substituting into equation (4.30) the values computed for the "excess" e in equation (4.25) and the "deficit" d in equation (4.27), we obtain, in matrix terms,

$$\begin{pmatrix} 1 & 1 \\ -e & d \end{pmatrix} \begin{pmatrix} x_1 \\ x_2 \end{pmatrix} = \begin{pmatrix} b_1 \\ 0 \end{pmatrix},$$

or, written as a system of equations,

$$\begin{aligned} x_1 + x_2 &= b_1 \\ -ex_1 + dx_2 &= 0. \end{aligned}$$

The solution is then easily calculated,

$$\begin{aligned} x_1 &= \frac{b_1 d}{d+e} \\ x_2 &= \frac{b_1 e}{d+e}, \end{aligned}$$

which is the formula given in the *Book of Computation*, in equations (4.28) and (4.29) above.

"Rice for Coins": Conclusions

In this problem, then, in contrast to the preceding "Dividing Coins," there is no mention of "cross multiplication," and in fact no mention of positions on a counting board. Instead, the given problem is transformed into an equivalent "excess and deficit" problem, and the latter is solved without further explanation. It is possible that solutions to this problem were calculated in the head, without using counting rods.

"Excess and Deficit," Chapter 7 of the *Nine Chapters*

In "Excess and Deficit," the absence of negative numbers results in the presentation of several related methods, such as "excess and deficit" and "double excess." In modern terms, we would describe these problems as similar, but with positive or negative results. Indeed, there appears to be no use of negative numbers in the "Excess and Deficit" chapter in the *Nine Chapters*. In the *Nine Chapters*, negative numbers are first introduced in chapter 8, "*Fangcheng*," which appears to have been compiled later than chapter 7, "Excess and Deficit." The earliest record we have of the use of *fangcheng* to solve systems of *n* conditions in *n* unknowns is in the *Nine Chapters*; as noted above, there is no mention of the term *fangcheng* in the *Book of Calculation*.[21]

The first four problems of "Excess and Deficit" are solved by the same method, together with two methods of solution. In the following translation I will (as I noted in chapter 3) use the earliest extant version of the "Excess and Deficit" chapter while noting the findings of Guo's important textual studies, since there are places in the text where I believe that the mathematical content may be inconsistent with Guo's reconstruction. The translation is as follows:

> Excess and Deficit: to solve the hidden complexity by comparison.
>
> 盈不足。以御隱[22]雜互見。[23]
>
> Now there are items to be purchased jointly: if each person contributes 8, then the excess is 3; if each person contributes 7, then the deficit is 4. Problem: What is the number of persons and the price of the items? Answer: 7 persons; the price of the items is 53.
>
> 今有共買物：人出八，盈三；人出七，不足四。問人數、物價各幾何？荅曰：七人，物價五十三。[24]
>
> Now there are chickens to be purchased jointly: if each person contributes 9, then the excess is 11; if each person contributes 6, then the deficit is 16. It is asked, what is the number of persons and what is the cost of the chickens? Answer: 9 persons; the cost of the chickens is 70.

---

[21] It should be noted that there is no reason to believe that the ordering of the chapters corresponds to their date of composition; in fact, early bibliographic records suggest that the chapters of the *Nine Chapters* were ordered differently in different editions.

[22] Substituted for a rare variant.

[23] Although the second phrase 以御隱雜互見 is not distinguished typographically as commentary by offsetting it by a blank space of one character from the top, it is clearly an annotation, and is attributed to Liu Hui.

[24] For the first several passages of chapter 7 "Excess and Deficit" translated here, in the critical edition Guo basically follows the version of the text preserved in Yang Hui's *Nine Chapters on the Mathematical Arts, with Detailed Explanations* (*Xiang jie jiuzhang suanfa* 詳解九章算法), placing much of the secondary annotations by Li Chunfeng et al. just after the initial passage 盈不足。以御隱雜互見。

今有共買雞[25]：人出九，盈一十一；人出六，不足十六。問人數、雞價各幾何？答曰：九人，雞價七十。

Now there is jade[26] to be purchased jointly: if each person contributes $\frac{1}{2}$, the excess is four; if each contributes $\frac{1}{3}$, the deficit is 3. It is asked, what is the number of persons and what is the cost of the jade? Answer: 42 persons; the cost of the jade is 17.

今有共買璡：人出半，盈四；人出少半，不足三。問人數、璡價各幾何？答曰：四十二人，璡價十七。

Now there are oxen to be purchased jointly: if [each group consisting of] seven families contributes 190,[27] the deficit is 330; if [each group of] nine families contributes 270,[28] the excess is 30. It is asked what is the number of families and what is the cost of the oxen. Answer: 126 families; the oxen cost 3750.

今有共買牛：七家共出一百九十，不足三百三十；九家共出二百七十，盈三十。問家數、牛價各幾何？答曰：一百二十六家，牛價三千七百五十。

Note:[29] This method adds together the excess and deficit [$330 + 30 = 360$], which [result] from the difference in the numbers of families grouped together. Therefore, [360] is taken as the "upper quantity" (*shi* 實). Place [on the counting-board] the [two] rates of contribution [190 and 270], and divide each by the number of families [7 and 9], obtaining the contribution for one family for each case [$\frac{190}{7}$ and $\frac{270}{9}$]. Subtract the lesser from the greater [$\frac{270}{9} - \frac{190}{7} = \frac{20}{7}$], obtaining the difference [between the excess and deficit] for one family, and divide [$\frac{360}{20/7} = 126$],[30] namely, the number of families. Multiply by the rate of

---

[25] CSJC uses the variant form *ji* 鷄 throughout this problem, but not in the secondary annotations by Li Chunfeng et al.

[26] *Jin* 璡 or 瑨 is a precious stone for which no modern Chinese equivalent is given in standard references; it was considered to be of lesser value than jade. *Nine Chapters on the Mathematical Arts, Pronunciations and Meanings* (*Jiuzhang suanshu yin yi* 九章算術音義) by Li Ji 李籍 states, "*jin*, pronounced with the initial of *jiang* and the final of *lin*, is a precious stone second to jade; one edition states a musical instrument [similar to a zither]" 璡：將鄰切美石次玉曰璡一本作準.

[27] That is, when the total number of families is divided into groups of seven, each group of seven families contributes 190. As the answer will show, there are 126 families, so there are $126 \div 7 = 18$ groups of seven families. More literally, "the seven-families together contribute 190."

[28] When the total number of families is divided into groups of nine, each group of nine families contributes 270; there are $126 \div 9 = 14$ groups of nine families. More literally, "the nine-families together contribute 270."

[29] Attributed to Liu Hui.

[30] It should be noted that whereas 360 is the numerator in a fraction, and thus the translation of *shi* 實 as "numerator" might also seem appropriate, it is not divided by something termed a "denominator" (*fa* 法).

contribution $[126 \times \frac{270}{9} = 3780]$, subtract the excess $[3780-30 = 3750]$, obtaining the cost of the oxen.

按：此術并盈不足者，為衆家之差，故以為實。置所出率，各以家數除之，各得一家所出率。以少減多者，得一家之差。以除，即家數。以出率乘之，減盈,[31]故得牛價也。

"Excess and Deficit" Problems in Modern Terms

After presenting the above four problems as exemplars, the *Nine Chapters* next presents the methods for solving "excess and deficit" problems. In order to analyze the methods of solution, we will again introduce modern terminology. Written in modern terms, the general form of an "excess and deficit" problem can be translated into a system of 2 conditions in 2 unknowns; we will denote the two unknowns here by x_1 and x_2. In each of the problems, the two conditions are presented in narrative form as follows:

1. A constant a_1 is given that results in an "excess" e, that is, $a_1x_1 - x_2 = e$ (where $e > 0$),
2. A second constant a_2 is given that results in a "deficit" d, that is, $a_2x_1 - x_2 = -d$ (where $d > 0$).

Note that these problems differ slightly from the problems in the *Book of Computation*: in "Dividing Coins," each person was receiving coins; in the first four problems from "Excess and Deficit" translated here, each person is contributing to a purchase. Thus the signs of the "excess" and "deficit" are reversed. This is then equivalent to a system of 2 equations in 2 unknowns x_1 and x_2,

$$a_1x_1 - x_2 = e \tag{4.31}$$

$$a_2x_1 - x_2 = -d \tag{4.32}$$

or in matrix form as

$$\begin{pmatrix} a_1 & -1 \\ a_2 & -1 \end{pmatrix} \begin{pmatrix} x_1 \\ x_2 \end{pmatrix} = \begin{pmatrix} e \\ -d \end{pmatrix}.$$

The solution, which is easily calculated, is as follows:

$$x_1 = \frac{e+d}{a_1 - a_2}, \tag{4.33}$$

$$x_2 = \frac{a_2e + a_1d}{a_1 - a_2}. \tag{4.34}$$

However, as we will see below, in the *Nine Chapters*, the solution is given not as a fraction but rather division, so perhaps it might be more precise to represent the

[31] Guo 1990 and Guo 1998 also insert the phrase "increase by the deficit" *zeng bu zu* 增不足.

solution as follows:

$$x_1 = (e + d) \div (a_1 - a_2) \tag{4.35}$$
$$x_2 = (a_2 e + a_1 d) \div (a_1 - a_2). \tag{4.36}$$

Solution to the First Four Problems

The first four "excess and deficit" problems, translated above, can be solved by a straightforward application of the formula given in equations (4.35) and (4.36). This, as we will see below, will turn out to be the approach given in the "excess and deficit method." The calculations below are presented here in some detail, to make the following analysis of the "excess and deficit method" easier to follow.

PROBLEM 1. The first problem, in modern terms, is equivalent to a system of 2 equations in 2 unknowns,

$$\begin{aligned} 8x_1 - x_2 &= 3 \\ 7x_1 - x_2 &= -4, \end{aligned}$$

which has the solution

$$\begin{aligned} x_1 &= (3 + 4) \div (8 - 7) = 7 \\ x_2 &= (7 \times 3 + 8 \times 4) \div (8 - 7) = 53. \end{aligned}$$

PROBLEM 2. The next problem is equivalent to the system

$$\begin{aligned} 9x_1 - x_2 &= 11 \\ 6x_1 - x_2 &= -16, \end{aligned}$$

which has the solution

$$\begin{aligned} x_1 &= (16 + 11) \div (9 - 6) = 9 \\ x_2 &= (6 \times 11 + 9 \times 16) \div (9 - 6) = 70. \end{aligned}$$

PROBLEM 3. The third, in modern terms, is equivalent to

$$\begin{aligned} \frac{1}{2}x_1 - x_2 &= 4 \\ \frac{1}{3}x_1 - x_2 &= -3, \end{aligned}$$

which has the solution

$$x_1 = (4 + 3) \div \left(\frac{1}{2} - \frac{1}{3}\right) = 7 \div \frac{1}{6} = 42$$

$$x_2 = \left(\frac{1}{3} \times 4 + \frac{1}{2} \times 3\right) \div \left(\frac{1}{2} - \frac{1}{3}\right) = \frac{17}{6} \div \frac{1}{6} = 17.$$

Though this problem is more complicated because of the appearance of fractions, the "Method for Excess and Deficit," which we will analyze following this section, provides explicit instructions on computations with fractions.

PROBLEM 4. The fourth problem is also slightly more complicated, and will require further explanation. In modern terms, we can write

$$\frac{270}{9} x_1 - x_2 = 30$$
$$\frac{190}{7} x_1 - x_2 = -330,$$

where the order of the equations has been switched from the order in which they are presented in the text, so that the equation with the "excess" is first, and the equation with the "deficit" is second. This system then has the solution

$$x_1 = (30 + 330) \div \left(\frac{270}{9} - \frac{190}{7}\right) = 360 \div \frac{20}{7} = 126$$
$$x_2 = \left(\frac{190}{7} \times 30 + \frac{270}{9} \times 330\right) \div \left(\frac{270}{9} - \frac{190}{7}\right) = \frac{75000}{7} \div \frac{20}{7} = 3750.$$

These calculations will then help us better understand the "excess and deficit method," presented below.

The "Excess and Deficit Method"

We are now ready to examine the method for solving these problems given in the *Nine Chapters*, using the modern notation and numbers calculated above. Chapter 7, "Excess and Deficit" continues as follows:

> 30 The "excess and deficit method" states: Place [on the counting-board] the contribution rates [a_1 and a_2]. The excess and the deficit [e and d] are each positioned below [the two contribution rates, respectively]. Then cross multiply the contribution rates [by the excess and deficit, yielding $a_1 d$ and $a_2 e$], and add [$a_1 d + a_2 e$], to become the upper quantity (*shi* 實).[32] Combine the
> 35 excess and deficit [$e + d$] to become the lower quantity (*fa* 法).[33]

---

[32] To be used to calculate the second unknown, x_2.

[33] To be used to calculate the first unknown, x_1. Note that the calculation of the first and second unknowns x_1 and x_2 is not yet finished at this point, and is continued following commentary that explains the calculations so far.

〔盈不足〕術曰：（盈不足相與同共買物者。）置所出率；盈、不足各居其下。令維乘所出率，并，以為實。并盈、不足為法。[34]

[Liu Hui's commentary] Note: "excess" (*ying* 盈) is what is called "waning crescent moon" (*tiao* 朓); "deficit" (*bu zu* 不足) is what is called "waxing crescent moon" (*nü* 朒).[35] The "rates of contribution" (*chu lü* 出率) [a_1 and a_2] are what are called "false assumptions" (*jia ling* 假令). The excess and deficit [e and d] are cross multiplied (*wei cheng* 維乘) by the two assumed values (*she zhe* 設者) [a_1 and a_2, with the result $a_2 e + a_1 d$], desiring to make them uniform to combine them (*tong qi* 同齊).[36] According to the problem "Jointly Purchasing an Object," "If each person contributes 8, then the excess is 3; if each person contributes 7, then the deficit is 4." Make the false suppositions uniform (*qi* 齊),[37] and combine (*tong* 同) the excess and deficit:[38] the excess and deficit [multiplied together] are 12 [$e \times d$, here 3×4]; completing calculation with the [quantities that are made] uniform is then the exact solution (*bu ying bu nü zhi zheng shu* 不盈不朒之正數).[39] Thus they [$a_2 e$ and $a_1 d$, here 21 and 32] are to be added [$a_2 e + a_1 d$, here 53] in [the calculation of] the upper quantity [in order to find x_1]. Add the excess and deficit [$e + d$, here $3 + 4$] in [the calculation of] the lower quantity [to find the first unknown, x_1]; 32 [$a_1 d$, here 4×8] is the result from making the assumed value of 4 uniform, giving an excess of 12; 21 [$a_2 e$, here 3×7] is the result from making the assumed value 3 uniform, giving a deficit of 12 [$a_2 e + a_1 d$, here $4 \times 8 + 3 \times 7$]. Add 7, the assumed values [$a_1 + a_2$, $4 + 3 = 7$], to become the upper quantity; thus add 3 [a_1] and 4 [a_2] to become the lower quantity [$a_1 + a_2$].

按：盈者，謂之朓；不足者，謂之朒；所出率謂之假令。盈、朒維乘兩設者，欲為同齊之意。據共買物：人出八，盈三；人出七，不足四。齊其假令，同其盈、朒。盈、朒俱十二，通計齊則不盈不朒之正數，故可并之為實；并盈、不足為法。齊之三十

---

[34] The SKQS edition has removed the phrase "and simplify the fraction" or more literally, "the numerator is divided by the denominator as one." Guo restores the phrase 實如法而一 to his editions. However, because this is not a fraction, and the *fa* is not divided into the *shi*, there is no "fraction" to reduce, and the phrase should be omitted. See JZSS, *juan* 7, 2a, and Chemla and Guo 2004, 562–63.

[35] The *Nine Chapters on the Mathematical Arts, Pronunciations and Meanings* (*Jiuzhang suanshu yin yi* 九章算術音義) by Li Ji 李籍 states "*Nü* is pronounced with the initial *nü* and final *liu*. One [edition] uses *fei* 朏, which is not correct" 朒：女六切不足也或作朏非是.

[36] The mathematical term *tong qi* 同齊 is usually used for fractions.

[37] The term *qi* 齊 is often used for operations on fractions, to mean finding a common denominator. Here, Liu makes an analogy to the cross multiplication in combining fractions.

[38] The term *tong* 同 is also used for operations on fractions, to mean combining the numerators after finding a common denominator. Here again, Liu makes an analogy to fractions.

[39] Literally, "neither excess nor deficit," here used in contrast with the "false assumptions" that result in the "excess" and "deficit."

二者，是四假令，有盈十二；齊之二十一者，是三假令，亦朒十二。并七假令合為一實，故并三、四為法。[40]

Those with fractions are made common.

有分者，通之。

60 [Liu Hui's commentary:] If the two assumed values [a_1 and a_2] have fractions, add the fractions.[41] In this [last] problem, both of the assumed values are seen to be fractions, therefore add the fractions.

若兩設有分者，齊其子，同其母。此問兩設俱見零分，故齊其子，同其母。

Again, place [on the counting board] the rates of contribution [a_1 and a_2], subtract the lesser from the greater, and use the remainder to reduce the 65 divisor and dividend. The "upper quantity" is the price of the object, and the "lower quantity" is the number of people.

副置所出率，以少減多，餘，以約法、實。實為物價，法為人數。

The "Method for Excess and Deficit" on a Counting Board

The "Method for Excess and Deficit," as presented in the *Nine Chapters*, is fairly straightforward, and intriguingly similar to the method presented in the *Book of Computation*:

STEP 1. The first instruction in the "method for excess and deficit" states, "Place [on the counting-board] the contribution rates [a_1 and a_2]" (lines 30 and 31 on page 61). On the counting board, this might be accomplished as follows, where the rates are placed in order from right to left,

a_2	a_1	"contribution rates" (*chu lü*)

STEP 2. The second instruction tells us, "The excess and the deficit [e and d] are each positioned below [the two contribution rates, respectively]" (lines 31 and 32 on page 61), which gives the following arrangement:

a_2	a_1	"contribution rates" (*chu lü*)
d	e	"excess and deficit" (*ying bu zu*)

[40] Guo places this passage after the title in Guo 1990, but places this passage here in Guo 1998.

[41] More literally, combine the numerators [by cross multiplying them by the denominators and adding the results together] and make the denominators uniform [by multiplying them together]. This is the usual way of adding fractions, but in Chinese mathematical treatises it is described as two operations, one on the numerators and one on the denominators.

STEP 3. The following instruction is arguably two steps: "Then cross multiply the contribution rates [by the excess and deficit, yielding $a_1 d$ and $a_2 e$], and add [$a_1 d + a_2 e$], to become the upper quantity (*shi* 實)" (lines 32 to 34 on page 61). That is, a step seems to have been omitted here, since the results of the "cross multiplication" must first be placed somewhere on the counting board before they are added together. In any case, the result is the following arrangement:

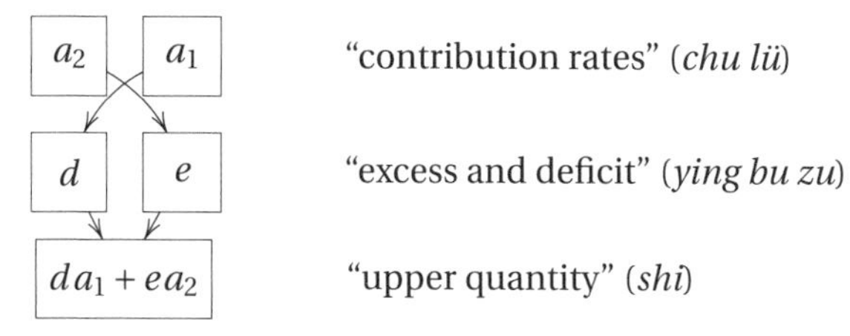

STEP 4. The next instruction, "Combine the excess and deficit [$e + d$] to become the lower quantity (*fa* 法)" (lines 34 and 35 on page 61), yields the following:

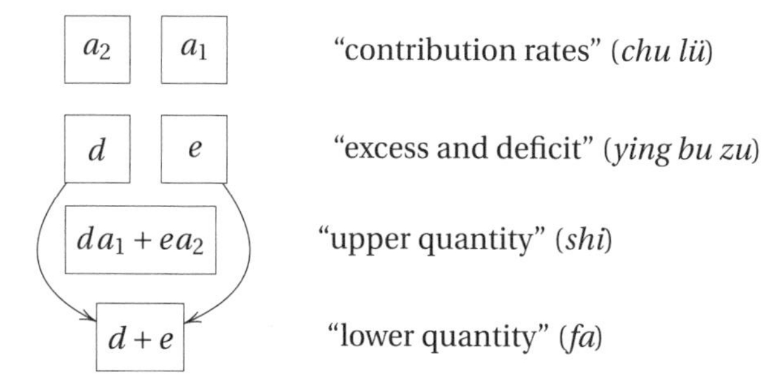

STEP 5. The following instruction, "Those with fractions are made common" (line 59 on the preceding page), asks us to follow the procedure, explained elsewhere in the *Nine Chapters*, for adding fractions, which, in modern terms, consists of finding a common denominator and then adding the numerators.

STEP 6. The next instruction, "Again, place [on the counting board] the rates of contribution [a_1 and a_2]" (line 63 on the previous page), seems perhaps spurious: the "contribution rates" have not been altered, and so there would seem to be no particular need to place them elsewhere on the counting board for comparison to determine which is greater;[42] also, no name or relative position is given where the "contribution rates" are to be placed. However, if we do follow the instruc-

[42] It should be noted that in some modern reconstructions of multiplication on the counting board, the multiplier or multiplicand may be removed in the process of multiplication; it is also possible that conventionally both were removed after completion of the multiplication. However, as Step 4 indicates, the "excess and deficit" were to remain on the counting board, so I see no particular reason why the "contribution rates," which are needed in this step, would have been removed.

tion given in the "method for excess and deficit," the counting board appears as follows:

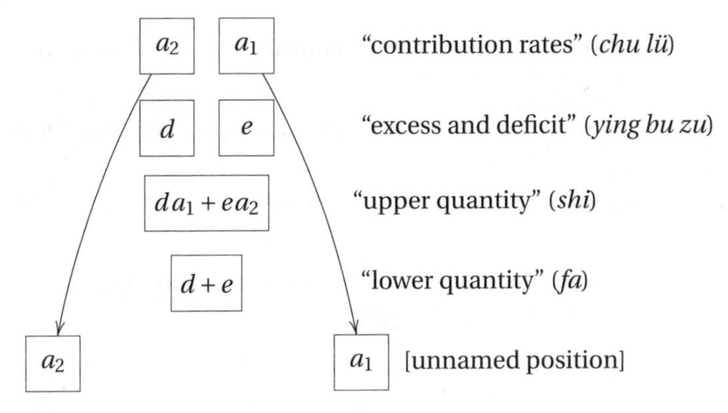

STEP 7. The following instruction, "subtract the lesser from the greater" (line 64 on page 63), represents a workaround to solve the problem without negative numbers, presumably at a time before negative numbers were understood. In modern terms, we might write this as $|a_1 - a_2|$, where, to be more specific,

$$|a_1 - a_2| = \begin{cases} a_1 - a_2 & \text{if } a_1 > a_2 \text{ (thus } a_1 - a_2 > 0) \\ a_2 - a_1 & \text{if } a_2 > a_1 \text{ (thus } a_2 - a_1 > 0). \end{cases}$$

The resulting arrangement on the counting board is the following:

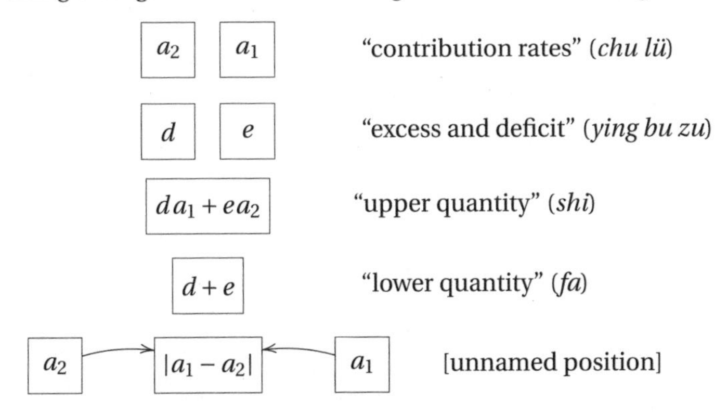

STEP 8. The final instruction, "use the remainder to reduce the divisor and dividend" (lines 64 and 65 on page 63), uses the terminology from operations with fractions, "reduce" (*yue* 約), which means to divide the numerator and denominator of a fraction by a common divisor (for example, dividing the numerator and denominator of $\frac{15}{21}$ by 3 to yield $\frac{5}{7}$). Here, however, we must note that the "upper quantity" and "lower quantity" do not represent the numerator and denominator

of a fraction. Instead, the instructions tell us to divide $|a_1 - a_2|$ into $da_1 + ea_2$ and $d + e$, as shown below:

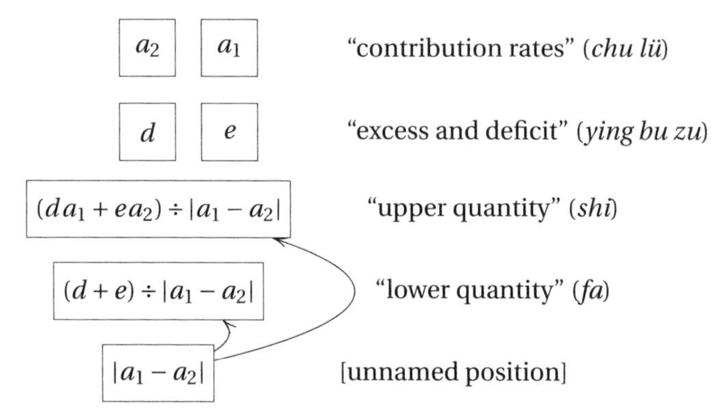

with the result that "the 'upper quantity' is the price of the object, and the 'lower quantity' is the number of people" (lines 65 and 66 on page 63), which is precisely the result we found above in equations (4.35) and (4.36).

Chapter 5
Fangcheng, Chapter 8 of the *Nine Chapters*

This chapter will focus on the *fangcheng* procedure, as presented in the *Nine Chapters*, to elucidate several key features: (1) *fangcheng* problems are displayed in two dimensions on the counting board; (2) entries are eliminated, in a manner similar to Gaussian elimination, by a form of "cross multiplication," *bian cheng* 徧乘, which involves "cross multiplying" an entire column by one entry; and (3) the approach to back substitution in the *fangcheng* procedure differs from the intuitive approach familiar from modern linear algebra. This chapter comprises two sections:

1. The first, "The *Fangcheng* Procedure," examines in detail how *fangcheng* problems were solved on the counting board, including both elimination and back substitution.
2. The second, and more brief, section, "Procedure for Positive and Negative Numbers," examines the rules presented in chapter 8 of the *Nine Chapters* for handling positive and negative numbers.

The *Fangcheng* Procedure

The original text of the *Nine Chapters* presents only this one "*fangcheng* procedure" for solving *all* the problems in Chapter 8, supplemented by the "procedure for positive and negative [numbers]" to deal with negative numbers.

Definitions of Fangcheng

Before proceeding, however, we must first examine the precise meaning of the term *fangcheng* 方程, which, as noted previously, is sometimes translated as "rectangular arrays" or "matrices."[1] Whereas the first character *fang*, which means

[1] For a discussion of this issue, see Martzloff [1987] 2006, 250–51; Chemla and Guo 2004, 922–23.

"rectangle" or "square," is unambiguous, early Chinese sources provide different interpretations of the second character *cheng*:

1. The earliest extant commentary, by Liu Hui, dated 263 C.E.—possibly two centuries after the date of the compilation of the *Nine Chapters*[2]—defines *cheng* as "measures," citing the nonmathematical term *kecheng* 課程, which means "collecting taxes according to tax rates." Liu then defines *fangcheng* as a "rectangle of measures." Liu's commentary states:

   "Measure" (*cheng*) means "to find the measure" (*kecheng*). The collection of objects combines the heterogeneous; each [object] is displayed as having a number, and the sum together is stated as the constant term. Each column is formulated as the terms of a ratio: two objects with two measures, three objects with three measures, all are given measures in accord with the number of objects, and placed side-by-side as columns. Therefore, it is called a "rectangular array" [lit., "rectangle of measures"].

   程，課程也。羣物總雜，各列有數，總言其實。令每行為率，二物者再程，三物者三程，皆如物數程之，并列為行，故謂之方程。(JZSS, *juan* 8, 1b).

   The term *kecheng* 課程, however, is not a mathematical term—it appears nowhere else in the *Nine Chapters*, and in fact does not appear in any of the other mathematical texts collected in the *Complete Collection of the Four Treasuries*; outside of mathematics, *kecheng* is a term most commonly used for collecting taxes.

2. Li Ji's 李籍 *Nine Chapters on the Mathematical Arts, Pronunciations and Meanings* (*Jiuzhang suanshu yin yi* 九章算術音義, 11th century C.E.), an early commentary that offers definitions of important terms in the *Nine Chapters*, also glosses *cheng* as "measure," again using a nonmathematical term, *kelü* 課率, commonly used for taxation. Li's definition of *fangcheng* states,

   Fang means [on the] left and right. *Cheng* means terms of a ratio. Terms of a ratio [on the] left and right, combining together numerous objects, therefore [it] is called a "rectangular array."

   方程，直成切。方者左右也。程者，課率也。左右課率，總統羣物故曰方程。(JZSS, *yin yi*, 21b).

   The term that Li uses to explain *cheng* 程 is *kelü* 課率, which again is not a technical mathematical term, but instead is most commonly used to mean "rates of taxation" (*shui lü* 稅率).

3. Yang Hui's 楊輝 *Nine Chapters on the Mathematical Arts, with Detailed Explanations* (*Xiang jie jiuzhang suanfa* 詳解九章算法, 1261), defines *cheng* as a general term for measuring weight, height, and length. *Detailed Explanations* states,

[2] On the dating of the *Nine Chapters*, see footnote 3 on page 30.

> What is called "rectangular" (*fang*) is the shape of the numbers; "measure" (*cheng*) is the general term for [all forms of] measurement, also a method for equating weights, lengths, and volumes, especially referring to measuring clearly and distinctly the greater and lesser.
>
> 謂方者，數之形也。程者，量度之總名，亦權衡丈尺斛之平法也，（尢）〔尤〕課分明多寡之義。(YHJZ, 21a).

In sum, Chinese explanations of the term *fangcheng*—ambiguous philological glosses from many centuries later—do not give us a definitive answer to the question of what the term meant at the time of the compilation of the *Nine Chapters.* For these reasons, *fangcheng* is sometimes left untranslated, as I have chosen to do in this book.[3] But as Martzloff notes, while "philological explanations leave room for doubt, the course of the calculations may be interpreted without ambiguity" (Martzloff [1987] 2006, 251).

The Procedure

The "*fangcheng* procedure" (supplemented by the "procedure for positive and negative [numbers]") is in many ways similar to Gaussian elimination as presented in modern textbooks on linear algebra.[4] There are, however, some important differences from the modern approach.

Fangcheng problems in the *Nine Chapters* are solved using counting rods. According to the "*fangcheng* procedure," the counting rods are placed in columns, where each condition is represented by a column. According to the "procedure for positive and negative numbers," red counting rods are used for positive numbers, and black counting rods for negative numbers. Throughout this book, however, because of typographic limitations, black will be used for positive rods, and bold black for negative rods (for example, || and **||** for 2 and −2, respectively).

The Chinese array, as placed on the counting board, is similar to the corresponding augmented matrix familiar from modern linear algebra, except for the system for writing numbers, and the orientation of the matrix. In the diagrams below, the position of each entry, when laid out using Chinese counting rods, is shown below on the left, and the augmented matrix, as written in modern linear algebra, is shown on the right:

---

[3] See for example Chemla and Guo 2004.

[4] The "*fangcheng* procedure" (*fangcheng shu* 方程術) is given in the *Nine Chapters* immediately after problem 1; the "procedure for positive and negative [numbers]" (*zheng fu shu* 正負術) is presented immediately after problem 3. Descriptions of the "*fangcheng* procedure" and "procedure for positive and negative [numbers]" are available in recent translations of the *Nine Chapters,* such as Chemla and Guo 2004, and Shen, Lun, and Crossley 1999, and in standard references on Chinese mathematics such as Martzloff [1987] 2006, and Li and Du 1987. There are of course many variations on Gaussian elimination. Standard introductions to modern linear algebra include Strang [1976] 1988; Hoffman and Kunze [1961] 1971; Horn and Johnson 1985; MacLane and Birkhoff 1979; and Hungerford 1974.

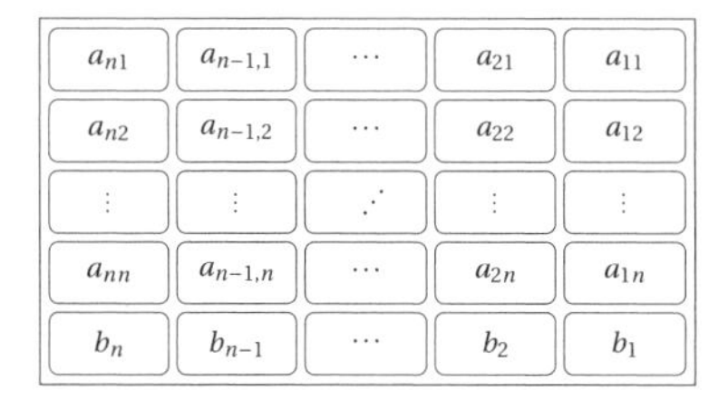

$$\begin{bmatrix} a_{11} & a_{12} & \cdots & a_{1n} & b_1 \\ a_{21} & a_{22} & \cdots & a_{2n} & b_2 \\ \vdots & \vdots & \ddots & \vdots & \vdots \\ a_{n1} & a_{n2} & \cdots & a_{nn} & b_n \end{bmatrix}.$$

Arrays of Chinese counting rods, when rotated counterclockwise 90 degrees, correspond to the modern orientation, and therefore statements about the counting board are equivalent to corresponding statements about the augmented matrix, and vice versa.

This difference in orientation, unfortunately, leads to difficulties with notation. In Chinese arrays, each condition is laid out as a column; each row consists of the coefficients for one of the unknowns, with the constant terms at the bottom. In modern matrix notation, each equation corresponds to a row, and each column consists of the coefficients of one unknown, with the constant terms in the right-hand column. For clarity, in this first problem, and where appropriate later in this book, I will note the corresponding position in the augmented matrix using brackets: for example, "the third entry in the fourth column [row]." In general, unless specified otherwise, the term "column" will always refer to a column in a Chinese array, where it represents one condition; the term "row" will refer to a row in the corresponding augmented matrix, where it represents the corresponding equation.[5] I will use the term "elimination" for both modern matrices and Chinese arrays (rather than, for example, "row reductions" in the former case) to refer to the process of elimination of entries by what is often now called "Gaussian elimination."

Problem 1

Here, we will analyze in detail Problem 1, the only passage in the *Nine Chapters* where the *fangcheng* procedure is explained. The 18 problems in chapter 8 all give the instruction "follow the *fangcheng* procedure" (*ru fangcheng* 如方程), sometimes followed by an instruction to use the "procedure for positive and negative [numbers]," and often with little else.

---

[5] Many recent studies have chosen to transcribe the equations as columns of an augmented matrix. Though this is convenient in the sense that columns on the Chinese counting board then correspond to columns in the matrix, it is contrary to standard practice in modern linear algebra, where rows always represent equations, and leads to confusion in translating these operations into modern mathematics. To avoid this, in this book, translations of Chinese diagrams will preserve the orientation in the Chinese original, but will not use brackets or parentheses, which might make them look like matrices; all matrices will be written in the standard manner. The purpose here is simply to present as accurate a translation as possible of Chinese mathematics into English and modern mathematics.

Translation of Problem 1

The following is a translation of the first problem presented in "*Fangcheng*," chapter 8 of the *Nine Chapters*, followed by the method of solution presented in the text.[6]

> Given 3 bundles[7] of superior paddy [unhusked rice],[8] 2 bundles of ordinary paddy, and 1 bundle of inferior paddy, [together they yield] 39 *dou*[9] of grain;
> 70 2 bundles of superior paddy, 3 bundles of ordinary paddy, and 1 bundle of inferior paddy [together yield] 34 *dou* of grain; 1 bundle of superior paddy, 2 bundles of ordinary paddy, and 3 bundles of inferior paddy [together yield] 26 *dou* of grain. Problem: 1 bundle of superior, ordinary, and inferior paddy each yield how much grain?
>
> 今有上禾三秉，中禾二秉，下禾一秉，實三十九斗；上禾二秉，中禾三秉，下禾一秉，實三十四斗；上禾一秉，中禾二秉，下禾三秉，實二十六斗。問上、中、下禾實一秉各幾何？
>
> 75 Answer: 1 bundle of superior paddy [yields] $9\frac{1}{4}$ *dou*,[10] 1 bundle of ordinary paddy [yields] $4\frac{1}{4}$ *dou*, and 1 bundle of inferior paddy [yields] $2\frac{3}{4}$ *dou*.
>
> 答曰：上禾一秉，九斗四分斗之一；中禾一秉，四斗四分斗之一；下禾一秉，二斗四分斗之三。
>
>> [Liu Hui's commentary:][11] The [columns to the] left and right of each column are not equal,[12] and furthermore the statement [of the problem] has some basis.[13] This is a general method,[14] and difficult to un-
>> 80 derstand through empty words, so it is especially important to use [as a concrete example] the paddy [in problem 1] to solve it. Also, place the middle and left columns just as the right column.[15]

---

[6] For a discussion in English, see Martzloff [1987] 2006, Li and Du 1987; in French, Chemla and Guo 2004; in Chinese, see Bai 1990 and Guo 1998.

[7] According to the earliest commentary, the term *bing* means bundle.

[8] Though the term "paddy" is more commonly used to mean "rice field," it also means "unhusked rice"; a single word seems preferable to two in the translation of this problem, so I have followed Shen, Lun, and Crossley 1999, 404, in translating *he* 禾 as "paddy."

[9] Preserved archaeological artifacts show that 1 *dou* was equal to approximately 2 dry liters, more precisely, between 1.98 and 2.10 liters. A helpful summary of the units of measure used in the *Nine Chapters* explanation can be found in Shen, Lun, and Crossley 1999, 6–12.

[10] Literally, "nine *dou* and one quarter–*dou*," which I have translated here using "mixed" fractions; improper fractions will always be used in mathematical calculations.

[11] Liu Hui's definition of *fangcheng* was translated above on page 68, so I have omitted it here.

[12] That is, no two columns [rows] are the same, a necessary (but not sufficient) condition for there to be a unique solution.

[13] This condition seems to express, albeit in a vague manner, an understanding that even if none of the columns are duplicates, there may be no solution.

[14] The term *dou shu* is used elsewhere in the *Nine Chapters*.

[15] This sentence seems to be out of place.

行之左右無所同存，且為有所據而言耳。此都術也，以空言難曉，故特繫之禾以決之。又列中〔、左〕[16]行如右行也。

Technique: In the right [column] of the array, place 3 for the [number of] bundles of superior paddy, 2 for the [number of] bundles of ordinary paddy,
85 1 for the [number of] bundles of inferior paddy, and 39 for the [number of] *dou* of grain. The middle and left [columns of numbers of bundles of] paddy are arranged [in a manner] similar to that for the right [column].[17] Take the number [of bundles] of superior paddy in the right column [namely, 3], and multiply term-by-term[18] the middle column $[3 \cdot (2,3,1,34) = (6,9,3,102)]$.[19]
90 Then subtract term-by-term[20] [the right column twice, giving for the middle column $(6,9,3,102) - (3,2,1,39) - (3,2,1,39) = (0,5,1,24)$].[21]

術曰：置上禾三秉，中禾二秉，下禾一秉，實三十九斗于右方。中、左（禾）〔方〕[22]列如右方。以右行上禾徧乘中行，而以直除。

[Liu Hui's commentary:] The purpose of using this method is to subtract columns with few [entries] from columns with many [entries],[23] repeatedly subtracting one from another,[24] thus the leading posi-
95 tion(s) must first be eliminated. Less one position, thus this column also is less one object. But using the ratios in order to subtract one from the other does not harm the measure of the remaining numbers. If the leading entry is eliminated, then below the [contribution of that entry to the] solution is eliminated. In this manner, repeat-
100 edly subtract the left and right columns from one another, taking into account the positive and negative, then [the result] can be obtained and known. First "take the superior paddy of the right column and multiply the middle,"[25] means to make uniform and combine.[26] To

---

[16] Inserted following Qian.

[17] For the initial layout of the counting rods, see diagram 5.2 on page 76.

[18] The term *biancheng* 徧乘 literally means "everywhere multiply"; Martzloff [1987] 2006 translates this as "multiply throughout" (253).

[19] See diagram 5.3 on page 76.

[20] The term *zhi chu* 直除 literally means "directly subtract"; Martzloff [1987] 2006 translates this as "direct reduction" (253).

[21] See diagram 5.4 on page 76.

[22] Qian emends 禾 to 行, but 方 is more precise here.

[23] Other translations have interpreted this as taking the columns with the smaller values to subtract from larger values (Chemla and Guo 2004; Shen, Lun, and Crossley 1999). This seems incorrect: *duo* and *shao* should be many and few, not large and small. But more important, it does not matter whether the individual entry is large or small, since they are cross multiplied. This appears to be a more general methodological reflection, rather than a specific instruction for this problem—indeed, in this example the column with the larger value, 3, is subtracted from the column with the smaller value, 2, after cross multiplication.

[24] That is, instead of multiplying by a constant, subtract multiple times.

[25] Quoted from above, but with *cheng* instead of *biancheng*.

[26] Terms used for fractions, here applied by analogy to *fangcheng*.

"make uniform and combine,"[27] means the middle column subtracts
105 term-by-term from the right column. Although for the sake of simplicity, "make uniform and combine" is not stated, looking at it from the meaning of "make uniform and combine," its purpose becomes clear.

為術之意，令少行減多行，反覆相減，則頭位必先盡。上無一位，則此行亦闕一物矣。然而舉率以相減，不害餘數之課也。若消去頭位，則下去一物之實。如是叠令左右行相減，審其正負，則可得而知。先令右行上禾乘中行，為齊同之意。為齊同者，謂中行直減[28]右行也。從簡易雖不言齊同，以齊同之意觀之，其義然矣。

Then multiply the next [namely, three times the left column, $3 \cdot (1,2,3,26) =$
110 $(3,6,9,78)$], and again subtract term-by-term [the right column, yielding in the left column $(0,4,8,39)$].[29]

又乘其次，亦以直除。

[Liu Hui's commentary:] Again eliminate the first [entry] of the left column.

復去左行首。

Then take [the number of] the ordinary paddy in the middle column, and
115 if it is not exhausted,[30] multiply [it] term-by-term with the left column [$5 \cdot (0,4,8,39) = (0,20,40,195)$], and with the result subtract term-by-term [four times the middle column, $4 \cdot (0,5,1,24) = (0,20,4,96)$, yielding in the left column $(0,0,36,99)$].[31]

然以中行中禾不盡者徧乘左行，而以直除。

[Liu Hui's commentary:] Also letting the two columns [middle and
120 left] multiply together to eliminate the ordinary paddy of the [left] column.

亦令兩行相（乘）[32]去行之中禾也。

[27] The term *qi tong* 齊同, a procedure to find a common denominator, is sometimes translated as "homogenize and uniformize."

[28] Dai suggests in a philological note emending 直減 to 上禾亦乘, so that the entire sentence reads 謂中行上禾亦乘右行也. Guo rejects this.

[29] See diagram 5.6 on page 77.

[30] Literally, *jin* means "exhausted" that is, non-zero—in this case it is 5. This seems to indicate a recognition that one could encounter zero pivots.

[31] See diagram 5.8 on page 77.

[32] Qian removes *cheng* 乘.

If the inferior paddy in the left [column] of the array is not exhausted,[33] take the upper [number, namely 36] as the divisor, and the lower [number, namely 99] as the dividend, namely the [total amount of] grain from the inferior paddy.[34]

左方下禾不盡者，上為法，下為實，實即下禾之實。

[Liu Hui's commentary:] The [entries for the] superior and ordinary paddies have both been eliminated [in the left column]. Then the number of the remainder [99] is correct for the [total amount of] grain [from 36 bundles] of inferior paddy, [but] not correct for [the amount of grain from] only one bundle. In order to reduce the [total amount of] grain from many bundles, [we] should take the number of bundles of paddy [36] as the divisor. Display this, and multiply the two columns [middle and right] by the number of bundles of inferior paddy [$36 \cdot (0,5,1,24) = (0,180,36,864)$ and $36 \cdot (3,2,1,39) = (108,72,36,1404)$]. Then subtract term-by-term [$(0,180,36,864)-(0,0,36,99) = (0,180,0,765)$ and $(108,72,36,1404)-(0,0,36,99) = (108,72,0,1305)$]. The position of the inferior paddy is then finished. If [the number of bundles in] the remaining one position [in the left column, 36] is divided into its [corresponding number for the total amount of] grain, then this is just the number of *dou* [from one bundle, $\frac{99}{36}$]. Using [this], the calculation is difficult and not efficient, so another method is presented to reduce it. But it is as if it is not as good as using the old "wide-difference" method.[35]

上、中禾皆去，故餘數是下禾實，非但一秉。欲約衆秉之實，當以禾秉數為法。列此，下[36]禾之秉（實）〔數〕[37]乘兩行，以直除，則下禾之位自[38]決矣。（若）〔各〕[39]以其餘一位之秉除其下實，即斗[40]數矣。用算繁而不省，所以別為法約也。然猶不如自用其舊，廣異法也。

[33] Again, non-zero, in this case it is 36.

[34] This final phrase clarifies the connection between the two distinct uses of the term *shi* (dividend and grain) here: "*shi* [the dividend] is just the [amount] of the *shi* [grain] from the inferior paddy." Note that this is not the amount of grain from one bundle of inferior paddy, but rather from 36 bundles. Also note that the value of the unknown, $\frac{11}{4}$, is not calculated at this point—that is, there is no instruction to divide 36 into 99, or to reduce the fraction (*shi ru fa de* 實如法得 or *shi ru fa de yi* 實如法得一). Note that Liu's approach differs from the original text.

[35] The meaning of the term *guang yi* 廣異 is not clear. The remainder of this translation will focus on the original text, and omit Liu Hui's commentary.

[36] Guo restores 以, which Dai removed, for 以下.

[37] Emended following Qian.

[38] In SKQS, ZKJD. Guo emends to 皆.

[39] Emended following Guo 1990.

[40] Guo rejects Dai's emendation of 斗 for 計.

To solve for ordinary paddy, take the “divisor” [36], and multiply [it] by the [number for] grain in the lower [position] of the middle column [36×24],[41] and subtract the [amount of] grain from the inferior paddy [$36 \times 24 - 99 = 765$].[42]

求中禾，以法乘中行下實，而除下禾之實。

Then take the remainder [765 as the numerator], with the number of bundles of ordinary paddy [5 as the denominator], and reduce the fraction [yielding 153], namely the [amount of] grain from the ordinary paddy [to be divided by 36 below].

餘，如中禾秉數而一，即中禾之實。

To solve for superior paddy, also multiply the divisor [36] by the lower [amount of grain] in the right column [39], and subtract the [amount of grain from] inferior and ordinary paddy, [following calculations similar to those above, yielding $(36 \times 39 - 2 \times 153 - 99) \div 3 = 333$ for] the remainder, namely the amount of grain from the superior paddy.

求上禾，亦以法乘右行下實，而除下禾、中禾之實。

The amounts of grain [99, 153, 333, respectively] are all divided by the divisor [36], giving the [amount of grain] of each as a mixed fraction [$2\frac{3}{4}$, $4\frac{1}{4}$, and $9\frac{1}{4}$, respectively].

餘，如上禾秉數而一，即上禾之實。實皆如法，各得一。

Elimination on the Counting Board

Problem 1, as it is presented in the *Nine Chapters*, is quite difficult to follow, even with the commentary of Liu Hui. To understand this problem would have required considerable training in *fangcheng* as a mathematical practice. We will see that the method of elimination is essentially the same as that presented in modern expositions of linear algebra.[43] Elimination proceeds as follows:

STEP 1. The initial placement of counting rods for problem 1, following the instructions in the text, is as follows:

𝍠	𝍡	𝍢
𝍡	𝍢	𝍡
𝍢	𝍠	𝍠
𝍪𝍥	𝍫𝍣	𝍫𝍨

[41] The term *xia shi* 下實 is specialized terminology for matrices, namely the constant terms.

[42] It should be noted that this instruction, though correct in this calculation, is incomplete in the sense that it is not in general true, if this problem is to serve as an exemplar for more general solutions. It should state that the amount of grain from inferior paddy [99] must be multiplied by the number of bundles of inferior paddy in the middle column [1], yielding $36 \times 24 - 99 \times 1 = 765$. Because the number of bundles of inferior paddy in the middle column is 1, this omission does not affect this calculation.

[43] For a standard account, see Strang [1976] 1988.

In modern notation, problem 1 can be written as follows:

$$\begin{pmatrix} 3 & 2 & 1 \\ 2 & 3 & 1 \\ 1 & 2 & 3 \end{pmatrix} \begin{pmatrix} x_1 \\ x_2 \\ x_3 \end{pmatrix} = \begin{pmatrix} 39 \\ 34 \\ 26 \end{pmatrix},$$

which is solved by writing the system in the form of an augmented matrix,

$$\begin{bmatrix} 3 & 2 & 1 & 39 \\ 2 & 3 & 1 & 34 \\ 1 & 2 & 3 & 26 \end{bmatrix}. \tag{5.1}$$

To show the similarity of the operations performed when using counting rods to those when using modern calculations, the first problem in Chapter 8, when laid out with Chinese counting rods, is shown below on the left; in modern linear algebra, the augmented matrix given by (5.1) above, is shown on the right. As noted above, they are essentially the same (except for the orientation and the system for writing numbers),

$$\begin{bmatrix} 3 & 2 & 1 & 39 \\ 2 & 3 & 1 & 34 \\ 1 & 2 & 3 & 26 \end{bmatrix}. \tag{5.2}$$

STEP 2. The second column of counting rods [that is, the second row in the augmented matrix] is multiplied term-by-term by 3, yielding for the second column [row] $3 \cdot (2,3,1,34) = (6,9,3,102)$,

$$\begin{bmatrix} 3 & 2 & 1 & 39 \\ 6 & 9 & 3 & 102 \\ 1 & 2 & 3 & 26 \end{bmatrix}. \tag{5.3}$$

STEP 3. Next, the first column [row] is subtracted twice from the second row, giving $(6,9,3,102) - (3,2,1,39) - (3,2,1,39) = (0,5,1,24)$,

$$\begin{bmatrix} 3 & 2 & 1 & 39 \\ 0 & 5 & 1 & 24 \\ 1 & 2 & 3 & 26 \end{bmatrix}. \tag{5.4}$$

STEP 4. The next step is to multiply the third column [row] by 3, $3 \cdot (1,2,3,26) = (3,6,9,78)$,

$$\begin{bmatrix} 3 & 2 & 1 & 39 \\ 0 & 5 & 1 & 24 \\ 3 & 6 & 9 & 78 \end{bmatrix}. \tag{5.5}$$

STEP 5. We then subtract the first column [row], yielding $(3,6,9,78)-(3,2,1,39) = (0,4,8,39)$,

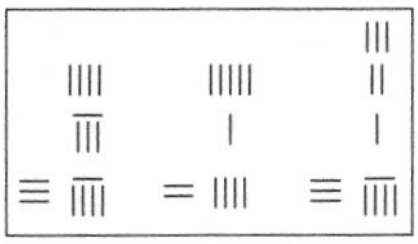

$$\begin{bmatrix} 3 & 2 & 1 & 39 \\ 0 & 5 & 1 & 24 \\ 0 & 4 & 8 & 39 \end{bmatrix}. \tag{5.6}$$

STEP 6. We multiply the third column [row] by 5, $5 \cdot (0,4,8,39) = (0,20,40,195)$,

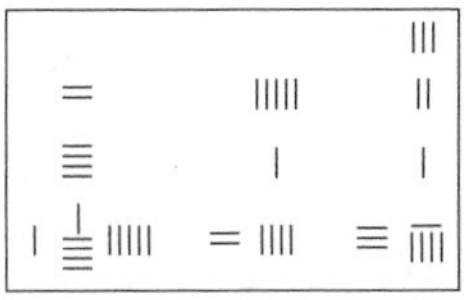

$$\begin{bmatrix} 3 & 2 & 1 & 39 \\ 0 & 5 & 1 & 24 \\ 0 & 20 & 40 & 195 \end{bmatrix}. \tag{5.7}$$

STEP 7. Finally, we subtract the second column [row] four times, $(0,20,40,195)-(0,5,1,24)-(0,5,1,24)-(0,5,1,24)-(0,5,1,24) = (0,0,36,99)$, arriving at the lower- [upper-] triangular form,

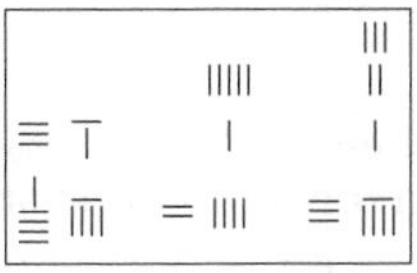

$$\begin{bmatrix} 3 & 2 & 1 & 39 \\ 0 & 5 & 1 & 24 \\ 0 & 0 & 36 & 99 \end{bmatrix}. \tag{5.8}$$

The counting board [augmented matrix] is now in lower- [upper-] triangular form, and we find the solution by back substitution.

Back Substitution in Modern Linear Algebra

The method for solution presented in the *fangcheng* procedure for solving (5.8) differs from the conventional approach familiar from modern mathematics texts. In the conventional modern approach, the augmented matrix in (5.8) represents the following equations,

$$3x_1 + 2x_2 + x_3 = 39, \tag{5.9}$$

$$5x_2 + x_3 = 24, \tag{5.10}$$

$$36x_3 = 99. \tag{5.11}$$

STEP 1. We first calculate the value of x_3 using equation (5.11), yielding

$$x_3 = \frac{99}{36} = \frac{11}{4}.$$

STEP 2. Then we substitute this value of x_3 into the equation immediately above, equation (5.10), giving

$$5x_2 + \frac{11}{4} = 24, \text{ or } x_2 = \frac{17}{4}.$$

Note that before the final division by 5 (the coefficient of x_2), this method requires calculations with fractions, namely,

$$24 - \frac{11}{4} = \frac{85}{4}.$$

STEP 3. The first unknown is then found from equation (5.9) by substituting in the values found for x_3 and x_2, yielding

$$3x_1 + 2 \times \frac{17}{4} + \frac{11}{4} = 39, \text{ or } x_1 = \frac{37}{4}.$$

Again, before the final division by 3 (the coefficient of x_1), this calculation requires subtraction using fractions,

$$39 - 2 \times \frac{17}{4} - \frac{11}{4} = \frac{111}{4}.$$

Note that in this approach, the quantities that are substituted back are easy to conceptualize—they are simply the values of the solution. That is, $\frac{11}{4}$, which is substituted back into equations (5.10) and (5.11), is just the value of x_3, the yield of 1 bundle of inferior paddy. When we solve equation (5.10), the result is $\frac{17}{4}$, the value of x_2, the yield of one bundle of ordinary paddy, which is substituted back into equation (5.11). Although all of this may seem obvious, it will not be the case for the *fangcheng* procedure, where the quantities used for back substitution do not correspond to any easily conceptualizable quantity.

Back Substitution Using the Fangcheng *Procedure*

The "*fangcheng* procedure," apparently in order to avoid the multiplication and addition of fractions, takes a different approach, which may at first seem perplexing. It differs from modern linear algebra, and previous studies have sometimes mistakenly assimilated this to modern methods. The *Nine Chapters* instructs us to solve for the three unknowns as follows:

STEP 1. To calculate the third unknown, instead of dividing 99 by 36, we keep 99 to use in our back substitutions. The value 99, unlike $\frac{11}{4}$, does not correspond to anything: it is $36 \times x_3$, but 36 is nothing more than the result of a calculation in the process of elimination.

STEP 2. To calculate the second unknown, we "cross multiply," subtract, and divide as follows:

$$(36 \times 24 - 99) \div 5 = 153.$$

STEP 3. Then we again "cross multiply," subtract, and divide, yielding

$$(36 \times 39 - 2 \times 153 - 99) \div 3 = 333.$$

STEP 4. Finally, we calculate the unknowns by dividing the results by 36,

$$x_1 = 333 \div 36 = \frac{37}{4},$$
$$x_2 = 153 \div 36 = \frac{17}{4},$$
$$x_3 = 99 \div 36 = \frac{11}{4}.$$

In sum, the *fangcheng* procedure avoids fractions until the very last step. A comparison of the approach to back substitution in modern linear algebra and the *fangcheng* procedure is given in Table 5.1.

Table 5.1: Comparison of the methods of back substitution.

Step	Unknown	*Fangcheng* procedure	Modern
1	3rd	99	$\frac{99}{36} = \frac{11}{4}$
2	2nd	$(36 \times 24 - 99) \div 5 = 153$	$\frac{24 - \frac{11}{4}}{5} = \frac{17}{4}$
3	1st	$(36 \times 39 - 2 \times 153 - 99) \div 3 = 333$	$\frac{39 - 2 \times \frac{17}{4} - \frac{11}{4}}{3} = \frac{37}{4}$

Back Substitution on the Counting Board

The *fangcheng* procedure provides only the instructions translated above; there are no explicit directions describing the placement of entries in the process of back substitution. And again, there are no diagrams in chapter 8 in the transmitted versions of the *Nine Chapters*. We can, however, attempt to reconstruct these calculations on the counting board. The computations turn out to be reasonably simple, and in fact it is easier to complete these calculations using diagrams than it is to explain the computations in modern mathematical terminology. As counterintuitive as back substitution in the *fangcheng* procedure may seem when it is translated into narrative form in classical Chinese, or into formulas using modern mathematical terminology, it is actually quite simple on the counting board, using "cross multiplications" to further "eliminate" entries.

STEP 1. The first step in this process of back substitution is a form of "cross multiplication": we multiply the second constant term, $b_2 = 24$, by the final pivot, $a_{33} = 36$; we then multiply the entry $a_{23} = 1$ by the final constant term, $b_3 = 99$,

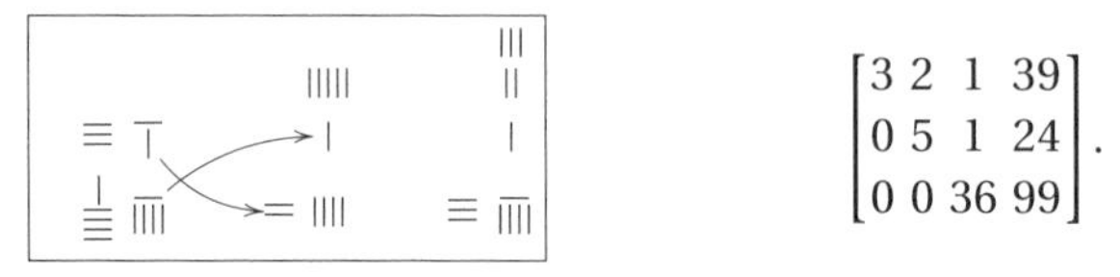

$$\begin{bmatrix} 3 & 2 & 1 & 39 \\ 0 & 5 & 1 & 24 \\ 0 & 0 & 36 & 99 \end{bmatrix}.$$

This gives us the following result on the counting board (note that from this point on, the matrices on the right-hand side no longer correspond to the original system of linear equations):

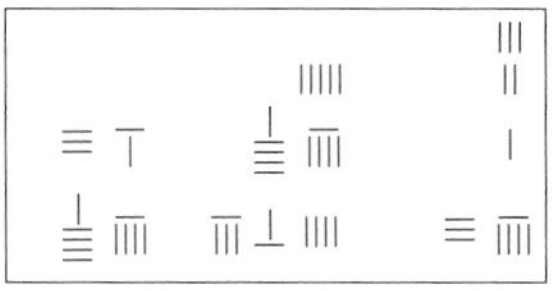

$$\begin{bmatrix} 3 & 2 & 1 & 39 \\ 0 & 5 & 99 & 864 \\ 0 & 0 & 36 & 99 \end{bmatrix}.$$

STEP 2. Next, we "eliminate" the entries in the second column [row], by subtracting 99 from 864, then dividing by 5, giving $(864 - 99) \div 5 = 153$,

$$\begin{bmatrix} 3 & 2 & 1 & 39 \\ 0 & 0 & 0 & 153 \\ 0 & 0 & 36 & 99 \end{bmatrix}.$$

STEP 3. We are now ready for back substitution into the final remaining column [row]. Again, we "cross multiply": we multiply $b_1 = 39$ by $a_{33} = 36$, $a_{13} = 1$ by $b_3 = 99$, and $a_{12} = 2$ by $b_2 = 153$,

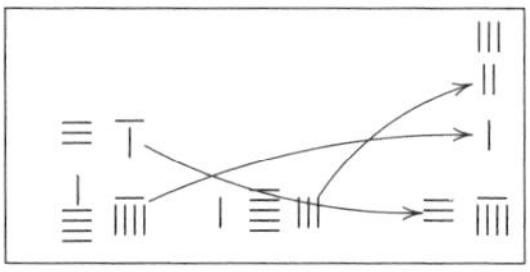

$$\begin{bmatrix} 3 & 2 & 1 & 39 \\ 0 & 0 & 0 & 153 \\ 0 & 0 & 36 & 99 \end{bmatrix}.$$

This gives the following result on the counting board:

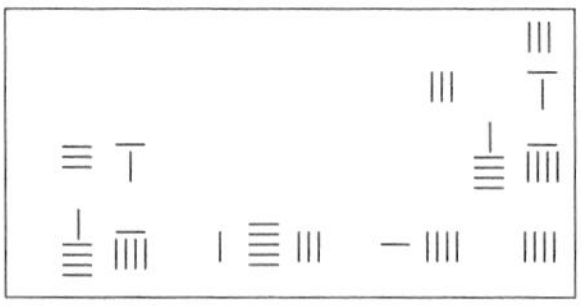

$$\begin{bmatrix} 3 & 306 & 99 & 1404 \\ 0 & 0 & 0 & 153 \\ 0 & 0 & 36 & 99 \end{bmatrix}.$$

STEP 4. Again, we "eliminate" the entries in the final column [row] by subtracting 99 and 306 from 1404 and dividing by 3, $(1404 - 99 - 306) \div 3 = 333$,

$$\begin{bmatrix} 0 & 0 & 0 & 333 \\ 0 & 0 & 0 & 153 \\ 0 & 0 & 36 & 99 \end{bmatrix}.$$

STEP 5. The next step is to divide 36 into the constant terms

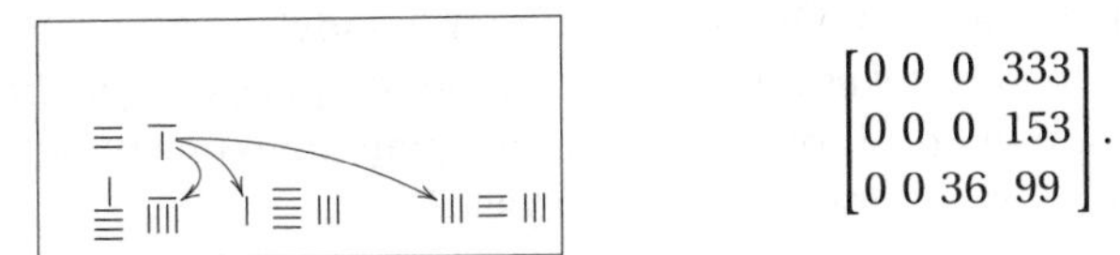

$$\begin{bmatrix} 0 & 0 & 0 & 333 \\ 0 & 0 & 0 & 153 \\ 0 & 0 & 36 & 99 \end{bmatrix}.$$

This might be arranged on the counting board as follows, by adding another entry below each column,

$$\begin{bmatrix} 0 & 0 & 0 & 333/36 \\ 0 & 0 & 0 & 153/36 \\ 0 & 0 & 0 & 99/36 \end{bmatrix}.$$

STEP 6. Finally, each of the fractions is reduced, giving the answer,

$$\begin{bmatrix} 0 & 0 & 0 & 37/4 \\ 0 & 0 & 0 & 17/4 \\ 0 & 0 & 0 & 11/4 \end{bmatrix}.$$

Procedure for Positive and Negative Numbers

The only other procedure given in "*Fangcheng*," chapter 8 of the *Nine Chapters*, is the "procedure for positive and negative [numbers]" (*zheng fu shu* 正負術), which follows problem 3.

Translation of Problem 3

In problem 3, a negative number emerges in the process of elimination, and the "procedure for positive and negative [numbers]" is presented. Problem 3, translated into English, is as follows:

> Given two bundles of superior paddy [unhusked rice],[44] three bundles of ordinary paddy, and four bundles of inferior paddy, the grain from each is less than one *dou* [1 *dou* ≈ 2 dry liters].[45] If [two bundles of] the superior

[44] See footnote 8 on page 71.

[45] For an explanation of Chinese units of dry measure capacity, see Shen, Lun, and Crossley 1999, 9-10: preserved archaeological artifacts show that 1 *dou* was equal to between 1.98 and 2.10 liters.

[paddy] takes one bundle of ordinary [paddy], [three bundles of] the ordinary [paddy] takes one bundle of inferior [paddy], and [four bundles of] the inferior [paddy] takes one bundle of superior [paddy], then the grain [in each case] is one *dou*. Problem: how much grain does one bundle of superior, ordinary, and inferior paddy each yield?

今有上禾二秉，中禾三秉，下禾四秉，實皆不滿斗。上取中，中取下，下取上各一秉而實滿斗。問上、中、下禾實一秉各幾何？

Solution: The superior paddy yields $\frac{9}{25}$ *dou* of grain, the ordinary paddy yields $\frac{7}{25}$ *dou* of grain, and the inferior paddy yields $\frac{4}{25}$ *dou* of grain.

荅曰：上禾一秉實二十五分斗之九，中禾一秉實二十五分斗之七，下禾一秉實二十五分斗之四。

Method: Use the *fangcheng* procedure; place what is taken in each case; apply the procedure for positive and negative [numbers].

術曰：如方程，各置所取，以正負術入之。

Elimination on the Counting Board

The calculations on the counting board are as follows:

STEP 1. The initial placement of the counting rods is as follows:

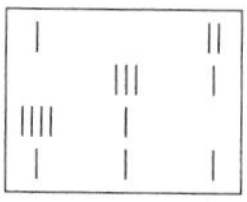

```
|          ||
     |||   |
||||  |
|     |    |
```

$$\begin{bmatrix} 2 & 1 & 0 & 1 \\ 0 & 3 & 1 & 1 \\ 1 & 0 & 4 & 1 \end{bmatrix}.$$

STEP 2. Next, we multiply the third column [row] by 2 and subtract the first column [row], $2 \cdot (1,0,4,1) - (2,1,0,1) = (0,-1,8,1)$. Note that when, following the "*fangcheng* procedure" outlined above, we eliminate the entry 1 in the upper left-hand corner of the counting rods (the lower left-hand entry in the augmented matrix), the result is a negative number, that is, the left-hand column (bottom row of the augmented matrix) becomes $(0,-1,8,1)$. It is immediately following this problem that negative numbers and the "procedure for positive and negative [numbers]" (*zheng fu shu* 正負術) are introduced. (In the following diagram, note that the negative value is denoted by a counting rod in bold black typeface.)

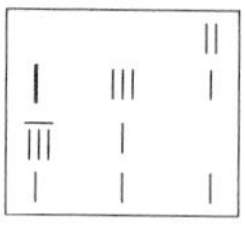

```
           ||
|    |||   |
T|||  |
|     |    |
```

$$\begin{bmatrix} 2 & 1 & 0 & 1 \\ 0 & 3 & 1 & 1 \\ 0 & -1 & 8 & 1 \end{bmatrix}.$$

STEP 3. Next, we multiply the third column [row] by 3 and add the second column [row], $3 \cdot (0,-1,8,1) + (0,3,1,1) = (0,0,25,4)$,

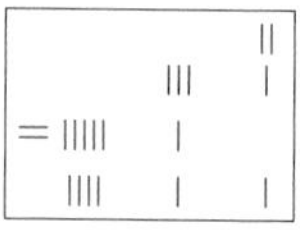

```
              ||
         |||  |
= |||||   |
  ||||    |   |
```

$$\begin{bmatrix} 2 & 1 & 0 & 1 \\ 0 & 3 & 1 & 1 \\ 0 & 0 & 25 & 4 \end{bmatrix}.$$

Back Substitution on the Counting Board

To solve by back substitution, we calculate the following quantities,

$$(25 \times 1 - 4) \div 3 = 7,$$
$$(25 \times 1 - 7) \div 2 = 9,$$

and the solution is then

$$x_1 = 9 \div 25 = \frac{9}{25}, \quad x_2 = 7 \div 25 = \frac{7}{25}, \quad x_3 = 4 \div 25 = \frac{4}{25}.$$

On the counting board, as I have explained above, the calculations were likely performed in the following manner:

STEP 1. As in the previous example, the first step in back substitution is to multiply the second constant term, $b_2 = 1$, by the final pivot, $a_{33} = 25$, and the entry $a_{23} = 1$ by the final constant term, $b_3 = 4$,

$$\begin{bmatrix} 2 & 1 & 0 & 1 \\ 0 & 3 & 1 & 1 \\ 0 & 0 & 25 & 4 \end{bmatrix}.$$

This yields the following result on the counting board (again, note that the matrices on the right-hand side do not correspond to the system of linear equations):

$$\begin{bmatrix} 2 & 1 & 0 & 1 \\ 0 & 3 & 4 & 25 \\ 0 & 0 & 25 & 4 \end{bmatrix}.$$

STEP 2. Next, we eliminate the entries in the second column [row] by subtracting 4 from 25 and then dividing by 3, giving $(25 - 4) \div 3 = 7$,

$$\begin{bmatrix} 2 & 1 & 0 & 1 \\ 0 & 0 & 0 & 7 \\ 0 & 0 & 25 & 4 \end{bmatrix}.$$

STEP 3. Next, we perform back substitution for the final remaining column [row]. We multiply $b_1 = 1$ by $a_{33} = 25$, $a_{13} = 0$ by $b_3 = 4$, and $a_{12} = 1$ by $b_2 = 7$,

$$\begin{bmatrix} 2 & 1 & 0 & 1 \\ 0 & 0 & 0 & 7 \\ 0 & 0 & 25 & 4 \end{bmatrix}.$$

This gives the following result on the counting board:

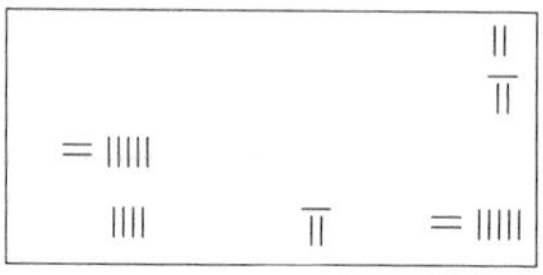

$$\begin{bmatrix} 2 & 7 & 0 & 25 \\ 0 & 0 & 0 & 7 \\ 0 & 0 & 25 & 4 \end{bmatrix}.$$

STEP 4. We then eliminate the entries in the final column [row] by subtracting 7 from 25 and dividing by 2, $(25 - 7) \div 2 = 9$,

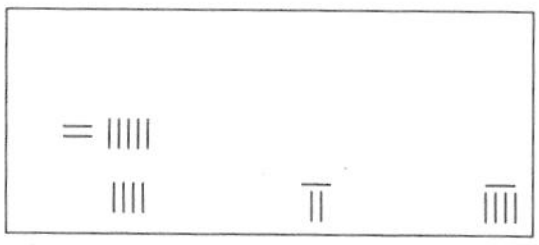

$$\begin{bmatrix} 0 & 0 & 0 & 9 \\ 0 & 0 & 0 & 7 \\ 0 & 0 & 25 & 4 \end{bmatrix}.$$

STEP 5. The last step is to divide 25 into the constant terms,

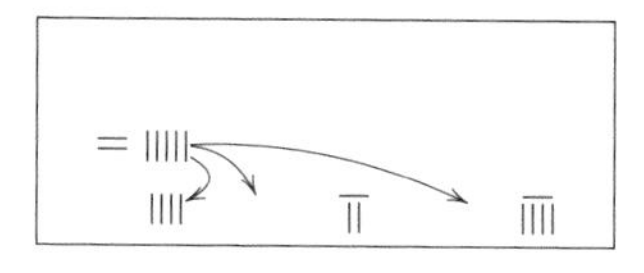

$$\begin{bmatrix} 0 & 0 & 0 & 9 \\ 0 & 0 & 0 & 7 \\ 0 & 0 & 25 & 4 \end{bmatrix}.$$

This might be arranged on the counting board as follows:

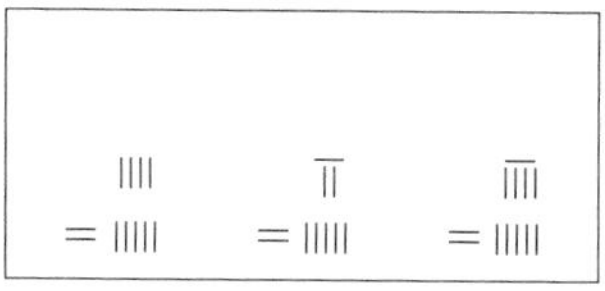

$$\begin{bmatrix} 0 & 0 & 0 & 9/25 \\ 0 & 0 & 0 & 7/25 \\ 0 & 0 & 0 & 4/25 \end{bmatrix}.$$

The Procedure for Positive and Negative

The "procedure for positive and negative [numbers]"
正負術曰：

> Now the two types of counting rods for gain and loss are opposites; in order to name them, let them be called positive and negative. Positive counting rods are red, negative rods are black; otherwise, use slanted and upright [counting rods] to differentiate [them]. In *fangcheng*, from the time there were red and black [counting rods] to take from each other, the procedure for mutually finding the solutions from the unknowns, the state of adding and subtracting did not obtain wide completion, and therefore red and black were used to cancel one another. In calculating, sometimes subtracting and sometimes adding, in the same column different positions represent two objects, each has adding or subtracting, and the difference is seen in

30 the low[est] column. In writing about these two passages, the example using paddy is given; this was specially provided in order to complete the meaning of these two passages. Therefore red and black, distinct from each other, are sufficient to establish the upper and lower; decreasing and increasing, although different, are sufficient to reduce
35 the numbers on the left and right; difference and constant terms, although different, are sufficient to bring into correspondence values (rates) that are similar and different. Thus if the positive has no corresponding entry, it is made negative, and its value is not lost.

今兩算得失相反，要令正負以名之。正算赤，負算黑，否則以邪正為異。方程自有赤、黑相取，（左、右）〔法、實〕數相推求之術，而其并減之勢不得（交）〔廣〕通，故使赤、黑相消奪之。于算或減或益，同行異位殊為二品，各有并、減，之差見于下焉。著此二條，特繫之禾以成此二條之意。故赤、黑相雜足以定上下之程；減、益雖殊足以通左右之數，差、實雖分足以應同異之率。然則其正無人以負之，〔負無入以正之，〕其率不妄也。

Same signs are subtracted,

同名相除，

40 Opposite signs are added [literally, increased],

異名相益，

[Subtracting,] Positive [numbers subtracted] from zero are made negative, negative [numbers subtracted] from zero are made positive.

正無入負之，負無入正之。

[Adding,] Opposite [literally, different] signs are subtracted, similar signs are added. Positive [and] zero [literally, nothing entered] is positive; neg-
45 ative [and] zero [literally, nothing entered] is negative.

其異名相除，同名相益，正無入正之，負無入負之

Conclusions

The *fangcheng* procedure, as recorded in the *Nine Chapters*, solves problems in the following manner: (1) entries are eliminated until the counting board is in lower triangular form; (2) a counterintuitive approach to back substitution is then used; (3) finally, the results are all divided by the final pivot, a''_{nn}. The *fangcheng* procedure, as presented in the *Nine Chapters*, is quite abbreviated. In the next chapter, we will analyze it in more detail and generality, using modern mathematical terminology.

Chapter 6
The *Fangcheng* Procedure in Modern Mathematical Terms

This chapter offers an analysis of the *fangcheng* procedure presented in chapter 8 of the *Nine Chapters,* which, as we saw in the preceding chapter, is presented there only in an abbreviated form. This chapter attempts to reconstruct the *fangcheng* procedure using modern mathematical terminology. The main goals here are the following: (1) to describe in modern mathematical terms what we know about the *fangcheng* procedure; (2) to offer reasonable hypotheses on the many questions left unanswered by the description of the *fangcheng* procedure; (3) to formalize the *fangcheng* procedure as an algorithm; and (4) to show that the counterintuitive approach to back substitution does, in most cases, avoid fractions. The chapter will be divided into the following sections:

1. The first section, "Conspectus of *Fangcheng* Problems in the *Nine Chapters,*" offers a compact summary of the 18 problems in *Fangcheng,* chapter 8 of the *Nine Chapters,* as grounding for our analysis of the problems that were recorded therein.
2. The second section, "Elimination," analyzes what we do know about elimination, the first half of the *fangcheng* procedure, as presented in chapter 8 of the *Nine Chapters*: the types of entries, the placement of the entries on the counting board, the order of elimination, and the resulting lower- [upper-] triangular form.
3. The third section, "Back Substitution," analyzes in general terms the approach to back substitution given in the *fangcheng* procedure as presented in chapter 8, again using modern mathematical terminology.
4. The final section, "Is the *Fangcheng* Procedure Integer-Preserving?" though somewhat technical, presents the key finding of this chapter—that by following the *fangcheng* procedure, in most cases, calculations with fractions are avoided.

Conspectus of *Fangcheng* Problems in the *Nine Chapters*

It may be helpful to present first a conspectus of these problems, toward the end of providing a concise overview for the analysis in the following sections (see Table 6.1 below). I have noted the conventional classification of these problems by their order, that is, by the number of conditions and unknowns. (In chapter 8 of this book I will present an alternative approach to classifying these problems.)

Table 6.1: The 18 problems of chapter 8 of the *Nine Chapters.*

Problem	*Augmented Matrix*	*Notes*
1	$\begin{bmatrix} 3 & 2 & 1 & 39 \\ 2 & 3 & 1 & 34 \\ 1 & 2 & 3 & 26 \end{bmatrix}$	Exemplar for *fangcheng* procedure. Three conditions in three unknowns.
2	$\begin{bmatrix} 8 & 2 & 9 \\ 2 & 7 & 11 \end{bmatrix}$	Two conditions in two unknowns.
3	$\begin{bmatrix} 2 & 1 & 0 & 1 \\ 0 & 3 & 1 & 1 \\ 1 & 0 & 4 & 1 \end{bmatrix}$	Exemplar for procedure for positive and negative numbers. Three conditions in three unknowns.
4	$\begin{bmatrix} 5 & -7 & 11 \\ 7 & -5 & 25 \end{bmatrix}$	Two conditions in two unknowns.
5	$\begin{bmatrix} 6 & -10 & 18 \\ -5 & 15 & 5 \end{bmatrix}$	Two conditions in two unknowns.
6	$\begin{bmatrix} 3 & -10 & -6 \\ -2 & 5 & -1 \end{bmatrix}$	Two conditions in two unknowns.
7	$\begin{bmatrix} 5 & 2 & 10 \\ 2 & 5 & 8 \end{bmatrix}$	Two conditions in two unknowns.
8	$\begin{bmatrix} 2 & 5 & -13 & 1000 \\ 3 & -9 & 3 & 0 \\ -5 & 6 & 8 & -600 \end{bmatrix}$	Three conditions in three unknowns.
9	$\begin{bmatrix} 4 & 1 & 8 \\ 1 & 5 & 8 \end{bmatrix}$	Two conditions in two unknowns.
10	$\begin{bmatrix} 1 & \frac{1}{2} & 50 \\ \frac{2}{3} & 1 & 50 \end{bmatrix}$	Two conditions in two unknowns.
11	$\begin{bmatrix} 1 & 2+\frac{1}{2} & 10000 \\ 2-\frac{1}{2} & 1 & 10000 \end{bmatrix}$	Two conditions in two unknowns.
12	$\begin{bmatrix} 1 & 1 & 0 & 40 \\ 0 & 2 & 1 & 40 \\ 1 & 0 & 3 & 40 \end{bmatrix}$	Three conditions in three unknowns.

Continued on next page

Problem	*Augmented Matrix*	*Notes*
13	$\begin{bmatrix} 2&1&0&0&0&y \\ 0&3&1&0&0&y \\ 0&0&4&1&0&y \\ 0&0&0&5&1&y \\ 1&0&0&0&6&y \end{bmatrix}$	Five conditions in six unknowns (x_1, x_2, x_3, x_4, x_5, and y).
14	$\begin{bmatrix} 2&1&1&0&1 \\ 0&3&1&1&1 \\ 1&0&4&1&1 \\ 1&1&0&5&1 \end{bmatrix}$	Four conditions in four unknowns.
15	$\begin{bmatrix} 2&-1&0&1 \\ 0&3&-1&1 \\ -1&0&4&1 \end{bmatrix}$	Three conditions in three unknowns.
16	$\begin{bmatrix} 1&5&10&10 \\ 10&1&5&8 \\ 5&10&1&6 \end{bmatrix}$	Three conditions in three unknowns.
17	$\begin{bmatrix} 5&4&3&2&1496 \\ 4&2&6&3&1175 \\ 3&1&7&5&958 \\ 2&3&5&1&861 \end{bmatrix}$	Four conditions in four unknowns.
18	$\begin{bmatrix} 9&7&3&2&5&140 \\ 7&6&4&5&3&128 \\ 3&5&7&6&4&116 \\ 2&5&3&9&4&112 \\ 1&3&2&8&5&95 \end{bmatrix}$	Five conditions in five unknowns.

Elimination

In this section we will examine how *fangcheng* were displayed using counting rods on a counting board, and the process of elimination.

Entries

On the basis of the eighteen exemplars recorded in chapter 8, the initial entries in *fangcheng* problems can be characterized as follows:

Positive integers: Most of the entries are positive integers, with the exceptions noted below. The coefficients are usually positive integers ranging from 1 to 9, though occasionally integers between 10 and 15 appear. The constant terms are usually between 1 and 150, though constant terms as large as 10000 are sometimes used.

Negative numbers: Negative integers appear in several problems—4, 5, 6, 8, and 15—usually expressly for the pedagogic purpose of showing how to handle negative numbers. Of course, negative numbers also emerge in the process of elimination, and are handled using the "procedure for positive and negative numbers."

Zeros: Zeros never appear as coefficients in chapter 8 of the *Nine Chapters*, except, as we will see in a later chapter, in the five problems that have determinantal solutions, namely problems 3 and 12–15; when these zeros do appear, they are placed symmetrically in such a manner as to allow determinantal solutions. Zeros appear only once in the constant terms, in problem 8, in which the statements for the three unknowns are formulated as "excess," "equality," and "deficit."

Fractions: Fractions appear in problems 10 and 11, and like negative numbers, they are there for pedagogic purposes, to explain how to handle fractions: problems 10 and 11 provide methods for transforming the fractions into integers by multiplying the entire columns [rows] by a suitable constant, usually the denominator of the fraction. Fractions presumably impose added complexity in problems in which two dimensions are already being used to display the problem as an array, and so are avoided in *fangcheng* computations.

In sum, the entries in *fangcheng* problems are limited to integers (positive, negative, and zero); any fractions that appear are transformed into integers before the process of elimination begins.

Placement and Orientation

The *Nine Chapters* offer only one exemplar, problem 1, for which $n = 3$, that is, 3 conditions in 3 unknowns. There is, however, little question about how to display *fangcheng* of a higher order. We can reconstruct the general form as follows:

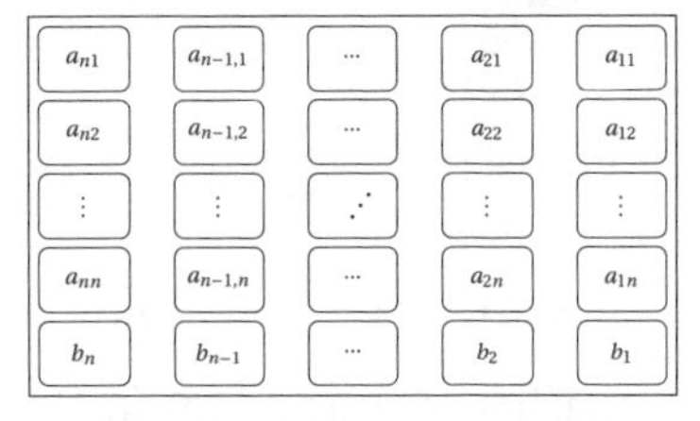

$$\begin{bmatrix} a_{11} & a_{12} & \cdots & a_{1n} & b_1 \\ a_{21} & a_{22} & \cdots & a_{2n} & b_2 \\ \vdots & \vdots & \ddots & \vdots & \vdots \\ a_{n1} & a_{n2} & \cdots & a_{nn} & b_n \end{bmatrix}.$$

That is, the counting rods are placed in columns, so that $a_{11}x_1 + a_{12}x_2 + \cdots + a_{1n}x_n = b_1$, which is the first row in the augmented matrix on the right, becomes the first column from the right in the diagram on the left.

There appears to be no variation in the arrangement or orientation of the placement of entries. That is, whereas chapter 8 of the *Nine Chapters* describes the counting board only in narrative form, providing no diagrams of any kind, later texts often record diagrams of the problems as they would be displayed on the counting board, albeit usually using Chinese characters for the entries rather than representations of the counting rods themselves. All of the texts I have seen follow the above conventions.

Order of Elimination

According to the *fangcheng* procedure as exemplified in the solution to problem 1, entries are eliminated from right to left, and from top to bottom, until reaching lower-triangular form. This corresponds to the order in the usual approach to Gaussian elimination in modern linear algebra—from top to bottom, and from left to right, until we reach upper-triangular form. That is, we proceed by column [row] operations, first eliminating the first entry in the second column [row], the first entry in the third column [row], continuing until we eliminate the first entry in the last column [row]. Similarly, we eliminate the second entry in the third column [row], on to the second entry in the final column [row].

It should be noted that the *fangcheng* procedure, presented in its very abbreviated form, makes no mention of the possibility of opportunistically eliminating entries in a different order, as proposed by Liu Hui in his commentary. Although eliminating entries in a different order may result in more efficiency at the elimination stage, such an approach might lead to a higher likelihood of fractions emerging in the process of back substitution, and hence be less efficient overall, as we will see later in this chapter.

Elimination of Entries

The first important difference between the *fangcheng* procedure and modern methods is in the column [row] reductions. In modern linear algebra, as in the *fangcheng* procedure, we first remove entry a_{21},

$$\begin{bmatrix} a_{11} & a_{12} & \cdots & a_{1n} & b_1 \\ a_{21} & a_{22} & \cdots & a_{2n} & b_2 \\ \vdots & \vdots & \ddots & \vdots & \vdots \\ a_{n1} & a_{n2} & \cdots & a_{nn} & b_n \end{bmatrix}.$$

In many treatments of modern linear algebra, however, a multiplier,

$$l_1 = \frac{a_{11}}{a_{21}},$$

is used to multiply the entire row, so that after subtraction, we have

$$\begin{aligned}&\frac{a_{11}}{a_{21}}(a_{21},a_{22},\cdots,a_{2n},b_2)-(a_{11},a_{12},\cdots,a_{1n},b_1)\\&=(0,\frac{a_{11}a_{22}-a_{12}a_{21}}{a_{21}},\cdots,\frac{a_{11}a_{2n}-a_{1n}a_{21}}{a_{21}},\frac{a_{11}b_2-b_1a_{21}}{a_{21}}).\end{aligned}$$

That is, after the elimination of the first element, we have

$$\begin{bmatrix} a_{11} & a_{12} & \cdots & a_{1n} & b_1 \\ 0 & \frac{a_{11}a_{22}-a_{12}a_{21}}{a_{21}} & \cdots & \frac{a_{11}a_{2n}-a_{1n}a_{21}}{a_{21}} & \frac{a_{11}b_2-b_1a_{21}}{a_{21}} \\ a_{31} & a_{32} & \cdots & a_{3n} & b_3 \\ \vdots & \vdots & \ddots & \vdots & \vdots \\ a_{n1} & a_{n2} & \cdots & a_{nn} & b_n \end{bmatrix}.$$

We continue until we have eliminated all of the entries above [below] the diagonal.

In contrast, in the *fangcheng* procedure, the column [row] with the entry to be eliminated is "multiplied term-by-term" (*bian cheng* 徧乘) by the pivot, a_{11}. That is, the column [row] with the entry to be eliminated is multiplied by a_{11}, from which the pivot row is subtracted a_{21} times (or, equivalently, a_{21} times the pivot row is subtracted), yielding

$$\begin{aligned}&a_{11}(a_{21},a_{22},\cdots,a_{2n},b_2)-a_{21}(a_{11},a_{12},\cdots,a_{1n},b_1)\\&=(0,a_{11}a_{22}-a_{21}a_{12},\cdots,a_{11}a_{2n}-a_{21}a_{1n},a_{11}b_2-a_{21}b_1).\end{aligned}$$

Thus, after the elimination of the first entry, the result is

$$\begin{bmatrix} a_{11} & a_{12} & \cdots & a_{1n} & b_1 \\ 0 & a_{11}a_{22}-a_{21}a_{12} & \cdots & a_{11}a_{2n}-a_{21}a_{1n} & a_{11}b_2-a_{21}b_1 \\ a_{31} & a_{32} & \cdots & a_{3n} & b_3 \\ \vdots & \vdots & \ddots & \vdots & \vdots \\ a_{n1} & a_{n2} & \cdots & a_{nn} & b_n \end{bmatrix}.$$

This is the approach used in Strang [1976] 1988, and also in what has been termed "fraction-free" or "integer-preserving" Gaussian elimination.

Lower- [Upper-] Triangular Form

We continue this process of elimination until the counting rods [augmented matrix] are [is] in lower- [upper-] triangular form. But contrary to some descriptions of the *fangcheng* procedure, the augmented matrix is not further transformed to row-echelon form with ones as leading entries, which could produce fractional entries; nor is the augmented matrix transformed into reduced row-echelon form of Gauss-Jordan elimination. (This will be discussed further in the following section.)

This procedure can be followed exactly for each of the problems in chapter 8 of the *Nine Chapters*, because in each problem the columns [rows] are linearly independent, that is, $\det A \neq 0$, so that we always can eliminate entries to arrive at the lower- [upper-] triangular form:

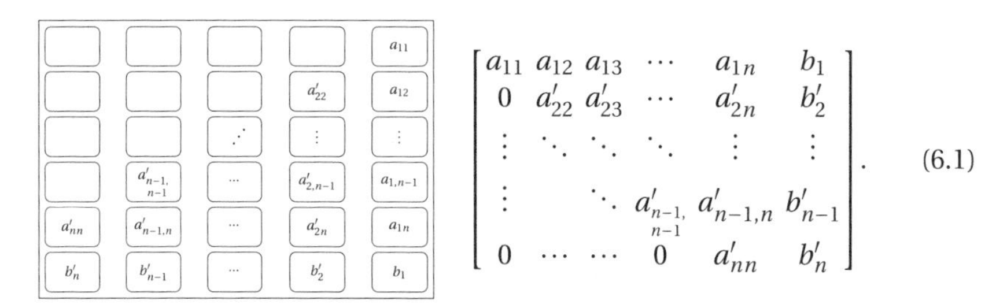

$$\begin{bmatrix} a_{11} & a_{12} & a_{13} & \cdots & a_{1n} & b_1 \\ 0 & a'_{22} & a'_{23} & \cdots & a'_{2n} & b'_2 \\ \vdots & \ddots & \ddots & \ddots & \vdots & \vdots \\ \vdots & & \ddots & a'_{n-1,n-1} & a'_{n-1,n} & b'_{n-1} \\ 0 & \cdots & \cdots & 0 & a'_{nn} & b'_n \end{bmatrix}. \quad (6.1)$$

In summary, when elementary row operations are completed following the *fangcheng* procedure, the transformed augmented matrix is of the above form, where primes denote the results after (possibly many) row reductions. All of the entries are now integers, and in general, a_{11}, a'_{22}, ..., a'_{nn} are not necessarily 1.

Algorithm for Elimination in the Fangcheng *Procedure*

We can translate the elimination stage in the *fangcheng* procedure into modern mathematical terminology as shown in Algorithm 1.

Algorithm 1 *Fangcheng* Procedure

1: **procedure** ELIMINATION(A, b)
2: **for** $k \leftarrow 1$ **to** $n-1$ **do** ▷ For columns 1 to $n-1$
3: **if** $A[k,k] = 0$ **then** ▷ If pivot a_{kk} is zero
4: [Not explicitly addressed] ▷ Problem 1 gives no instructions
5: **else** ▷ Pivot $a_{kk} \neq 0$
6: **for** $i \leftarrow k+1$ **to** n **do** ▷ For each row below the pivot a_{kk}
7: $a[i] \leftarrow A[i,k]$ ▷ Save the value of the entry a_{ik}
8: **for** $j \leftarrow k$ **to** $n+1$ **do** ▷ For the remaining entries in the row
9: $A[i,j] \leftarrow A[k,k] * A[i,j]$ ▷ Multiply the row by the pivot a_{kk}
10: **end for** j
11: **for** $j \leftarrow k$ **to** $n+1$ **do** ▷ For the remaining entries in the row
12: $A[i,j] \leftarrow A[i,j] - a[i] * A[k,j]$ ▷ Subtract entries multiplied by a_{ik}
13: **end for** j
14: **end for** i
15: **end if**
16: **end for** k
17: **return** A', b' ▷ Transformed augmented matrix (in upper-triangular form)
18: **end procedure**

Back Substitution

Now we are in a position to analyze the approach taken to back substitution in the *fangcheng* procedure. Though the solution by back substitution in modern linear algebra is well known, the approach in the *fangcheng* procedure differs, apparently in an attempt to avoid calculations with fractions, and may seem counterintuitive to the modern reader, and so requires at least some explanation.

Conflation with Solutions from Modern Linear Algebra

Perhaps because of the similarity between elimination in the *fangcheng* procedure and that in modern linear algebra, and perhaps because the instructions of the *fangcheng* procedure are so counterintuitive, historians of Chinese mathematics have often misunderstood the remaining steps, and instead explain the remaining steps by borrowing from approaches familiar from modern linear algebra (important exceptions include Lam and Shen 1989; Shen, Lun, and Crossley 1999):

1. Some historians, perhaps in an effort to follow Liu Hui's commentary, have mistakenly used Gauss-Jordan elimination to explain the *fangcheng* procedure. For example, Chemla and Guo (2004) eliminate the entries above the diagonal to give, for problem 1,

$$\begin{bmatrix} 4 & 0 & 0 & 37 \\ 0 & 4 & 0 & 17 \\ 0 & 0 & 4 & 11 \end{bmatrix},$$

   arguing "ce procédé est de fait identique à la méthode de séparation des coefficients des mathématiques modernes" (Chemla and Guo 2004, 604). As we have seen above, this is not the instruction given in the *fangcheng* procedure, and indeed, Gauss-Jordan elimination is less efficient than Gaussian elimination.[1]
2. Other modern historians of Chinese mathematics reduce the problem to the row-echelon form, giving

$$\begin{bmatrix} 1 & a'_{12} & a'_{13} & \cdots & a'_{1n} & b'_1 \\ 0 & 1 & a'_{23} & \cdots & a'_{2n} & b'_2 \\ 0 & 0 & 1 & \ddots & \vdots & \vdots \\ \vdots & \ddots & \ddots & \ddots & a'_{n-1,n} & b'_{n-1} \\ 0 & \cdots & 0 & 0 & 1 & b'_n \end{bmatrix}, \tag{6.2}$$

[1] "Gauss-Jordan is slower in practical calculations" and "requires too many operations to be the first choice on a computer" (Strang [1976] 1988, 77), which is also true when calculating with counting rods.

where the entries on the diagonal are $a_{ii} = 1$ for $1 \leq i \leq n$.

3. Still other historians have used the conventional approach to back substitution familiar from modern linear algebra, whereby the augmented matrix in equation (5.8), which represents the equations

$$36x_3 = 99, \quad 5x_2 + x_3 = 24, \quad 3x_1 + 2x_2 + x_1 = 39,$$

is solved by back substitution. However, this involves calculations with fractions: since $x_3 = \frac{99}{36} = \frac{11}{4}$, we must solve $5x_2 + \frac{11}{4} = 24$, giving $x_2 = \frac{17}{4}$, and then solve $3x_1 + 2 \times \frac{17}{4} + \frac{11}{4} = 39$, which gives $x_1 = \frac{37}{4}$. In this approach, the quantities that are substituted back are easy to conceptualize—they are simply the values of the solution.

Back Substitution in Modern Linear Algebra

The augmented matrix in equation (6.1) is equivalent to the following associated system of linear equations:

$$a_{11}x_1 + a_{12}x_2 + a_{13}x_3 + \cdots + a_{1,n-1}x_{n-1} + a_{1n}x_n = b_1, \tag{6.3}$$

$$a'_{22}x_2 + a'_{23}x_3 + \cdots + a'_{2,n-1}x_{n-1} + a'_{2n}x_n = b'_2, \tag{6.4}$$

$$\vdots$$

$$a'_{n-2,n-2}x_{n-2} + a'_{n-2,n-1}x_{n-1} + a'_{n-2,n}x_n = b'_{n-2}, \tag{6.5}$$

$$a'_{n-1,n-1}x_{n-1} + a'_{n-1,n}x_n = b'_{n-1}, \tag{6.6}$$

$$a'_{nn}x_n = b'_n. \tag{6.7}$$

We then solve them in sequence, working backward from x_n to x_1, as follows:

STEP 1. First, we solve for x_n,

$$x_n = \frac{b'_n}{a'_{nn}}. \tag{6.8}$$

STEP 2. Next, we can solve for x_{n-1} by substituting the value $\frac{b'_n}{a'_{nn}}$ that we found for x_n in equation (6.8) back into equation (6.6), yielding

$$x_{n-1} = \frac{b'_{n-1} - a'_{n-1,n}\left(\frac{b'_n}{a'_{nn}}\right)}{a'_{n-1,n-1}}. \tag{6.9}$$

STEP 3. Next, substituting the values computed for x_n and x_{n-1} in equations (6.8) and (6.9) into equation (6.5), we can solve for x_{n-2}, as follows:

$$x_{n-2} = \frac{b'_{n-2} - a'_{n-2,n-1}\left(\frac{b'_{n-1} - a'_{n-1,n}\left(\frac{b'_n}{a'_{nn}}\right)}{a'_{n-1,n-1}}\right) - a'_{n-2,n}\left(\frac{b'_n}{a'_{nn}}\right)}{a'_{n-2,n-2}}. \tag{6.10}$$

STEPS 4–N. The values of x_n, x_{n-1}, x_{n-2} found in equations (6.8), (6.9), and (6.10) are then substituted back into the equation for x_{n-3}, and so on.

It should be noted that this process is greatly simplified in practice, since numbers calculated at each step, rather than complicated symbolic expressions, are used for the substitutions.

Back Substitution in the Fangcheng *Procedure with Three Unknowns*

Undoubtedly, the reason the method of back substitution in the *fangcheng* procedure has been misunderstood is that it is so counterintuitive. It is conceptually much simpler to understand back substitution in modern linear algebra, where the value is substituted back. The key difference in the *fangcheng* procedure is that division by a'_{nn} is postponed until the end. That is, in the modern approach to back substitution above, the first step is to divide b'_n by a'_{nn} in equation (6.8); in contrast, in the *fangcheng* procedure, this is the last step.

STEP 1. We are instructed to calculate first the following quantity:

$$(a'_{33}b'_2 - a'_{23}b'_3) \div a'_{22}. \tag{6.11}$$

STEP 2. The *fangcheng* procedure next instructs us to calculate the following quantity, using the result previously calculated in equation (6.11):

$$\left(a'_{33}b_1 - a_{12}((a'_{33}b'_2 - a'_{23}b'_3) \div a'_{22}) - a_{13}b'_3\right) \div a_{11}. \tag{6.12}$$

STEP 3. The example in the *fangcheng* procedure has only three conditions and three unknowns, and so at this point the "back substitutions" are complete, and we are instructed to divide the results obtained in (6.11) and (6.12) by a_{33}, yielding

$$\begin{aligned}
x_1 &= \left(\left(a'_{33}b_1 - a_{12}((a'_{33}b'_2 - a'_{23}b'_3) \div a'_{22}) - a_{13}b'_3\right) \div a_{11}\right) \div a'_{33}\\
x_2 &= \left((a'_{33}b'_2 - a'_{23}b'_3) \div a'_{22}\right) \div a'_{33}\\
x_3 &= b'_3 \div a'_{33}.
\end{aligned}$$

Back Substitution in the Fangcheng *Procedure in General*

Problem 1 is meant to be an exemplar, and it is not difficult to find how to extend this approach beyond three unknowns simply by trial and error, but we will need to use modern terminology to explain this approach and justify its validity. The easiest way to understand this approach is to see that it is equivalent to transforming the system of equations (6.3–6.7) by multiplying both sides of the $n-1$

equations (6.3–6.6) by a'_{nn}, thus

$$\begin{aligned}
a_{11}a'_{nn}x_1 + a_{12}a'_{nn}x_2 + \cdots + a_{1,n-1}a'_{nn}x_{n-1} + a_{1n}a'_{nn}x_n &= a'_{nn}b_1 \\
a'_{22}a'_{nn}x_2 + \cdots + a'_{2,n-1}a'_{nn}x_{n-1} + a'_{2n}a'_{nn}x_n &= a'_{nn}b'_2 \\
&\vdots \\
a'_{n-1,n-1}a'_{nn}x_{n-1} + a'_{n-1,n}a'_{nn}x_n &= a'_{nn}b'_{n-1} \\
a'_{nn}x_n &= b'_n.
\end{aligned}$$

Then, substituting $z_j = a'_{nn}x_j$ $(1 \le j \le n)$ throughout yields

$$z_n = b'_n, \tag{6.13}$$

$$a'_{n-1,n-1}z_{n-1} + a'_{n-1,n}z_n = a'_{nn}b'_{n-1} \tag{6.14}$$

$$\vdots \tag{6.15}$$

$$a'_{22}z_2 + a'_{23}z_3 + \cdots + a'_{2,n-1}z_{n-1} + a'_{2n}z_n = a'_{nn}b'_2 \tag{6.16}$$

$$a_{11}z_1 + a_{12}z_2 + a_{13}z_3 + \cdots + a_{1,n-1}z_{n-1} + a_{1n}z_n = a'_{nn}b_1. \tag{6.17}$$

The approach presented in *fangcheng* procedure is then the equivalent to solving this transformed system of n equations in n unknowns:

$$z_n = b'_n, \tag{6.18}$$

$$z_{n-1} = (a'_{nn}b'_{n-1} - a'_{n-1,n}z_n) \div a'_{n-1,n-1}, \tag{6.19}$$

$$z_{n-2} = (a'_{nn}b'_{n-2} - a'_{n-2,n-1}z_{n-1} - a'_{n-2,n}z_n) \div a'_{n-2,n-2}, \tag{6.20}$$

$$\vdots \tag{6.21}$$

$$z_2 = (a'_{nn}b'_2 - a'_{23}z_3 - \cdots - a'_{2,n-1}z_{n-1} - a'_{2n}z_n) \div a'_{22}, \tag{6.22}$$

$$z_1 = (a'_{nn}b_1 - a_{12}z_2 - a_{13}z_3 - \cdots - a_{1,n-1}z_{n-1} - a_{1n}z_n) \div a_{11}. \tag{6.23}$$

Then, after all the back substitutions are complete, and we have solved for all z_j, we can find x_j simply by dividing each z_j by a'_{nn}, thus

$$x_j = z_j \div a'_{nn},$$

yielding the results.

Back Substitution on the Counting Board

What is probably most important here is to reconstruct how this computation was most likely performed on a counting board with counting rods. The process is simply an extension to n unknowns of the computation described in the *fangcheng* procedure following problem 1. This procedure is as follows:

STEP 1. Beginning with the lower- [upper-] diagonal form, we first "cross multiply" b'_{n-1} by a'_{nn} and $a'_{n-1,n}$ by b'_n,

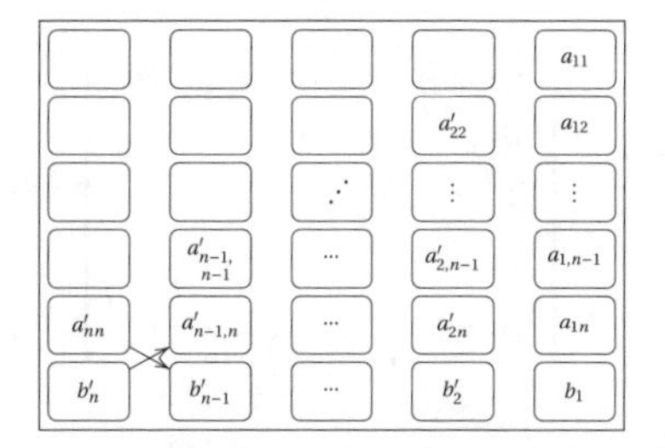

$$\begin{bmatrix} a_{11} & a_{12} & a_{13} & \cdots & a_{1n} & b_1 \\ 0 & a'_{22} & a'_{23} & \cdots & a'_{2n} & b'_2 \\ \vdots & \ddots & \ddots & \ddots & \vdots & \vdots \\ \vdots & & \ddots & a'_{n-1,n-1} & a'_{n-1,n} & b'_{n-1} \\ 0 & \cdots & \cdots & 0 & a'_{nn} & b'_n \end{bmatrix}.$$

This gives the following (again, it should be noted that throughout back substitution, the matrices no longer correspond to the system of linear equations):

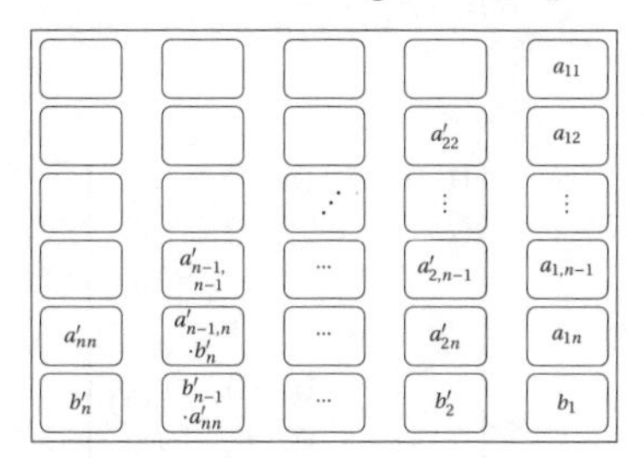

$$\begin{bmatrix} a_{11} & a_{12} & a_{13} & \cdots & a_{1n} & b_1 \\ 0 & a'_{22} & a'_{23} & \cdots & a'_{2n} & b'_2 \\ \vdots & \ddots & \ddots & \ddots & \vdots & \vdots \\ \vdots & & \ddots & a'_{n-1,n-1} & a'_{n-1,n}\cdot b'_n & b'_{n-1}\cdot a'_{nn} \\ 0 & \cdots & \cdots & 0 & a'_{nn} & b'_n \end{bmatrix}.$$

STEP 2. We then subtract $b'_{n-1}\cdot a'_{nn} - a'_{n-1,n}\cdot b'_n$, and divide the result by $a'_{n-1,n-1}$, giving the result $b''_{n-1} = (b'_{n-1}\cdot a'_{nn} - a'_{n-1,n}\cdot b'_n) \div a'_{n-1,n-1}$,

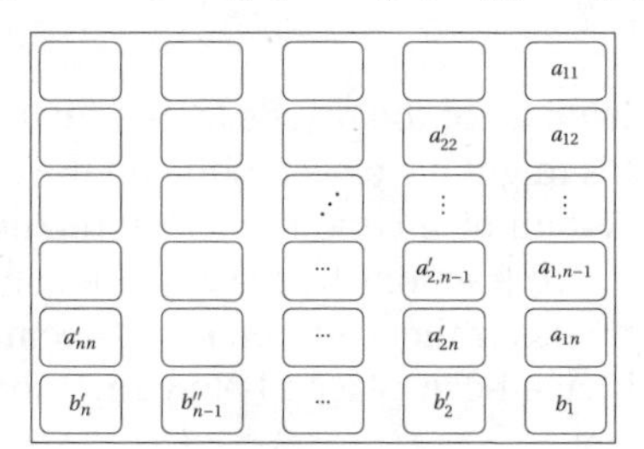

$$\begin{bmatrix} a_{11} & a_{12} & a_{13} & \cdots & a_{1n} & b_1 \\ 0 & a'_{22} & a'_{23} & \cdots & a'_{2n} & b'_2 \\ \vdots & \ddots & \ddots & \ddots & \vdots & \vdots \\ \vdots & & \ddots & 0 & 0 & b''_{n-1} \\ 0 & \cdots & \cdots & 0 & a'_{nn} & b'_n \end{bmatrix}.$$

Here we have cleared away the entries $a'_{n-1,n-1}$ and $a'_{n-1,n}\cdot b'_n$, now that they have been used in the calculations.

STEP $(n-1)$. We continue in precisely this manner until we reach the last step in the back substitution, shown as

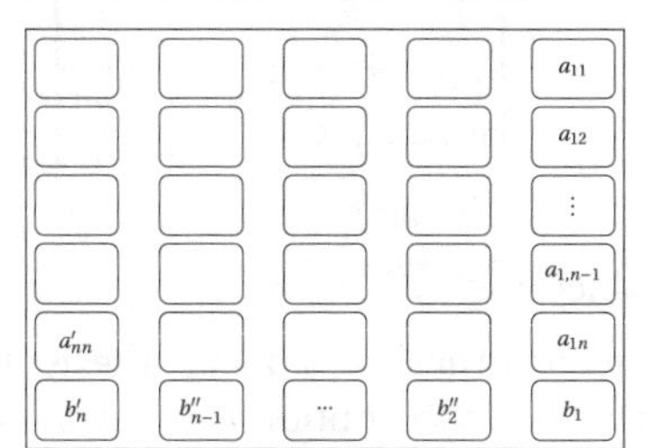

$$\begin{bmatrix} a_{11} & a_{12} & a_{13} & \cdots & a_{1n} & b_1 \\ 0 & \cdots & \cdots & \cdots & 0 & b''_2 \\ \vdots & & & & \vdots & \vdots \\ \vdots & & & 0 & 0 & b''_{n-1} \\ 0 & \cdots & \cdots & 0 & a'_{nn} & b'_n \end{bmatrix}.$$

STEP n. Here again, we "cross multiply" the entries, $a'_{nn} \cdot b_1$ and then $b'_n \cdot a_{1n}$, $b''_{n-1} \cdot a_{1,n-1}, \ldots, b''_2 \cdot a_{12}$, which is quite simple on the counting board,

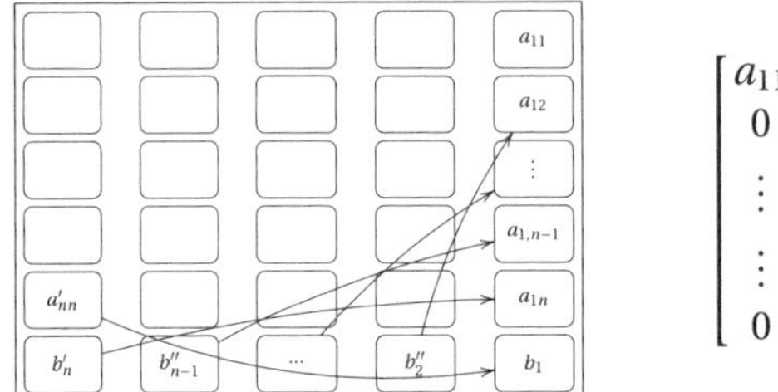

$$\begin{bmatrix} a_{11} & a_{12} & a_{13} & \cdots & a_{1n} & b_1 \\ 0 & \cdots & \cdots & \cdots & 0 & b''_2 \\ \vdots & & & & \vdots & \vdots \\ \vdots & & & 0 & 0 & b''_{n-1} \\ 0 & \cdots & \cdots & 0 & a'_{nn} & b'_n \end{bmatrix}.$$

We then simply subtract from $a'_{nn} \cdot b_1$ all of the entries above it, $b'_n \cdot a_{1n}$, $b''_{n-1} \cdot a_{1,n-1}, \ldots, b''_2 \cdot a_{12}$, giving $a'_{nn} \cdot b_1 - b'_n \cdot a_{1n} - b''_{n-1} \cdot a_{1,n-1} - \ldots - b''_2 \cdot a_{12}$, and divide the result by a_{11}, giving the result $b'_1 = (a'_{nn} \cdot b_1 - b'_n \cdot a_{1n} - b''_{n-1} \cdot a_{1,n-1} - \ldots - b''_2 \cdot a_{12}) \div a_{11}$,

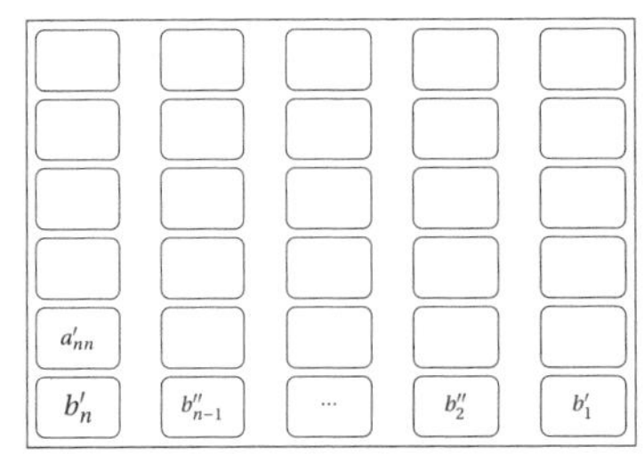

$$\begin{bmatrix} 0 & 0 & 0 & 0 & 0 & b'_1 \\ 0 & \cdots & \cdots & \cdots & 0 & b''_2 \\ \vdots & & & & \vdots & \vdots \\ \vdots & & & 0 & 0 & b''_{n-1} \\ 0 & \cdots & \cdots & 0 & a'_{nn} & b'_n \end{bmatrix}.$$

Here again, we can clear all the entries after they have been used in the calculation.

FINAL STEP. The final step is then to divide all of the results by the final pivot, a'_{nn}. Again, we can only speculate how this might have been done on the counting board that was used to display the system of n conditions in n unknowns. Assuming again that the Chinese represented fractions by placing the counting rods for the numerator over the counting rods for the denominator, it seems reasonable that the final pivot, a'_{nn}, would have been placed below the modified constant terms, b'_n, in the following manner:

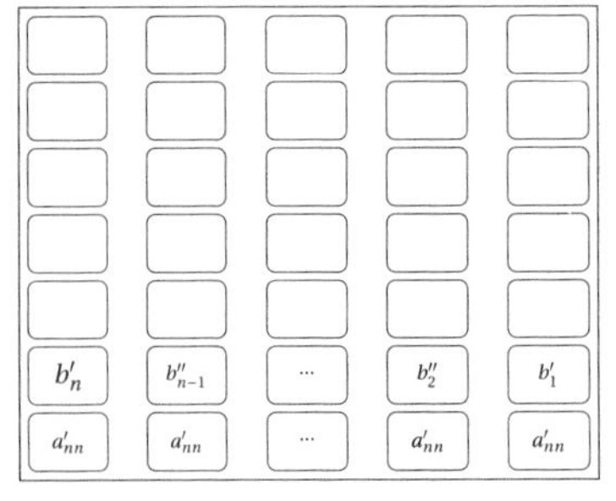

$$\begin{bmatrix} 0 & \cdots & \cdots & \cdots & 0 & b'_1/a'_{nn} \\ \vdots & & & & \vdots & b''_2/a'_{nn} \\ \vdots & & & & \vdots & \vdots \\ \vdots & & & & \vdots & b''_{n-1}/a'_{nn} \\ 0 & \cdots & \cdots & \cdots & 0 & b'_n/a'_{nn} \end{bmatrix}.$$

Of course, the fractions would then be reduced.

Again, I should note that the above reconstruction is a hypothesis: I have not seen any diagrams that illustrate how back substitution was calculated on a counting board; nor have I seen any descriptions that specify the positions for various entries in the process of back substitution. Still, it seems reasonable to assume that the back substitution was calculated on the same counting board that was used to calculate the eliminations; the only question is precisely how this was done, and the above seems to be a reasonable reconstruction. Whether or not this reconstruction is in fact how back substitution was calculated, what should be apparent is that computations with fractions in the process of back substitution would be quite complicated. With integers, each position on the counting board corresponds to a single entry, that is, the coefficient of one unknown for one of the conditions. Fractions, however, would require two entries for each coefficient, in which case the two dimensions of the counting board would be used to display both the n conditions in n unknowns, and the fractions. This would likely increase the possibility of error, and in addition considerably increase the complexity of the calculations, requiring the use of common denominators throughout.

Algorithm for Substitution in the **Fangcheng** *Procedure*

Back substitution in the *fangcheng* procedure, as analyzed above, can be written in the form shown in Algorithm 2.

Algorithm 2 Back Substitution

```
1: procedure SUBSTITUTION(A, b, x)
2:     z[n] = A[n,n+1]                          ▷ z_n = b'_n
3:     for k ← n − 1 to 1 do                    ▷ Calculate each z_k, from z_{n−1} to z_1
4:         z[k] ← A[n,n] ∗ A[k,n+1]             ▷ First multiply a'_nn · b'_k
5:         for i ← n to k+1 do
6:             z[k] ← z[k] − A[k,i]z[i]         ▷ Subtract a'_kn · z_n + ... + a'_k,k+1 · z_k+1
7:         end for
8:         z[k] ← z[k]/A[k,k]                   ▷ Divide by a'_kk
9:     end for
10:    for k ← n to 1 do                        ▷ For each k find x_k
11:        x[k] = z[k]/A[n,n]                   ▷ Divide by a'_nn
12:    end for
13:    return x                                 ▷ x = (x_1, x_2, ..., x_n)
14: end procedure
```

Is the *Fangcheng* Procedure Integer-Preserving?

As Table 6.2 on the following page shows, for the 18 problems in *Fangcheng*, in chapter 8 of the *Nine Chapters*, fractions never emerge in the process of back substitution. The central point of this section is to show that this is true not just for the exemplars selected for inclusion in the *Nine Chapters*, but is true in most cases—that is, in all but a small percentage of cases, the *fangcheng* procedure avoids calculations with fractions. To demonstrate this, however, requires some work: we must calculate symbolically the results of following the *fangcheng* procedure.[2] This section, then, is somewhat technical.

Table 6.2: Appearance of fractions in back substitution and in the solutions for the 18 problems in chapter 8 of the *Nine Chapters.*

Problem	1	2	3	4	5	6	7	8	9	10	11	12	13	14	15	16	17	18
Solutions:																		
Fractions	+	+	+	−	−	−	+	−	+	+	+	+	−	+	+	+	−	−
Integers	−	−	−	+	+	+	−	+	−	+	−	−	+	−	−	−	+	+
Back substitution:																		
Reducible $\frac{b'_i}{a'_{nn}}$	−	+	−	+	+	+	−	+	−	+	−	−	+	+	−	+	+	+
Fractions	−	−	−	−	−	−	−	−	−	−	−	−	−	−	−	−	−	−

Calculation of Results of the Row Reductions

In the *fangcheng* procedure, as we have seen above, we first form an augmented matrix,

$$\begin{bmatrix} a_{11} & a_{12} & a_{13} & \cdots & a_{1n} & b_1 \\ a_{21} & a_{22} & a_{23} & \cdots & a_{2n} & b_2 \\ a_{31} & a_{32} & a_{33} & \cdots & a_{3n} & b_3 \\ \vdots & \vdots & \vdots & & \vdots & \vdots \\ a_{n1} & a_{n2} & a_{n3} & \cdots & a_{nn} & b_n \end{bmatrix}.$$

We then successively eliminate nondiagonal entries. To accomplish this, we first multiply the second row by the first entry in the first row, a_{11}, yielding $a_{11} \cdot (a_{21}, a_{22}, a_{23}, \ldots, a_{2n}, b_2)$. We then multiply the first row [column] by the first entry in the second row [column], a_{21}, with the result $a_{21} \cdot (a_{11}, a_{12}, a_{13}, \ldots, a_{1n}, b_1)$, and subtract, giving $(0, a_{11}a_{22} - a_{12}a_{21}, a_{11}a_{23} - a_{13}a_{21}, \ldots, a_{11}a_{2n} - a_{1n}a_{21}, a_{11}b_2 - b_1a_{21})$:

$$\begin{bmatrix} a_{11} & a_{12} & a_{13} & \cdots & a_{1n} & b_1 \\ 0 & \begin{matrix} a_{11}a_{22} \\ -a_{12}a_{21} \end{matrix} & \begin{matrix} a_{11}a_{23} \\ -a_{13}a_{21} \end{matrix} & \cdots & \begin{matrix} a_{11}a_{2n} \\ -a_{1n}a_{21} \end{matrix} & \begin{matrix} a_{11}b_2 \\ -b_1a_{21} \end{matrix} \\ a_{31} & a_{32} & a_{33} & \cdots & a_{3n} & b_3 \\ \vdots & \vdots & \vdots & & \vdots & \vdots \\ a_{n1} & a_{n2} & a_{n3} & \cdots & a_{nn} & b_n \end{bmatrix}.$$

[2] In the following analysis we will assume that a unique solution exists, that is, in modern mathematical terms, that the equations are linearly independent.

Note that using determinants, we can write this in a more compact manner, as

$$\begin{bmatrix} a_{11} & a_{12} & a_{13} & \cdots & a_{1n} & b_1 \\ 0 & \begin{vmatrix} a_{11} & a_{12} \\ a_{21} & a_{22} \end{vmatrix} & \begin{vmatrix} a_{11} & a_{13} \\ a_{21} & a_{23} \end{vmatrix} & \cdots & \begin{vmatrix} a_{11} & a_{1n} \\ a_{21} & a_{2n} \end{vmatrix} & \begin{vmatrix} a_{11} & b_1 \\ a_{21} & b_2 \end{vmatrix} \\ a_{31} & a_{32} & a_{33} & \cdots & a_{3n} & b_3 \\ \vdots & \vdots & \vdots & & \vdots & \vdots \\ a_{n1} & a_{n2} & a_{n3} & \cdots & a_{nn} & b_n \end{bmatrix}.$$

But if we consider *fangcheng* problems in general, this is actually not quite correct. Let us write $a_{2j}^{(1)}$ for the actual result after this first set of row reductions in the second row and j^{th} column. In practice, we have to consider the three following cases:

1. If $a_{21} = 0$, then there is no subtraction, and the row is left untouched, so that we must multiply $\begin{vmatrix} a_{11} & a_{1j} \\ a_{21} & a_{2j} \end{vmatrix}$ by $\frac{1}{a_{11}}$ to achieve the correct result:

$$a_{2j}^{(1)} = a_{2j} = \frac{1}{a_{11}}(a_{11}a_{2j} - 0 \cdot a_{1j}) = \frac{1}{a_{11}}\begin{vmatrix} a_{11} & a_{1j} \\ a_{21} & a_{2j} \end{vmatrix}.$$

2. Similarly, if $a_{21} = a_{11}$, then we simply subtract the first row from the second row (instead, that is, of multiplying the second row by a_{11} and subtracting the first row multiplied by a_{21}), so we again must correct by a factor of $\frac{1}{a_{11}}$,

$$a_{2j}^{(1)} = a_{2j} - a_{1j} = \frac{1}{a_{11}}(a_{11}a_{2j} - a_{21}a_{1j}) = \frac{1}{a_{11}}\begin{vmatrix} a_{11} & a_{1j} \\ a_{21} & a_{2j} \end{vmatrix}.$$

3. In general, if the greatest common divisor $(a_{11}, a_{21}) \neq 1$, then we *could* multiply the second row by $a_{11}/(a_{11}, a_{21})$ and subtract the first row multiplied by $a_{21}/(a_{11}, a_{21})$ (instead, that is, of multiplying by a_{11} and a_{21}, respectively), so we *could* have

$$a_{2j}^{(1)} = \frac{a_{11}}{(a_{11}, a_{21})}a_{2j} - \frac{a_{21}}{(a_{11}, a_{21})}a_{1j} = \frac{1}{(a_{11}, a_{21})}\begin{vmatrix} a_{11} & a_{1j} \\ a_{21} & a_{2j} \end{vmatrix}.$$

Unfortunately, problem 1, our exemplar, does not address any of these three cases. For case 1, $a_{21} = 0$, the text does note this exception, but implies only that we should move to the next row. For case 2, if $a_{21} = a_{11}$, it seems only reasonable to assume that we would directly subtract (instead, that is, of first multiplying the second row by a_{11} and subtracting the first row multiplied by a_{21}, in order to subtract a_{11}^2 from a_{11}^2). Case 3 is the most ambiguous, and above I have used the phrase "could have" to indicate that the text of problem 1 of chapter 8 gives us no explicit instructions on what to do in case the greatest common divisor $(a_{11}, a_{21}) \neq 1$; furthermore, in practice, for large numbers, one might not recognize the fact that there is a common divisor, and finding the greatest common divisor may be difficult.

Therefore, to correct these calculations, we must multiply the entries $\begin{vmatrix} a_{11} & a_{1j} \\ a_{21} & a_{2j} \end{vmatrix}$ by the constant $\frac{1}{k_{21}}$, which will be defined as follows:

$$k_{21} = \begin{cases} 1 \text{ if } (a_{11}, a_{21}) = 1, \\ a_{11} \text{ if } a_{21} = 0, \\ a_{11} \text{ if } a_{21} = \pm a_{11}, \\ d \text{ if } (a_{11}, a_{21}) \neq 1, \text{ where } d \text{ is an integer that divides } (a_{11}, a_{21}). \end{cases}$$

That is, we must assume that the greatest common divisor might not always be found, or might not be used, so that we will have to use a more vague but more general definition of k_{21}: for the case $(a_{11}, a_{21}) \neq 1$, we can only assume that $k_{21} = d$ for one of possibly many $d \in \mathbb{N}$ such that d divides the greatest common divisor, $d|(a_{11}, a_{21})$, where possibly $d = 1$. Then we have the desired result

$$a_{2j}^{(1)} = \begin{cases} \begin{vmatrix} a_{11} & a_{1j} \\ a_{21} & a_{2j} \end{vmatrix} = a_{11}a_{2j} - a_{21}a_{1j} \text{ if } (a_{11}, a_{21}) = 1, \\ \frac{1}{a_{11}} \begin{vmatrix} a_{11} & a_{1j} \\ 0 & a_{2j} \end{vmatrix} = a_{2j} \text{ if } a_{21} = 0, \\ \frac{1}{a_{11}} \begin{vmatrix} a_{11} & a_{1j} \\ a_{21} & a_{2j} \end{vmatrix} = a_{2j} \pm a_{1j} \text{ if } a_{21} = \pm a_{11}, \\ \frac{1}{d} \begin{vmatrix} a_{11} & a_{1j} \\ a_{21} & a_{2j} \end{vmatrix} = \frac{a_{11}}{d} a_{2j} - \frac{a_{21}}{d} a_{1j} \text{ if } (a_{11}, a_{21}) \neq 1, \text{ where } d|(a_{11}, a_{21}). \end{cases}$$

Next we eliminate the first entry in the third, fourth, and following rows in succession. As above, in general we multiply each row by the first entry in the first row, a_{11}, and subtract from it the first row multiplied by the first entry in that row, a_{i1}. But again, we must consider other possibilities, and so for each row i we must include a distinct constant

$$k_{i1} = \begin{cases} 1 \text{ if } (a_{11}, a_{i1}) = 1, \\ a_{11} \text{ if } a_{i1} = 0, \\ a_{11} \text{ if } a_{i1} = \pm a_{11}, \\ d_i \text{ if } (a_{11}, a_{i1}) \neq 1, \text{ where } d_i \in \mathbb{N} \text{ and } d_i|(a_{11}, a_{i1}). \end{cases}$$

We can then rewrite the result after the first set of row reductions as follows:

$$\begin{bmatrix} a_{11} & a_{12} & a_{13} & \cdots & a_{1n} & b_1 \\ 0 & \frac{1}{k_{21}} \begin{vmatrix} a_{11} & a_{12} \\ a_{21} & a_{22} \end{vmatrix} & \frac{1}{k_{21}} \begin{vmatrix} a_{11} & a_{13} \\ a_{21} & a_{23} \end{vmatrix} & \cdots & \frac{1}{k_{21}} \begin{vmatrix} a_{11} & a_{1n} \\ a_{21} & a_{2n} \end{vmatrix} & \frac{1}{k_{21}} \begin{vmatrix} a_{11} & b_1 \\ a_{21} & b_2 \end{vmatrix} \\ 0 & \frac{1}{k_{31}} \begin{vmatrix} a_{11} & a_{12} \\ a_{31} & a_{32} \end{vmatrix} & \frac{1}{k_{31}} \begin{vmatrix} a_{11} & a_{13} \\ a_{31} & a_{33} \end{vmatrix} & \cdots & \frac{1}{k_{31}} \begin{vmatrix} a_{11} & a_{1n} \\ a_{31} & a_{3n} \end{vmatrix} & \frac{1}{k_{31}} \begin{vmatrix} a_{11} & b_1 \\ a_{31} & b_3 \end{vmatrix} \\ \vdots & \vdots & \vdots & & \vdots & \vdots \\ 0 & \frac{1}{k_{n1}} \begin{vmatrix} a_{11} & a_{12} \\ a_{n1} & a_{n2} \end{vmatrix} & \frac{1}{k_{n1}} \begin{vmatrix} a_{11} & a_{13} \\ a_{n1} & a_{n3} \end{vmatrix} & \cdots & \frac{1}{k_{n1}} \begin{vmatrix} a_{11} & a_{1n} \\ a_{n1} & a_{nn} \end{vmatrix} & \frac{1}{k_{n1}} \begin{vmatrix} a_{11} & b_1 \\ a_{n1} & b_n \end{vmatrix} \end{bmatrix}.$$

For simplicity, we will sometimes write the result after this first set of row reductions as follows:

$$\begin{bmatrix} a_{11} & a_{12} & a_{13} & \cdots & a_{1n} & b_1 \\ 0 & a_{22}^{(1)} & a_{23}^{(1)} & \cdots & a_{2n}^{(1)} & b_2^{(1)} \\ 0 & a_{32}^{(1)} & a_{33}^{(1)} & \cdots & a_{3n}^{(1)} & b_3^{(1)} \\ \vdots & \vdots & \vdots & & \vdots & \vdots \\ 0 & a_{n2}^{(1)} & a_{n3}^{(1)} & \cdots & a_{nn}^{(1)} & b_n^{(1)} \end{bmatrix}, \tag{6.24}$$

where we have written the results of the computation in the previous step as $(a_{i2}^{(1)}, a_{i3}^{(1)}, \ldots, a_{in}^{(1)}, b_i^{(1)})$.

Following the example, problem 1, we now eliminate the second entries in rows 3 to n. As above, we first multiply the ith row by $a_{22}^{(1)}$, and subtract from the result the second row multiplied by $a_{i2}^{(1)}$, if the greatest common divisor $(a_{22}^{(1)}, a_{i2}^{(1)}) = 1$. But in case $(a_{22}^{(1)}, a_{i2}^{(1)}) \neq 1$, we have again to define k_{i2}, thus

$$k_{i2} = \begin{cases} 1 \text{ if } (a_{22}^{(1)}, a_{i2}^{(1)}) = 1, \\ a_{22}^{(1)} \text{ if } a_{i2}^{(1)} = 0, \\ a_{22}^{(1)} \text{ if } a_{i2}^{(1)} = \pm a_{22}^{(1)}, \\ d \text{ if } (a_{22}^{(1)}, a_{i2}^{(1)}) \neq 1, \text{ where } d \in \mathbb{N} \text{ and } d | (a_{22}^{(1)}, a_{i2}^{(1)}), \end{cases}$$

where again d is not necessarily $(a_{22}^{(1)}, a_{i2}^{(1)})$. Thus we must multiply the ith row by

$$\frac{1}{k_{i2}} a_{22}^{(1)} = \frac{1}{k_{i2}} \frac{1}{k_{21}} \begin{vmatrix} a_{11} & a_{12} \\ a_{21} & a_{22} \end{vmatrix}, \tag{6.25}$$

and subtract from that the second row multiplied by

$$\frac{1}{k_{i2}} a_{i2}^{(1)} = \frac{1}{k_{i2}} \frac{1}{k_{i1}} \begin{vmatrix} a_{11} & a_{12} \\ a_{i1} & a_{i2} \end{vmatrix},$$

so that entries in the ith row become

$$a_{ij}^{(2)} = \frac{1}{k_{i2}} \frac{1}{k_{21}} \begin{vmatrix} a_{11} & a_{12} \\ a_{21} & a_{22} \end{vmatrix} \cdot \frac{1}{k_{i1}} \begin{vmatrix} a_{11} & a_{1j} \\ a_{i1} & a_{ij} \end{vmatrix} - \frac{1}{k_{i2}} \frac{1}{k_{i1}} \begin{vmatrix} a_{11} & a_{12} \\ a_{i1} & a_{i2} \end{vmatrix} \cdot \frac{1}{k_{21}} \begin{vmatrix} a_{11} & a_{1j} \\ a_{21} & a_{2j} \end{vmatrix}.$$

We can then simplify this, as can be shown by direct calculation, as follows,

$$a_{ij}^{(2)} = \frac{a_{11}}{k_{21} k_{i1} k_{i2}} \begin{vmatrix} a_{11} & a_{12} & a_{1j} \\ a_{21} & a_{22} & a_{2j} \\ a_{i1} & a_{i2} & a_{ij} \end{vmatrix}.$$

We continue this process until we have eliminated all of the entries, and we can write the result after the second set of row reductions, as follows:

$$\begin{bmatrix} a_{11} & a_{12} & a_{13} & \cdots & b_1 \\ 0 & \frac{1}{k_{21}}\begin{vmatrix} a_{11} & a_{12} \\ a_{21} & a_{22} \end{vmatrix} & \frac{1}{k_{21}}\begin{vmatrix} a_{11} & a_{13} \\ a_{21} & a_{23} \end{vmatrix} & \cdots & \frac{1}{k_{21}}\begin{vmatrix} a_{11} & b_1 \\ a_{21} & b_2 \end{vmatrix} \\ 0 & 0 & \frac{a_{11}}{k_{21}k_{31}k_{32}}\begin{vmatrix} a_{11} & a_{12} & a_{13} \\ a_{21} & a_{22} & a_{23} \\ a_{31} & a_{32} & a_{33} \end{vmatrix} & \cdots & \frac{a_{11}}{k_{21}k_{31}k_{32}}\begin{vmatrix} a_{11} & a_{12} & b_1 \\ a_{21} & a_{22} & b_2 \\ a_{31} & a_{32} & b_3 \end{vmatrix} \\ \vdots & \vdots & \vdots & & \vdots \\ 0 & 0 & \frac{a_{11}}{k_{21}k_{n1}k_{n2}}\begin{vmatrix} a_{11} & a_{12} & a_{13} \\ a_{21} & a_{22} & a_{23} \\ a_{n1} & a_{n2} & a_{n3} \end{vmatrix} & \cdots & \frac{a_{11}}{k_{21}k_{n1}k_{n2}}\begin{vmatrix} a_{11} & a_{12} & b_1 \\ a_{21} & a_{22} & b_2 \\ a_{n1} & a_{n2} & b_n \end{vmatrix} \end{bmatrix}. \tag{6.26}$$

We must now introduce some simplifying notation. Let $\{i_1, i_2, \ldots, i_r\}$ and $\{j_1, j_2, \ldots, j_r\}$ be subsequences of $\{1, 2, \ldots, n\}$ with r elements. Then a *minor of order r* is the determinant of the $r \times r$ matrix formed by the entries that lie in the intersection of rows $\{i_1, i_2, \ldots, i_r\}$ and columns $\{j_1, j_2, \ldots, j_r\}$:

$$A\begin{pmatrix} i_1, i_2, \ldots, i_r \\ j_1, j_2, \ldots, j_r \end{pmatrix} = \begin{vmatrix} a_{i_1,j_1} & a_{i_1,j_2} & \cdots & a_{i_1,j_r} \\ a_{i_2,j_1} & a_{i_2,j_2} & \cdots & a_{i_2,j_r} \\ \vdots & \vdots & & \vdots \\ a_{i_r,j_1} & a_{i_r,j_2} & \cdots & a_{i_r,j_r} \end{vmatrix}.$$

More concretely, we can write, for example,

$$A\begin{pmatrix} 1 \\ 1 \end{pmatrix} = a_{11} \qquad A\begin{pmatrix} 1,2,3 \\ 1,2,3 \end{pmatrix} = \begin{vmatrix} a_{11} & a_{12} & a_{13} \\ a_{21} & a_{22} & a_{23} \\ a_{31} & a_{32} & a_{33} \end{vmatrix} \qquad A\begin{pmatrix} 1,\ldots,r,i \\ 1,\ldots,r,j \end{pmatrix} = \begin{vmatrix} a_{11} & \cdots & a_{1r} & a_{1j} \\ \vdots & & \vdots & \vdots \\ a_{r1} & \cdots & a_{rr} & a_{rj} \\ a_{i1} & \cdots & a_{ir} & a_{ij} \end{vmatrix},$$

where $i, j > r$.

Calculation of $a_{ij}^{(r)}$

Using this notation, Proposition C.1 (see page 258, in Appendix C) allows us to write $a_{ij}^{(r)}$, the entry in the ith row and j^{th} column after r row reductions $(i, j < r)$ as the following product of r terms, beginning with the $r-1$ leading principal minors $A\binom{1}{1}$ up to $A\binom{1,\ldots,r-1}{1,\ldots,r-1}$, and the final r^{th} term being the leading principal minor $A\binom{1,\ldots,r}{1,\ldots,r}$, with the first r terms of the ith row and j^{th} column appended to form $A\binom{1,\ldots,r,i}{1,\ldots,r,j}$, as shown in the following equation:

$$a_{ij}^{(r)} = \left(\frac{A\binom{1}{1}}{k_{21}}\right)^{2^{r-2}} \left(\frac{A\binom{1,2}{1,2}}{k_{31}k_{32}}\right)^{2^{r-3}} \left(\frac{A\binom{1,2,3}{1,2,3}}{k_{41}k_{42}k_{43}}\right)^{2^{r-4}} \cdots \left(\frac{A\binom{1,\ldots,r-1}{1,\ldots,r-1}}{k_{r1}k_{r2}\cdots k_{r,r-1}}\right)\left(\frac{A\binom{1,\ldots,r,i}{1,\ldots,r,j}}{k_{i1}k_{i2}\cdots k_{ir}}\right). \quad (6.27)$$

And after completing $r-1$ row reductions in the manner outlined above, we arrive at the row-reduced form, and can write the result as follows:

$$\begin{bmatrix} a_{11} & a_{12} & a_{13} & \cdots & a_{1,n-1} & a_{1n} & b_1 \\ 0 & a_{22}^{(1)} & a_{23}^{(1)} & \cdots & a_{2,n-1}^{(1)} & a_{2n}^{(1)} & b_2^{(1)} \\ 0 & 0 & a_{33}^{(2)} & \cdots & a_{3,n-1}^{(2)} & a_{3n}^{(2)} & b_3^{(2)} \\ \vdots & & \ddots & \ddots & \vdots & \vdots & \vdots \\ 0 & \cdots & \cdots & 0 & a_{n-1,n-1}^{(n-2)} & a_{n-1,n}^{(n-2)} & b_{n-1}^{(n-2)} \\ 0 & \cdots & \cdots & 0 & 0 & a_{nn}^{(n-1)} & b_n^{(n-1)} \end{bmatrix}, \quad (6.28)$$

where for $j \geq i$, $a_{ij}^{(i-1)}$ is as given in equation (6.27). Now, in order to check whether back substitution will be free of fractions, we must first explicitly calculate $a_{nn}^{(n-1)}$.

Calculation of $a_{nn}^{(n-1)}$

From the above analysis, we can explicitly calculate the term $a_{nn}^{(n-1)}$ as follows:

$$a_{nn}^{(n-1)} = \left(\frac{A\binom{1}{1}}{k_{21}}\right)^{2^{n-3}} \left(\frac{A\binom{1,2}{1,2}}{k_{31}k_{32}}\right)^{2^{n-4}} \left(\frac{A\binom{1,2,3}{1,2,3}}{k_{41}k_{42}k_{43}}\right)^{2^{n-5}} \cdots \left(\frac{A\binom{1,\ldots,n-2}{1,\ldots,n-2}}{k_{n-1,1}k_{n-1,2}\cdots k_{n-1,n-2}}\right)\left(\frac{A\binom{1,\ldots,n}{1,\ldots,n}}{k_{n1}k_{n2}\cdots k_{n,n-1}}\right).$$

Calculation of z_j

Now, by Cramer's Rule,

$$x_j = \frac{\det B_j}{\det A},$$

and since by definition $A\binom{1,\ldots,n}{1,\ldots,n} = \det A$, we have the following equation:

$$z_j = a_{nn}^{(n-1)} x_j = \left(\frac{A\binom{1}{1}}{k_{21}}\right)^{2^{n-3}} \left(\frac{A\binom{1,2}{1,2}}{k_{31}k_{32}}\right)^{2^{n-4}} \left(\frac{A\binom{1,2,3}{1,2,3}}{k_{41}k_{42}k_{43}}\right)^{2^{n-5}} \cdots \left(\frac{A\binom{1,\dots,n-2}{1,\dots,n-2}}{k_{n-1,1}k_{n-1,2}\cdots k_{n-1,n-2}}\right)\left(\frac{\det B_j}{k_{n1}k_{n2}\cdots k_{n,n-1}}\right).$$

Thus we have the immediate result that if at each step r of the row reductions we multiply each row by $a_{rr}^{(r-1)}$, that is, if $k_{ij} = 1$ for all i, j (recall that $j < i \le n$), then

$$z_j = a_{nn}^{(n-1)} x_j = (a_{11})^{2^{n-3}} \left(A\binom{1,2}{1,2}\right)^{2^{n-4}} \left(A\binom{1,2,3}{1,2,3}\right)^{2^{n-5}} \cdots A\binom{1,\dots,n-2}{1,\dots,n-2} \det B_j$$

is always an integer, for all j $(1 \le j \le n)$. That is, if at every stage we cross multiply according to the *fangcheng* procedure, if at no stage we use a reduced multiplier, then all the z_j are integers, that is, our calculations are fraction-free until the final division.

However, this makes several assumptions that need not be true: (1) least common multiples are never used; (2) even if the entry to be eliminated is already 0, we still cross multiply; (3) even if the entry to be eliminated is equal to the pivot (in absolute value), we still cross multiply. Although (1) is plausible, (2) and (3) are not. So we must seek a more precise estimate. The problem is that we do not know what relationship holds between the terms k_{ij} in the denominator and the minors $A\binom{1,\dots,r}{1,\dots,r}$—that is, the k_{ij} need not divide $A\binom{1,\dots,r}{1,\dots,r}$. But we do know, by definition, that each k_{ij} divides the pivot $a_{jj}^{(j-1)}$. So we wish to write $a_{nn}^{(n-1)}$ in terms of the pivots $a_{rr}^{(r-1)}$, and then rearrange the k_{ij} so that each k_{ij} appears under the $a_{jj}^{(j-1)}$ that it divides. That is, we must rewrite the $A\binom{1}{1}$ to $A\binom{1,\dots,n-1}{1,\dots,n-1}$ in terms of the $a_{rr}^{(r-1)}$.

Rewriting $a_{nn}^{(n-1)}$ *in Terms of* $a_{rr}^{(r-1)}$

We can obtain a value that more accurately reflects the mathematical practice, but it requires some work. Using Proposition C.2 (see page 259, in Appendix C), for $n \ge 3$, we can write $z_j = a_{nn}^{(n-1)} x_j$ as follows:

$$z_j = a_{nn}^{(n-1)} x_j = \left(\prod_{r=1}^{n-2} \left(\frac{\left(a_{rr}^{(r-1)}\right)^{n-r-1}}{\prod_{i=r+1}^{n} k_{ir}}\right)\right) \frac{\det B_j}{k_{n,n-1}}. \tag{6.29}$$

Expanding the product, this is equivalent to the following:

$$z_j = \left(\frac{\det B_j}{k_{n,n-1}}\right)\left(\frac{a_{n-2,n-2}^{(n-3)}}{k_{n-1,n-2}k_{n,n-2}}\right)\left(\frac{(a_{n-3,n-3}^{(n-4)})^2}{k_{n-2,n-3}k_{n-1,n-3}k_{n,n-3}}\right)\cdots$$

$$\cdot\left(\frac{(a_{33}^{(2)})^{n-4}}{k_{43}k_{53}\cdots k_{n3}}\right)\left(\frac{(a_{22}^{(1)})^{n-3}}{k_{32}k_{42}\cdots k_{n-1,2}k_{n2}}\right)\left(\frac{(a_{11}^{(0)})^{n-2}}{k_{21}k_{31}k_{41}\cdots k_{n-1,1}k_{n1}}\right).$$

Note that each of the terms on the right-hand side, with the exception of the first, is $(a_{rr}^{(r-1)})^{n-r-1}$ divided by $n-r$ terms k_{ir} $(r+1 \le i \le n)$. By definition, each of the terms k_{ir} divides $a_{rr}^{(r-1)}$, so each grouping

$$\frac{(a_{rr}^{(r-1)})^{n-r-1}}{k_{r+1,r}k_{r+2,r}\cdots k_{n-1,r}k_{nr}}$$

will be an integer if at least one of the $k_{ir}=1$. That is, we can state as a sufficient (but not necessary) condition that each grouping will be an integer if

$$k_{ir}=1 \text{ for at least one } i \text{ with } (r+1\le i\le n).$$

This is equivalent to stating that a sufficient (but not necessary) condition that each z_j is an integer is that for each $r=1,\ldots,n-1$, $k_{ir}=1$ for at least one i with $(r+1\le i\le n)$. That is, each z_j is an integer if each pivot $a_{rr}^{(r-1)}$ is used in at least one of the "cross multiplications" to eliminate the entries below it. In other words, in almost all calculations following the *fangcheng* procedure, fractions will be avoided.

Frequency Tables

We can give a more specific answer to what it means to say that "almost all" calculations are fraction-free by generating matrices at random and solving them following the *fangcheng* procedure. Table 6.3 shows the percentage of matrices for which the procedure for back substitution used in the *fangcheng* procedure fails to be integer-preserving, that is, a fraction emerges in the process before the final division by $a_{nn}^{(n-1)}$.

Table 6.3: Percentage of matrices of order 2 to 5 for which the *fangcheng* procedure fails to be integer-preserving.

Order		2	3	4	5
Matrices		1000000	1000000	1000000	1000000
$\det A=0$		3.2%	1.5%	0.5%	0.12%
x_j all integers		12.0%	3.9%	1.0%	0.27%
Non-integer:					
	z_4				0.70%
	z_3			2.3%	0.12%
	z_2		5.8%	0.6%	0.06%
	z_1	6.4%	0.8%	0.1%	0.01%
Totals:		6.4%	6.6%	3.0%	0.91%

For each order n ($2 \le n \le 5$), 1,000,000 matrices were generated by (pseudo-) randomly assigning integers 1–9 as coefficients of the unknowns a_{ij}, and integers 1–150 for the constants b_i. The row labeled "$\det A = 0$" gives the percentage of cases with no unique solution. If there was a unique solution (that is, if $\det A \neq 0$), it was then calculated following the *fangcheng* procedure, allowing partial pivoting, using the pivot as the multiplier unless the element to be eliminated was 0 or equal to plus or minus the pivot. The row labeled "x_j all integers" gives the percentage of cases in which the solution consisted only of integers, and no fractions. The rows labeled "Non-integer: z_j" give the percentage of cases in which a fraction appeared in the process of back substitution in the calculation of z_j; in most cases, the remaining z_j were also fractions.

From this table, we can observe several preliminary points regarding these matrices:

1. The percentage of systems of equations with no unique solution is small, and for $n \ge 4$, quite small. That is, contrary to the impression from modern mathematical texts, which emphasize proofs and counter-examples, the actual number of cases of inconsistent equations is small.
2. The percentage of solutions in which the unknowns are all integers is also small, and for $n \ge 4$, very small. (The integral solutions to problems 17 and 18 shows the highly contrived nature of the exemplars selected.)

Still, the most important result for this study is the demonstration that the *fangcheng* procedure is almost always integer-preserving.[3]

Table 6.4 shows that if in the process of row reductions we use the greatest common divisor to find the least common multiple of the pivot $a_{rr}^{(r-1)}$ and the element to be eliminated, $a_{ir}^{(r-1)}$, then the calculations are often not fraction-free. That is, in about one fifth of the cases, fractions emerge in the process of back substitution.

Table 6.4: Reduced using $(a_{rr}^{(r-1)}, a_{ir}^{(r-1)})$

Order		2	3	4	5
Matrices		1000000	1000000	1000000	1000000
Non-integer:					
	z_4				14.3%
	z_3			17.3%	3.6%
	z_2		19.1%	4.4%	1.2%
	z_1	17.2%	4.2%	1.2%	0.36%
Total		17.2%	23.4%	22.9%	19.6%

[3] Also note that if the last unknown z_1 is a fraction, this is of little consequence, since back substitutions are finished, and the only remaining calculation is to divide all the results, z_j, by $a_{nn}^{(n-1)}$.

Finally, Table 6.5 shows that using the greatest common divisor of $a_{nn}^{(n-1)}$ and $b_n^{(n-1)}$ to reduce them both before the process of back substitution results in the appearance of fractions in about one of three cases. Thus, while this reduction seems very natural in modern mathematics, there is a good reason for not doing so on the counting board.

Table 6.5: Reduced using $(a_{nn}^{(n-1)}, b_n^{(n-1)})$

Order		2	3	4	5
Matrices		1000000	1000000	1000000	1000000
Non-integer:					
	z_4				26.5%
	z_3			25.7%	7.0%
	z_2		24.0%	6.7%	2.4%
	z_1	19.6%	5.7%	1.9%	0.8%
Total		19.6%	29.7%	34.4%	36.7%

Conclusions

The following points should be noted about the approach to elimination in the *fangcheng* procedure in chapter 8:

1. In general, only integers are used in these problems as coefficients of the unknowns and constant terms; when fractions are introduced, as in problems 10 and 11, it is done just to explain how to transform the problem into one containing only integers.
2. Entries are eliminated from right to left, and from top to bottom, until reaching triangular form. This corresponds to the order in the usual approach to Gaussian elimination in modern linear algebra—from top to bottom, and from left to right, until reaching upper-triangular form. However, contrary to some descriptions of the *fangcheng* procedure, the augmented matrix is not further transformed to row-echelon form with ones as leading entries, which could produce fractional entries; nor is the augmented matrix transformed into reduced row-echelon form of Gauss-Jordan elimination.
3. The problems in chapter 8 are exemplars, and none are in any way "pathological": all of the problems have unique nonzero solutions; none of the problems require pivoting (row or column exchanges) for their solution.
4. The elementary row operations performed are either multiplication of a row by a nonzero integer or the addition of an integral multiple of one row to an integral multiple of another. Fractional multipliers are not used.
5. Elimination of entries is "integer-preserving": to remove an element, the entire row is multiplied by the pivot, from which is subtracted the pivot row

multiplied by the element. In contrast, in modern linear algebra, Gaussian elimination usually calls for multiplying the entire row by the fraction consisting of the pivot as the numerator and the element as the denominator.

The most important results in this chapter, however, lie not in the analysis of elimination, but in that of back substitution. We found that back substitution in the *fangcheng* procedure differs from that of modern linear algebra. Its counter-intuitive approach avoids the emergence of fractions in almost all cases, which demonstrates the sophistication of the *fangcheng* procedure.

Chapter 7
The Well Problem

This chapter focuses on the "well problem,"[1] problem 13 from *Fangcheng*, chapter 8 of the *Nine Chapters*. This is arguably the most important of all the *fangcheng* problems. We will see that it is in commentaries on the well problem that we find the earliest written records—from anywhere in the world—of what we now call determinantal calculations and determinantal solutions. The chapter will be divided into three sections:

1. "Traditional Solutions to the Well Problem" presents a survey of various interpretations and solutions that have been offered for the well problem: (i) the problem as presented in the *Nine Chapters*, and as solved using the *fangcheng* procedure; (ii) Liu Hui's commentary (c. 263 C.E.); (iii) Dai Zhen's notes (c. 1774 C.E.); and (iv) interpretations of the well problem as "indeterminate," as presented by modern historians of Chinese mathematics.
2. "The Earliest Extant Record of a Determinantal Calculation" presents the earliest known surviving written record of such a calculation, which is preserved in a commentary written in 1025 C.E. to the solution of the well problem.
3. "The Earliest Extant Record of a Determinantal Solution" presents the earliest known surviving written record of a determinantal solution, which is preserved in a text compiled in 1661 C.E. as an alternative solution to the well problem.

Translated into modern notation, the well problem is an exemplar for a system of 5 equations in 6 unknowns, x_1, x_2, x_3, x_4, x_5, and y,[2] represented by the following augmented matrix:

$$\begin{bmatrix} k_1 & 1 & 0 & 0 & 0 & y \\ 0 & k_2 & 1 & 0 & 0 & y \\ 0 & 0 & k_3 & 1 & 0 & y \\ 0 & 0 & 0 & k_4 & 1 & y \\ 1 & 0 & 0 & 0 & k_5 & y \end{bmatrix}. \tag{7.1}$$

[1] I would like to thank Jean-Claude Martzloff for suggesting this name.

[2] It is because the well problem has more unknowns than equations that modern studies have classified it as "indeterminate" (or "underdetermined").

Later commentaries to the *Nine Chapters* present a slightly more general form than that given in equation (7.1), which is equivalent to

$$\begin{bmatrix} k_1 & l_1 & 0 & \cdots & 0 & b_1 \\ 0 & k_2 & l_2 & \ddots & \vdots & b_2 \\ \vdots & \ddots & \ddots & \ddots & 0 & \vdots \\ 0 & \cdots & 0 & k_{n-1} & l_{n-1} & b_{n-1} \\ l_n & 0 & \cdots & 0 & k_n & b_n \end{bmatrix}. \tag{7.2}$$

Traditional Solutions to the Well Problem

We begin with a translation of the well problem, as recorded in the *Nine Chapters*, together with commentaries, in order to establish three important results: (1) We will calculate the solution to the well problem following the *fangcheng* procedure given in the *Nine Chapters*, which will serve as the basis for comparisons with alternative methods of solution. (2) The commentaries, I will argue, should not be viewed as mere explications of the original text of the *Nine Chapters*; instead, each presents a different approach to solving the well problem. (3) The modern view of the well problem as indeterminate, I will argue, is a rather recent interpretation: there is no clear evidence from either the *Nine Chapters* or the early commentaries that it was considered indeterminate at that time. Dai Zhen, for his part, clearly considers the problem to be indeterminate; and the earliest statements I have found in extant treatises that the well problem is indeterminate date from 1674, about 1400 years after Liu's commentary.

Translation of the Well Problem

The format of the following translation follows that found in the *Hall of Martial Eminence Collectanea, Movable Type Edition* (*Wuyingdian juzhen ban congshu* 武英殿聚珍版叢書, 1774), reprinted in ZKJD (see Figure 7.1 on the facing page). The format requires some explanation. First, in full-size characters, is the original text of the *Nine Chapters*, as reconstructed by Dai. Following the statement of the problem, the solution, and the method (that is, immediately after the sentence "Method: ... positive and negative [numbers]") is the commentary by Liu Hui, which in this edition is offset by the space of one character from the top; in my translation, Liu's commentary is offset as a quotation within the translation, and for added clarity preceded by the interpolation "[Liu Hui:]."[3] Interspersed throughout this text are notes by Dai, which are in half-size characters, and are

[3] Again, it should be noted that there is no subcommentary by Li Chunfeng et al. in chapter 8.

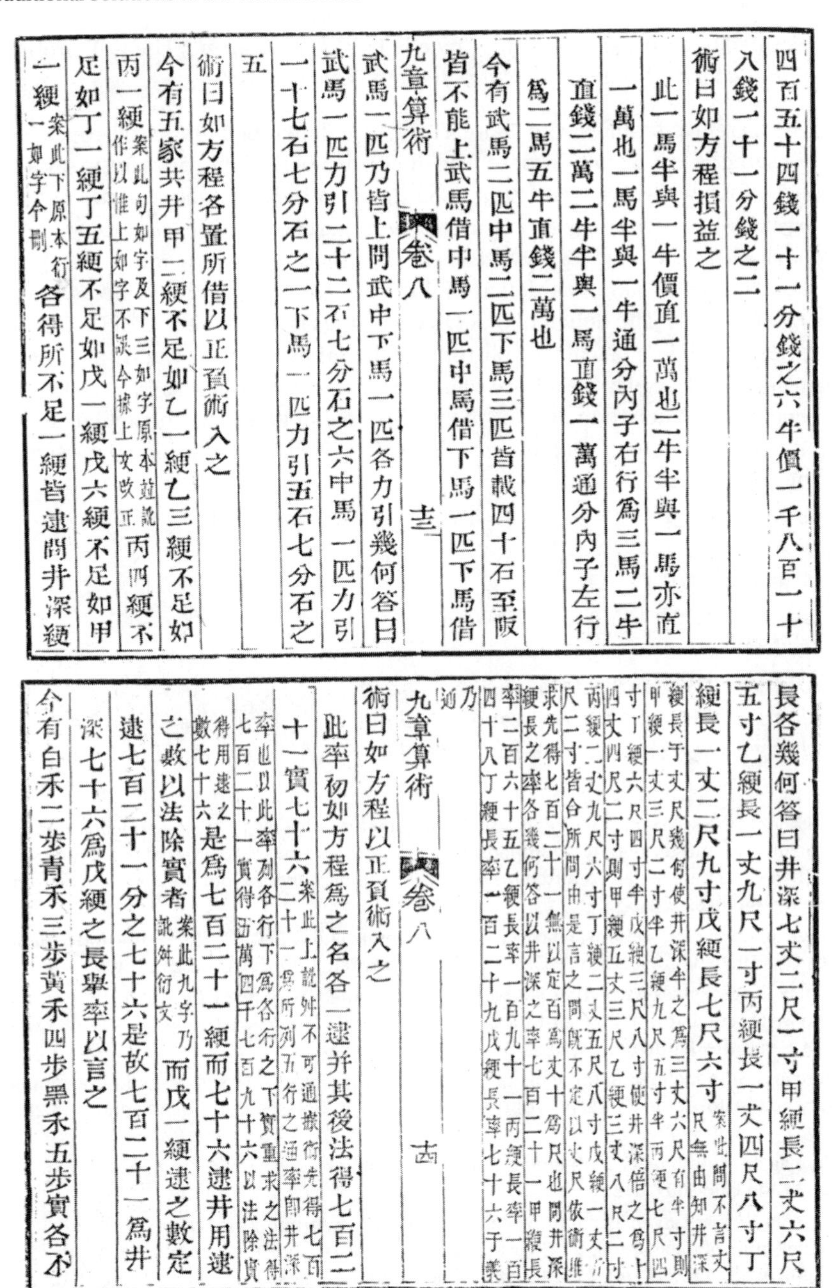

四百五十四錢一十一分錢之六牛價一千八百一十
八錢一十一分錢之二
術曰如方程損益之
此一馬半與一牛價直一萬也二牛半與一馬亦直
一萬也一馬半與一牛通分內子右行爲三馬二牛
直錢二萬二牛半與一馬直錢一萬通分內子左行
爲二馬五牛直錢二萬也
今有武馬二匹中馬二匹下馬三匹皆載四十石至阪
皆不能上武馬借中馬一匹中馬借下馬一匹下馬借
九章算術　卷八　十三
武馬一匹乃皆上問武中下馬一匹各力引幾何答曰
武馬一匹力引二十二石七分石之六中馬一匹力引
一十七石七分石之一下馬一匹力引五石七分石之
五
術曰如方程各置所借以正負術入之
今有五家共井甲二綆不足如乙一綆乙三綆不足如
丙一綆案此句如字及下三如字原本並譌作以惟上如字不誤今據上文改正丙四綆不
足如丁一綆丁五綆不足如戊一綆戊六綆不足如甲
一綆案此下原本衍一如字今刪各得所不足一綆皆逮問井深綆

長各幾何答曰井深七丈二尺一寸甲綆長二丈六尺
五寸乙綆長一丈九尺一寸丙綆長一丈四尺八寸丁
綆長一丈二尺九寸戊綆長七尺六寸案此問不言丈尺無由知井深
綆長于丈尺幾何使井深半之爲三丈六尺有半寸則甲綆一丈三尺二寸半乙綆九尺五寸半丙綆七尺四
寸丁綆六尺四寸半戊綆三尺八寸使井深倍之爲十四丈四尺二寸則甲綆五丈三尺乙綆三丈八尺二寸
丙綆二丈九尺六寸丁綆二丈五尺八寸戊綆一丈五尺二寸皆合所問由是言之問既不定以丈尺依術推
求先得七百二十一無以定百爲丈十爲尺也問井深綆長之率各幾何答以井深之率七百二十一甲綆長
率二百六十五乙綆長率一百九十一丙綆長率一百四十八丁綆長率一百二十九戊綆長率七十六于義
乃通
九章算術　卷八　十四
術曰如方程以正負術入之
此率初如方程爲之名各一逮并其後法得七百二
十一實七十六案此上譌舛不可通據術先得七百二十一爲所列五行之通率即井深
率也以此率列各行下爲各行之下實重求之法得七百二十一實得五萬四千七百九十六以法除實
得用逮之數七十六是爲七百二十一綆而七十六逮井用逮
之數以法除實者案此九字乃譌舛衍文而戊一綆逮之數定
逮七百二十一分之七十六是故七百二十一爲井
深七十六爲戊綆之長舉率以言之
今有白禾二步青禾三步黃禾四步黑禾五步實各不

Fig. 7.1: The well problem, photolithographic reproduction, *Hall of Martial Eminence Collectanea, Movable Type Edition* (*Wuyingdian juzhen ban congshu* 武英殿聚珍版叢書, 1774), reprinted in ZKJD. Full-size characters are the text of the *Nine Chapters*, together with the commentary by Liu Hui, which is indented a space of one character from the top, as can be seen in the middle of the lower page. Li Chunfeng did not offer any subcommentary for chapter 8. Half-size text is Dai Zhen's notes on the text and on Liu's commentaries.

transcribed here in a smaller font, preceded by "[Dai Zhen:]."[4] The well problem from the *Juzhen* edition of the *Nine Chapters* is as follows:[5]

> Given are five families who share a well. The deficit of 2 of A's well-ropes is the same as 1 of B's well-ropes.[6] The deficit of 3 of B's well-ropes is the same as 1 of C's well-ropes. [Dai Zhen:] Note: In this sentence, for the character *ru* 如, together with the following three characters *ru* 如, the original edition incorrectly used the character *yi* 以; only the above character *ru* 如 is not mistaken. Here it is emended according to the above. The deficit of 4 of C's well-ropes is the same as 1 of D's well-ropes. The deficit of 5 of D's well-ropes is the same as 1 of E's well-ropes. The deficit of 6 of E's well-ropes is the same as 1 of A's well-ropes. [Dai Zhen:] Note: Below this, the original edition has an extraneous character *ru* 如, which is removed here. Each then obtains the well-rope that makes up the deficit, and all reach the water. Problem: how deep is the well, and how long is each of the well-ropes?
>
> 今有五家共井，甲二綆不足，如乙一綆；乙三綆不足，如[7]丙一綆；案：此句如字及下三如字，原本並訛作以。惟上如字不誤。今據上文改正。丙四綆不足，如丁一綆；丁五綆不足，如戊一綆；戊六綆不足，如甲一綆。案：此下原本衍一如字，今删。各[8]得所不足一綆，皆逮。問井深、綆長各幾何。
>
> Solution: The depth of the well is 7 *zhang* 2 *chi* 1 *cun*,[9] A's well-rope is 2 *zhang* 6 *chi* 5 *cun*, B's well-rope is 1 *zhang* 9 *chi* 1 *cun*, C's well-rope is 1 *zhang* 4 *chi* 8 *cun*, D's well-rope is 1 *zhang* 2 *chi* 9 *cun*, and E's well-rope

---

[4] Dai Zhen's notes, written 1500 years later, have sometimes been viewed as erroneous, a source of corruption, and ultimately an impediment to understanding the "original" *Nine Chapters*. Dai Zhen's notes are usually omitted from critical editions, and have not previously been translated into English.

[5] In the following translations, I have consulted Chemla and Guo 2004; Shen, Lun, and Crossley 1999; Guo 1990; Bai 1990; Guo 1985; and Bai 1983. Any remaining errors are my own.

[6] That is, the distance by which 2 of A's well-ropes combined fail to reach the water at the bottom of the well is equal to 1 of B's well-ropes, or in other words, 2 of A's well-ropes plus 1 of B's well-ropes equals the depth of the well.

[7] Here and in the following three instances of *ru* 如, Guo rejects Dai's emendations as unnecessary and restores the character *yi* 以 (Guo 1990, 411 n. 133; see also Chemla and Guo 2004, 642).

[8] Here again, Guo rejects Dai's emendations and restores *ru* 如 before *ge* 各 (Guo 1990, 390; Chemla and Guo 2004, 642).

[9] The solutions are given as metrological numbers, that is, numbers in which places are designated by units of measurement rather than decimal places. Here, for example, the depth of the well is expressed as 7 *zhang* 2 *chi* 1 *cun*. The terms *zhang* 丈, *chi* 尺, and *cun* 寸 are all units of length; 1 *zhang* = 10 *chi* = 100 *cun*, so that in this case 7 *zhang* 2 *chi* 1 *cun* = 721 *cun*. However, such units of measure are sometimes related by multiples other than 10: for example, 1 *jin* = 16 *liang*, so 1 *jin* 4 *liang* = 20 *liang*, not 14 *liang*. Preserved archaeological artifacts show that 1 *cun* was approximately 2.3 cm. For detailed studies of Chinese units of measure, see Yang 1938 and Wu 1937. For helpful summaries of Chinese units of measure as used in the *Nine Chapters*, see Shen, Lun, and Crossley 1999, 9–10; Chemla and Guo 2004, inside of front cover. For a discussion of metrological numbers, see Martzloff [1987] 2006, 197–200.

is 7 *chi* 6 *cun*. [Dai Zhen:] Note: This problem does not state the measurements [lit., *zhang* and *chi*], and there is no basis from which one can ascertain the measurements of the depth of the well and the lengths of the well-ropes. Halve the depth of the well, 3 *zhang* 6 *chi* $\frac{1}{2}$ *cun*, then A's well-rope is 1 *zhang* 3 *chi* 2 *cun* and $\frac{1}{2}$ [*cun*], B's well-rope is 9 *zhang* 5 *chi* $\frac{1}{2}$ *cun*, C's well-rope is 7 *chi* 4 *cun*, D's well-rope is 6 *chi* 4 *cun* and $\frac{1}{2}$ [*cun*], and E's well-rope is 3 *chi* 8 *cun*. Double the depth of the well, 14 *zhang* 4 *chi* 2 *cun*, then A's well-rope is 5 *zhang* 3 *chi*, B's well-rope is 3 *zhang* 8 *chi* 2 *cun*, C's well-rope is 2 *zhang* 9 *chi* 6 *cun*, D's well-rope is 2 *zhang* 5 *chi* 8 *cun*, and E's well-rope is 1 *zhang* 5 *chi* 2 *cun*. Both [sets of solutions also] fit the problem. From this we can state that the problem then does not determine the measurements, which are deduced following the method.[10] First obtain 721,[11] and there is no reason to use *zhang* for hundreds and *chi* for tens. What are the "relative values" [*lü*][12] of the depth of the well and the length of each of the well-ropes? The answer takes the relative value of the depth of the well to be 721, the relative value of A's well-rope is then 265, the relative value of the length of B's well-rope is 191, the relative value of the length of C's well-rope is 148, the relative value of the length of D's well-rope is 129, and the relative value of the length of E's well-rope is 76. This is compatible with the meaning [of the problem].

答[13]曰：井深七丈二尺一寸，甲綆長二丈六尺五寸，乙綆長一丈九尺一寸，丙綆長一丈四尺八寸，丁綆長一丈二尺九寸，戊綆長七尺六寸。

案：此問不言丈尺，無由知井深、綆長于丈尺幾何？使井深半之，為三丈六尺有半寸。則甲綆一丈三尺二寸半。乙綆九尺五寸半。丙綆七尺四寸。丁綆六尺四寸半。戊綆三尺八寸。使井深倍之，為十四丈四尺二寸。則甲綆五丈三尺。乙綆三丈八尺二寸。丙綆二丈九尺六寸。丁綆二丈五尺八寸。戊綆一丈五尺二寸。皆合所問。由是言之。問既不定以丈尺。依術推求。先得七百二十一。無以定百為丈，十為尺也。問井深綆長之率各幾何。答以井深之率七百二十一。甲綆長率二百六十五。乙綆長率一百九十一。丙綆長率一百四十八。丁綆長率一百二十九。戊綆長率七十六。于義乃通。

Method: Follow [the procedure for] *fangcheng*; apply the method for positive and negative [numbers].

術曰：如方程，以正負術入之。

---

[10] Dai is presumably referring to the *fangcheng* procedure, but he does not explicitly describe how it is to be used to determine these measurements, nor does the *fangcheng* procedure, as preserved in the *Nine Chapters*, provide such a method.

[11] Dai does not explicitly state the calculations that give this result. A reconstruction of the method he likely used is presented below, beginning on page 124.

[12] Dai's use of the term *lü* 率 seems to convey the sense that the values of the lengths of the well-ropes depend on the assignment of the value for the depth of the well (also note that units of measure are not used here). Accordingly, I have translated *lü* here as "relative value," as in "relative value of the depth" (*shen zhi lü* 深之率) and "relative value of the length" (*chang lü* 長率). Dai seems to use the term *lü* in a manner different from that used by Liu (see footnote 14 on page 116).

[13] Guo has the variant form *da* 荅 (Guo 1990, 390; Chemla and Guo 2004, 642).

[Liu Hui:] The terms of the ratios [*lü*][14] are initially formulated according to [the procedure for] *fangcheng*; it is stipulated that each one [of the combinations of well-ropes] reaches the [depth of the] well.[15] Subsequently, obtain 721 as the divisor, 76 as the dividend.[16] [Dai Zhen:] Note: The above is corrupt and contradictory, and cannot be followed. Following the procedure [for *fangcheng*], first obtain 721 as the relative value to be placed uniformly below all five rows, that is, the relative value of the depth of the well. With this relative value placed beneath each row, as the constant term in each row, again find it [the solution using the *fangcheng* procedure], obtaining 721 as the divisor, and obtaining 54796 as the dividend,[17] then divide the divisor into the dividend, obtaining 76 for the length [of E's well-rope that was] used to reach [the bottom of the well]. This means 721 [of E's] well-ropes are equal to 76 depths of the well, so use the number for the depth, the result of dividing the divisor [721] into the dividend [76]. [Dai Zhen:] Note: These nine characters [immediately above, namely "so use the number for the depth, the result of dividing the divisor into the dividend"] are corrupt, contradictory, and spurious. The number for the length of E's well-rope is determined, reaching $\frac{76}{721}$. For this reason 721 is the depth of the well, and 76 is the length of E's well-rope. Use the terms of the ratios to state it.

此率初如方程為之，名各一逮井。其後，法得七百二十一，實七十六，　案：此上訛舛不可通。據術先得七百二十一，為所列五行之通率，即井深率也。以此率列各行下，為各行之下實，重求之，法得七百二十一，實得五萬四千七百九十六，以法除實得用逮之數七十六。是為七百二十一綆而七十六逮井，用[18]逮之數，以法除實者，　案：此九字乃訛舛衍

---

[14] *Lü* 率 is one of Liu's central concepts throughout his commentary to the *Nine Chapters*, and one he uses in a variety of ways in different contexts, as has been analyzed in detail in Chemla and Guo 2004, 956–59. Chemla notes that Liu introduces *lü* with the following explanation: "Whenever quantities are given in relation to one another, they are called *lü*" 凡數相與者謂之率 (Chemla and Guo 2004, 167–68). In this general case, it seems best to translate *lü* as "term(s) in the ratio(s)" (that is, as "one term in a ratio" for the singular, and "terms in a ratio" or "terms in the ratios" for the plural). "Ratio," "proportion," or "rate" are arguably less precise—I thank Donald Wagner for his suggestions on this point.

[15] The exact meaning of this sentence, and indeed of Liu's entire commentary to this problem, is difficult to ascertain. Shen, Lun, and Crossley translate: "First lay out the array according to the condition that [the well-ropes] reach the depth of the well in each case" (Shen, Lun, and Crossley 1999, 413). Guo translates into modern Chinese: "将这些率最初是如方程术那样求解出来的，指的是各达到一次井深" [That these rates are initially found by solving according to the *fangcheng* procedure, what is being pointed out is that each reaches one depth of the well] (Guo 1998, 175). Guo and Chemla translate: "Si ces *lü* sont dans un premier temps traités en suivant *fangcheng*, ils renvoient à chaque fois au fait qu'on atteint le fond du puits une fois" (Chemla and Guo 2004, 643).

[16] It is not clear how Liu obtains these results—he provides no explanation or even indication of how they are to be calculated.

[17] These numbers, 721 and 54796, result from the row reductions using the *fangcheng* procedure (see equation 7.7 on page 118).

[18] Guo inserts *bing* 并 before *yong* 用 (Guo 1990, 390; Chemla and Guo 2004, 642).

75 文。而戊一綆逮之數定，逮七百二十一分之七十六。是故七百二十一為井深，七十六為戊綆之長，舉率以言之。

(JZSS, *juan* 8, 13b–14b, in ZKJD, v. 1, 184–85)

Elimination Using the Fangcheng *Procedure*

In order to solve this problem using the *fangcheng* procedure, we must first perform a series of operations to eliminate entries.

STEP 1. The initial placement of the counting rods for the well problem is as follows:

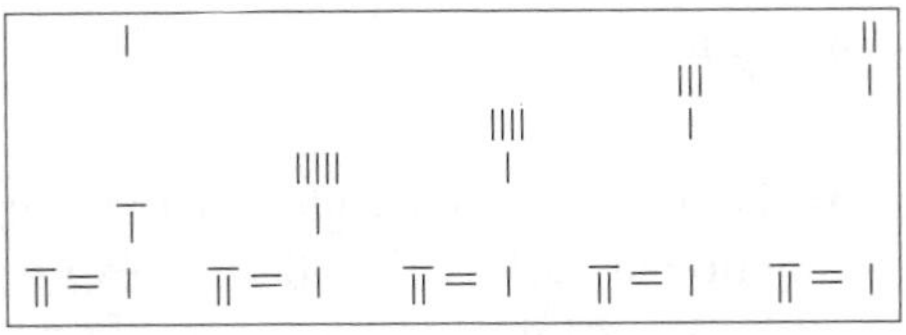

$$\begin{bmatrix} 2 & 1 & 0 & 0 & 0 & 721 \\ 0 & 3 & 1 & 0 & 0 & 721 \\ 0 & 0 & 4 & 1 & 0 & 721 \\ 0 & 0 & 0 & 5 & 1 & 721 \\ 1 & 0 & 0 & 0 & 6 & 721 \end{bmatrix}. \tag{7.3}$$

STEP 2. The only entry to be removed is the first entry in the fifth column [row, namely $a_{51} = 1$]. First we multiply the fifth column [row] by 2, and subtract the first column [row], giving $2 \cdot (1,0,0,0,6,721) - (2,1,0,0,0,721) = (0,-1,0,0,12,721)$,

$$\begin{bmatrix} 2 & 1 & 0 & 0 & 0 & 721 \\ 0 & 3 & 1 & 0 & 0 & 721 \\ 0 & 0 & 4 & 1 & 0 & 721 \\ 0 & 0 & 0 & 5 & 1 & 721 \\ 0 & -1 & 0 & 0 & 12 & 721 \end{bmatrix}. \tag{7.4}$$

STEP 3. We then multiply the fifth column [row] by 3, and add the second column [row], $3 \cdot (0,-1,0,0,12,721) + (0,3,1,0,0,721) = (0,0,1,0,36,2884)$,

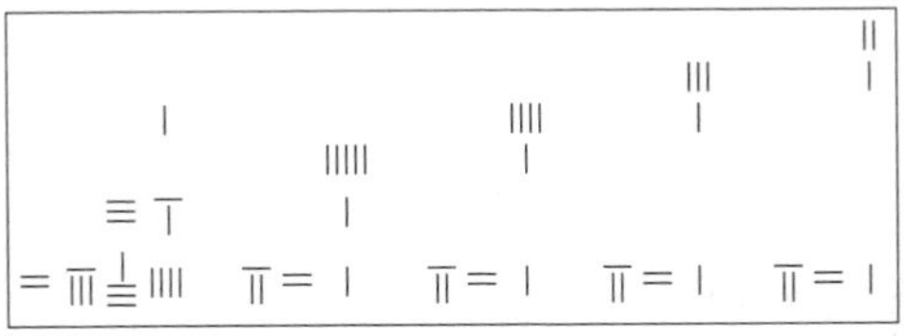

$$\begin{bmatrix} 2 & 1 & 0 & 0 & 0 & 721 \\ 0 & 3 & 1 & 0 & 0 & 721 \\ 0 & 0 & 4 & 1 & 0 & 721 \\ 0 & 0 & 0 & 5 & 1 & 721 \\ 0 & 0 & 1 & 0 & 36 & 2884 \end{bmatrix}. \tag{7.5}$$

STEP 4. We multiply the fifth column [row] by 4, and subtract the third column [row], $4 \cdot (0,0,1,0,36,2884) - (0,0,4,1,0,721) = (0,0,0,-1,144,10815)$,

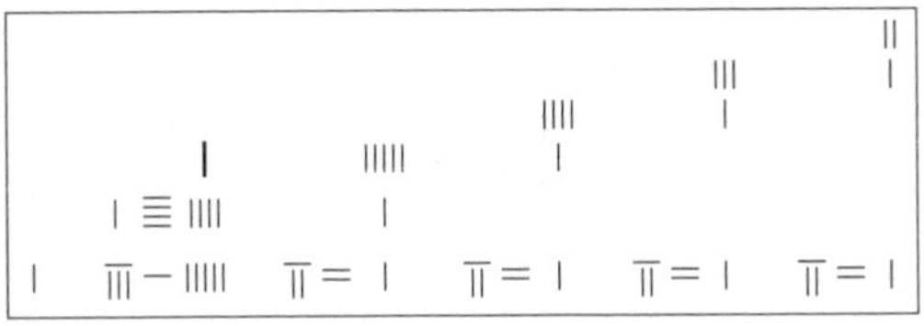

$$\begin{bmatrix} 2 & 1 & 0 & 0 & 0 & 721 \\ 0 & 3 & 1 & 0 & 0 & 721 \\ 0 & 0 & 4 & 1 & 0 & 721 \\ 0 & 0 & 0 & 5 & 1 & 721 \\ 0 & 0 & 0 & -1 & 144 & 10815 \end{bmatrix}. \tag{7.6}$$

STEP 5. Finally, we multiply the fifth column [row] by 5, and add the fourth column [row], $5 \cdot (0,0,0,-1,144,10815) + (0,0,0,5,1,721) = (0,0,0,0,721,54796)$,

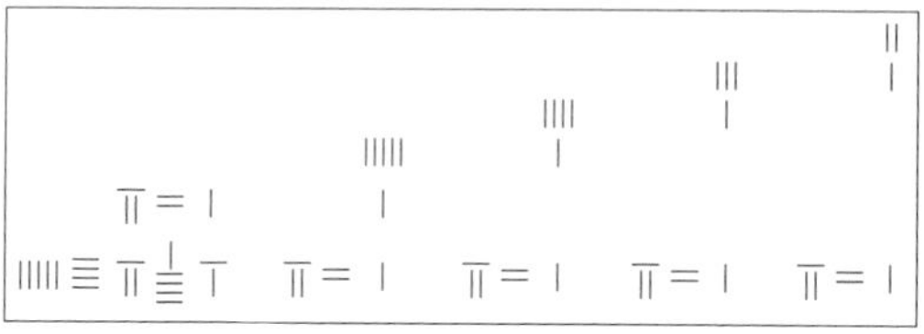

$$\begin{bmatrix} 2 & 1 & 0 & 0 & 0 & 721 \\ 0 & 3 & 1 & 0 & 0 & 721 \\ 0 & 0 & 4 & 1 & 0 & 721 \\ 0 & 0 & 0 & 5 & 1 & 721 \\ 0 & 0 & 0 & 0 & 721 & 54796 \end{bmatrix}. \tag{7.7}$$

The counting board [augmented matrix] is now in lower- [upper-] triangular form. We can now find the unknowns by back substitution.

Back Substitution Using the Fangcheng *Procedure*

We must now solve for the unknowns. Again, the approach given in the *fangcheng* procedure differs from the modern method of back substitution. As explained in chapter 6, the result from each step is back substituted into the next by a kind of "cross multiplication."

STEP 6. First, we "cross multiply" the final entry in the fourth column [row] by the fifth entry in the final column [row], giving $721 \times 721 = 519841$. We then "cross multiply" the fifth entry in the fourth column [row] by the final entry in the final column [row], $54796 \times 1 = 54796$, giving the result

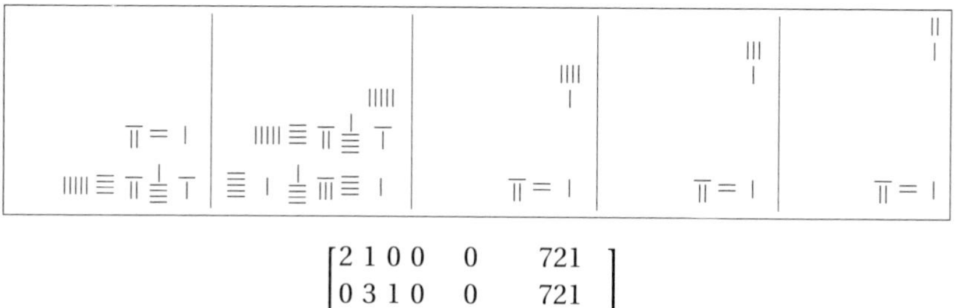

$$\begin{bmatrix} 2 & 1 & 0 & 0 & 0 & 721 \\ 0 & 3 & 1 & 0 & 0 & 721 \\ 0 & 0 & 4 & 1 & 0 & 721 \\ 0 & 0 & 0 & 5 & 54796 & 519841 \\ 0 & 0 & 0 & 0 & 721 & 54796 \end{bmatrix}.$$

STEP 7. We subtract the fifth entry in the fourth column [row] from the final entry in the fourth column [row], giving $519841 - 54796 = 465045$, and then divide this result by the fourth entry in the fourth column [row], $465045 \div 5 = 93009$, giving

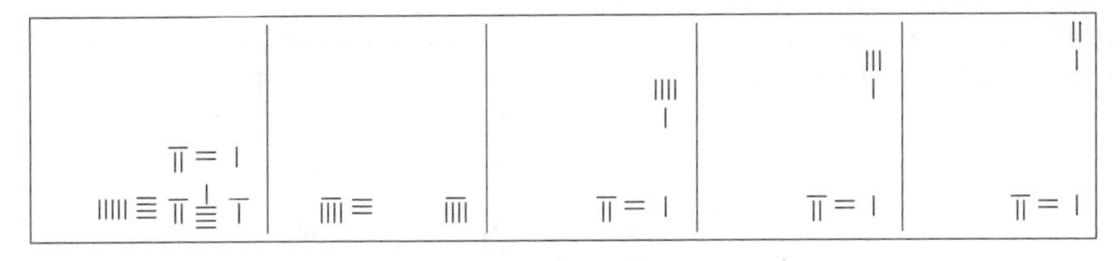

$$\begin{bmatrix} 2 & 1 & 0 & 0 & 0 & 721 \\ 0 & 3 & 1 & 0 & 0 & 721 \\ 0 & 0 & 4 & 1 & 0 & 721 \\ 0 & 0 & 0 & 0 & 0 & 93009 \\ 0 & 0 & 0 & 0 & 721 & 54796 \end{bmatrix}.$$

STEP 8. Next, we simplify the third column. We "cross multiply" the final entry in the third column [row] by the fifth entry in the final column [row], again giving $721 \times 721 = 519841$. We "cross multiply" the fourth entry in the third column [row] by the final entry in the fourth column [row], giving $93009 \times 1 = 93009$:

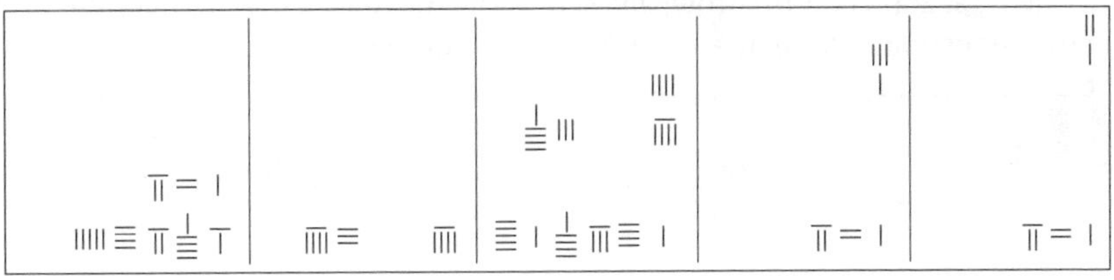

$$\begin{bmatrix} 2 & 1 & 0 & 0 & 0 & 721 \\ 0 & 3 & 1 & 0 & 0 & 721 \\ 0 & 0 & 4 & 93009 & 0 & 519841 \\ 0 & 0 & 0 & 0 & 0 & 93009 \\ 0 & 0 & 0 & 0 & 721 & 54796 \end{bmatrix}.$$

STEP 9. Subtracting the fourth entry in the third column [row] from the final entry in the third column [row] gives $519841 - 93009 = 426832$. We then divide the result by the third entry in the third column [row], giving $426832 \div 4 = 106708$,

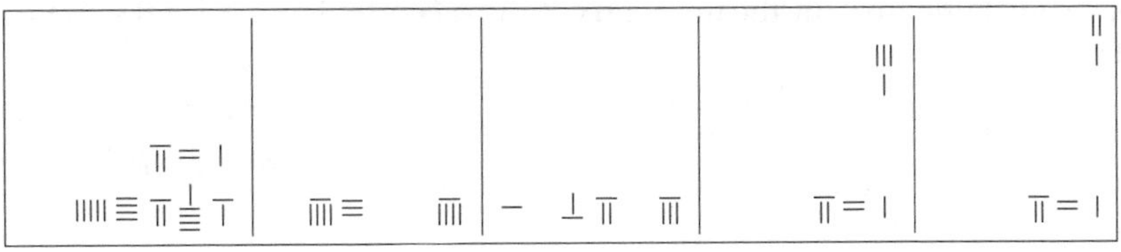

$$\begin{bmatrix} 2 & 1 & 0 & 0 & 0 & 721 \\ 0 & 3 & 1 & 0 & 0 & 721 \\ 0 & 0 & 0 & 0 & 0 & 106708 \\ 0 & 0 & 0 & 0 & 0 & 93009 \\ 0 & 0 & 0 & 0 & 721 & 54796 \end{bmatrix}.$$

STEP 10. The next step simplifies the second column [row]. We "cross multiply" the final entry in the second column [row] by the fifth entry in the final column

[row], $721 \times 721 = 519841$. We "cross multiply" the third entry in the second column [row] by the final entry in the third column [row], $106708 \times 1 = 106708$,

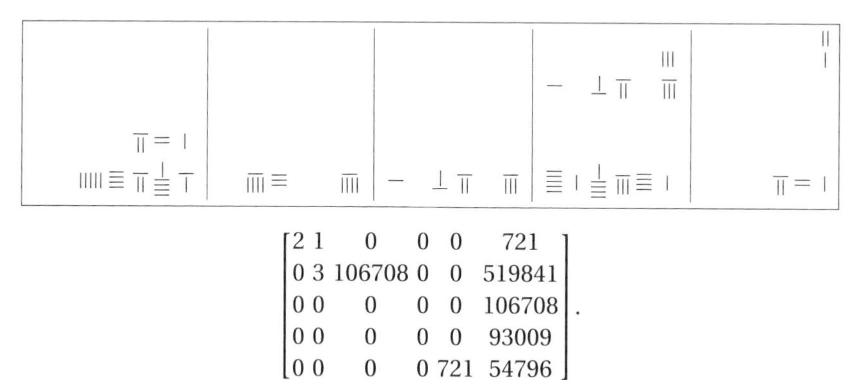

$$\begin{bmatrix} 2 & 1 & 0 & 0 & 0 & 721 \\ 0 & 3 & 106708 & 0 & 0 & 519841 \\ 0 & 0 & 0 & 0 & 0 & 106708 \\ 0 & 0 & 0 & 0 & 0 & 93009 \\ 0 & 0 & 0 & 0 & 721 & 54796 \end{bmatrix}.$$

STEP 11. Subtracting the third entry in the second column [row] from the final entry in the second column [row] gives $519841 - 106708 = 413133$. Dividing the result by the second entry in the second column [row] yields $413133 \div 3 = 137711$,

$$\begin{bmatrix} 2 & 1 & 0 & 0 & 0 & 721 \\ 0 & 0 & 0 & 0 & 0 & 137711 \\ 0 & 0 & 0 & 0 & 0 & 106708 \\ 0 & 0 & 0 & 0 & 0 & 93009 \\ 0 & 0 & 0 & 0 & 721 & 54796 \end{bmatrix}.$$

STEP 12. We then "cross multiply" the final entry in the first column [row] by the fifth entry in the final column [row], $721 \times 721 = 519841$, and the second entry in the first column [row] by the final entry in the second column [row], $137711 \times 1 = 137711$,

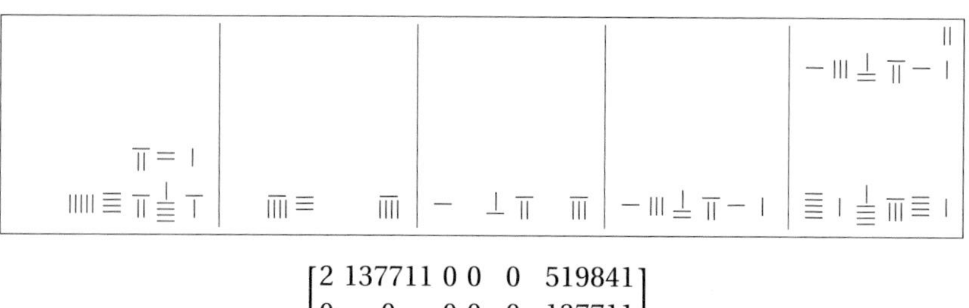

$$\begin{bmatrix} 2 & 137711 & 0 & 0 & 0 & 519841 \\ 0 & 0 & 0 & 0 & 0 & 137711 \\ 0 & 0 & 0 & 0 & 0 & 106708 \\ 0 & 0 & 0 & 0 & 0 & 93009 \\ 0 & 0 & 0 & 0 & 721 & 54796 \end{bmatrix}.$$

STEP 13. We subtract the second entry in the first column [row] from the final entry in the first column [row], 519841 − 137711 = 382130, and divide the result by the first entry in the first column [row], 382130 ÷ 2 = 191065,

$$\begin{bmatrix} 0\,0\,0\,0 & 0 & 191065 \\ 0\,0\,0\,0 & 0 & 137711 \\ 0\,0\,0\,0 & 0 & 106708 \\ 0\,0\,0\,0 & 0 & 93009 \\ 0\,0\,0\,0 & 721 & 54796 \end{bmatrix}.$$

STEP 14. Finally, we divide the final entry of each column [row] by the fifth entry in the final column [row], 721, to give the following answers:

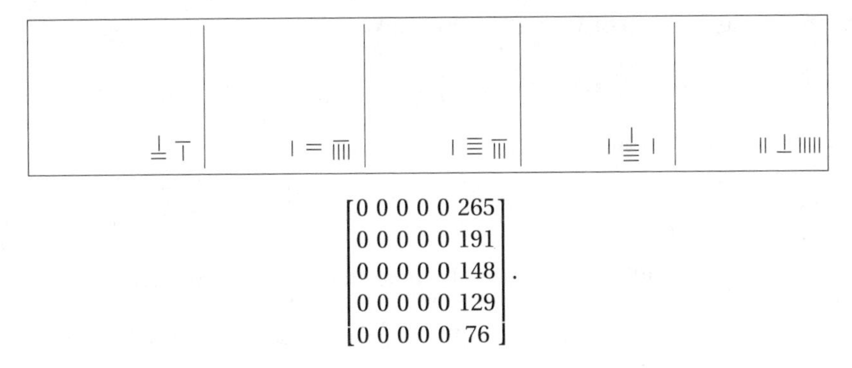

$$\begin{bmatrix} 0\,0\,0\,0\,0 & 265 \\ 0\,0\,0\,0\,0 & 191 \\ 0\,0\,0\,0\,0 & 148 \\ 0\,0\,0\,0\,0 & 129 \\ 0\,0\,0\,0\,0 & 76 \end{bmatrix}.$$

Conclusions

What this entire calculation shows is that there is no pedagogical reason whatsoever to include the well problem in chapter 8 of the *Nine Chapters*. That is, neither in the process of elimination, nor in the process of back substitution, was there anything noteworthy about this problem. In particular, the symmetric placement of zeros serves no purpose whatsoever in the application of the *fangcheng* procedure.

The Well Problem as "Indeterminate"

The previous section, "Elimination Using the *Fangcheng* Procedure," showed how the well problem is solved once a value for the sixth unknown y, the depth of the well, has been assigned. Now we must analyze how the sixth unknown is found. From the point of view of modern linear algebra, the well problem, which is equivalent to a system of 5 equations in 6 unknowns, does not have a unique solution, and is therefore indeterminate. Modern analyses of the *Nine Chapters*—without exception, as far as I have been able to determine—all classify the well problem as indeterminate, and solve it using row reductions in a manner that can be roughly summarized as follows.[19] We first write this problem as a system of 5 equations in 6 unknowns,

$$2x_1 + x_2 = y, \quad 3x_2 + x_3 = y, \quad 4x_3 + x_4 = y, \quad 5x_4 + x_5 = y, \quad x_1 + 6x_5 = y,$$

where x_1, x_2, x_3, x_4, x_5, and y represent the six unknowns. The trick used in modern interpretations to solve this problem is to transform it into a system of 5 equations in 5 unknowns, which is solvable, by dividing both sides of all five equations by the unknown y. If we rewrite the unknowns as

$$z_j = \frac{x_j}{y} \quad (1 \le j \le 5), \tag{7.8}$$

we then have a system of 5 equations in 5 unknowns,

$$2z_1 + z_2 = 1, \quad 3z_2 + z_3 = 1, \quad 4z_3 + z_4 = 1, \quad 5z_4 + z_5 = 1, \quad z_1 + 6z_5 = 1,$$

which does have a unique solution,

$$z_1 = \frac{265}{721}, \quad z_2 = \frac{191}{721}, \quad z_3 = \frac{148}{721}, \quad z_4 = \frac{129}{721}, \quad z_5 = \frac{76}{721}. \tag{7.9}$$

In these modern studies, this result is then taken to represent the solution,

$$x_1 = 265, \quad x_2 = 191, \quad x_3 = 148, \quad x_4 = 129, \quad x_5 = 76, \quad y = 721.$$

This interpretation of the well problem as indeterminate is, however, relatively recent.[20] The original text of the *Nine Chapters*, Liu's commentary, and Dai's notes represent not one but three different approaches to solving the well problem, and only in Dai's notes, written 1500 years after the previous two, is there an explicit statement that the problem is indeterminate.

There is no evidence in the original text of the *Nine Chapters* that this problem was viewed as indeterminate. The original text simply states the depth of the well without any explanation whatsoever: "The depth of the well is 7 *zhang* 2 *chi* 1 *cun*" (line 18 on page 114). As noted above, the explanation that the original text offers for finding the solution is just the same two laconic instructions that are

---

[19] See, for example, Chemla and Guo 2004, 612–13; Guo 1998, 422; Guo 1990, 395; Shen, Lun, and Crossley 1999, 413–23; Martzloff [1987] 2006, 257; Ma 1994, 90–92; Bai 1983, 281; Libbrecht 1973, 267–68; and Xu 1952, 10–12.

[20] Dai is not the first to assert that the well problem is indeterminate—the earliest statement of this claim that I have found is Mei Wending's *On* Fangcheng, c. 1674 (FCL, *juan* 4, 43a–45b).

offered for most of the problems in chapter 8: "Follow [the procedure for] *fangcheng*; apply the method for positive and negative [numbers]" (lines 48 and 49 on page 115). It is possible that the compiler of the original text did not know how the depth of the well was found in practice, just as later writers seem to have merely copied the solution to this problem without understanding how it was to be solved.

Liu Hui's Commentary

Recent studies have often imputed to Liu Hui a recognition that the well problem is indeterminate. It is true that Liu does formulate the problem using "terms of the ratios" (*lü* 率, sometimes translated as "proportion," "ratio," or "rate," line 51 on page 116), which has seemed to some historians to suggest an understanding of the problem similar to modern analyses, and in particular to setting $z_i = \frac{x_i}{y}$ (see equation 7.8). Furthermore, Liu states "obtain 721 as the divisor, 76 as the dividend" (line 54 on page 116); he notes "721 [of E's] well-ropes are equal to 76 depths of the well" (lines 62 and 63 on page 116); and he arrives at the result $\frac{76}{721}$ (line 68 on page 116), as do modern analyses (see equation 7.9).

However, a closer examination of Liu's commentary shows that there is in fact little evidence from his commentary to support this interpretation:

1. In contrast to Dai Zhen (as we will see below), Liu never makes any explicit statements that indicate that he viewed the problem as indeterminate: he never states that the depth of the well is not determined by the problem; he never offers any possible alternative values for the depth of the well; and he never states that assigning an alternative value yields an alternative set of solutions.
2. Liu's use of "terms of the ratios" (*lü*) does not indicate that he viewed this problem as indeterminate. This is simply the approach he proposes for solving *all fangcheng*. For example, in his commentary to the *fangcheng* procedure, used for solving all the *fangcheng* problems in chapter 8 of the *Nine Chapters*, Liu stipulates that "each column is formulated as the terms of a ratio" 令每行為率; this wording is very similar to Liu's statement here, that the "terms of the ratios are initially formulated according to [the procedure for] *fangcheng*" 此率初如方程為之 (lines 51 and 52 on page 116). In his commentary to problem 18, Liu criticizes what he pejoratively terms the "old method" (*jiu shu* 舊術) together with those who use it—he seems not to have understood that the purpose of the *fangcheng* procedure is to avoid calculations with fractions until the final step. He then presents a "new *fangcheng* procedure" (*fangcheng xin shu* 方程新術), together with an "alternative procedure" (*qi yi shu* 其一術). His "new procedure" differs from the *fangcheng* procedure in that he uses column [row] reductions to eliminate the constant terms b_i in all but one of the columns [rows]. As a result, the remaining transformed entries in the other rows of the coefficient matrix a'_{ij} can be consid-

ered as related to one another without a constant term as a "ratio," and therefore in this manner fit the definition of "terms of the ratios."

3. In fact, in his commentary to the 18 problems of "*Fangcheng*," chapter 8 of the *Nine Chapters*, Liu uses the term *lü* ("terms of the ratios") 40 times; and yet only one problem—the well problem—can be viewed as in any way "indeterminate" and here Liu uses the term *lü* only twice. Indeed, the example Liu chooses to illustrate the application of his "new method" and "alternative method" using "terms of the ratios" to solve *fangcheng* is problem 18, which is equivalent to a system of 5 equations in 5 unknowns, and is determinate.
4. Liu seems to take the depth of the well as determined. Though the exact mathematical sense of his commentary to the well problem is difficult to ascertain, and he never explains the calculations that lead to the statement "obtain 721 as the divisor, 76 as the dividend" (line 54 on page 116), he does state that "721 is the depth of the well, and 76 is the length of E's well-rope" (lines 68 and 69 on page 116). His statement "use the terms of the ratios to state it" (line 69 on page 116) seems to be an instruction to use these results to calculate the lengths of the remaining well-ropes.

Dai Zhen's Notes

Dai's notes, in contrast, show that he clearly regarded the problem as indeterminate. Though Dai's first two notes are philological (lines 3 to 6 on page 114, and lines 9 and 10 on page 114), most of the remainder of his notes on the well problem consist of his analysis showing that the problem is indeterminate. Dai observes: "This problem does not state the measurements, and there is no basis from which one can ascertain the measurements of the depth of the well and the lengths of the well-ropes" (lines 21 to 23 on page 115). Dai presents two different choices for the depth of the well, and then calculates the length of each of the well-ropes for both choices: "Halve the depth of the well, 3 *zhang* 6 *chi* $\frac{1}{2}$ *cun*, then A's well-rope is 1 *zhang* 3 *chi* 2 *cun* and $\frac{1}{2}$ [*cun*], B's well-rope is 9 *zhang* 5 *chi* $\frac{1}{2}$ *cun*, C's well-rope is 7 *chi* 4 *cun*, D's well-rope is 6 *chi* 4 *cun* and $\frac{1}{2}$ [*cun*], and E's well-rope is 3 *chi* 8 *cun*" (lines 23 to 26 on page 115); "Double the depth of the well, 14 *zhang* 4 *chi* 2 *cun*, then A's well-rope is 5 *zhang* 3 *chi*, B's well-rope is 3 *zhang* 8 *chi* 2 *cun*, C's well-rope is 2 *zhang* 9 *chi* 6 *cun*, D's well-rope is 2 *zhang* 5 *chi* 8 *cun*, and E's well-rope is 1 *zhang* 5 *chi* 2 *cun*" (lines 26 to 29 on page 115). Dai concludes that these lead to two different sets of solutions: "Both [sets of solutions also] fit the problem. From this we can state that the problem then does not determine the measurements" (lines 29 and 30 on page 115).

Although Dai does not provide any step-by-step instructions for how he calculates 721 to be the depth of the well, we might speculate that his approach is the following. It seems that Dai first performs column [row] reductions on the well problem before any value has been assigned to the depth of the well: "First obtain 721" (line 31 on page 115); "Following the procedure [for *fangcheng*], first obtain 721" (line 56 on page 116). Thus it seems he might first start with the following diagram:

$$\begin{bmatrix} 2&1&0&0&0 \\ 0&3&1&0&0 \\ 0&0&4&1&0 \\ 0&0&0&5&1 \\ 1&0&0&0&6 \end{bmatrix}. \tag{7.10}$$

If he then followed the *fangcheng* procedure, just as in the row reductions given in equations (7.4–7.7), but without the constant terms, Dai would then arrive at the following result:

$$\begin{bmatrix} 2&1&0&0&0 \\ 0&3&1&0&0 \\ 0&0&4&1&0 \\ 0&0&0&5&1 \\ 0&0&0&0&721 \end{bmatrix}. \tag{7.11}$$

The value 721 would then be assigned as the depth of the well: 721 is "placed uniformly below all five rows" (line 57 on page 116). The *fangcheng* procedure is then applied to the entire array: "again find it" (line 59 on page 116).

Dai does not seem to interpret Liu's commentary as viewing the problem as indeterminate, and is in fact highly critical of it. Following Liu's statement, "obtain 721 as the divisor, 76 as the dividend," Dai states "the above is corrupt and contradictory, and cannot be followed" (line 55 on page 116), apparently because, according to Dai's calculations, the "dividend" should be 54796 and not 76. Dai further argues that Liu's statement "so use the number for the depth, the result of dividing the divisor into the dividend" is "corrupt, contradictory, and spurious" (lines 66 and 67 on page 116), apparently because, again following Dai's approach, the depth of the well should be 721 rather than 721 divided by 76.

The Earliest Extant Record of a Determinantal Calculation

This section presents the earliest known surviving record of a calculation of a determinant,[21] which is preserved in Yang Hui's *Nine Chapters on the Mathematical Arts, with Detailed Explanations* (c. 1261). There, a commentary to the well problem provides a method for assigning a value for the depth of the well. This method instructs us to calculate what in modern terminology is the determinant of the coefficient matrix. This value is then assigned to the sixth unknown, y, transforming the problem into a system of 5 conditions in 5 unknowns that does

[21] Excluding, again, ambiguous cases of solutions to systems of 2 conditions in 2 unknowns.

have a unique solution. Thus, contrary to more recent interpretations, the well problem was not considered to be indeterminate.

Though the compilation of *Detailed Explanations* is attributed to Yang Hui, Guo Shuchun has shown that much of the material recorded in this work is from Jia Xian's 賈憲 (fl. 1023–1063) *"Yellow Emperor's Mathematical Classic in Nine Chapters," with Detailed Notes* (*Huangdi jiuzhang suan jing xi cao* 黃帝九章算經細草, c. 1025), which is no longer extant. Guo argues that in fact Yang's *Detailed Explanations* is a commentary on Jia Xian's *Detailed Notes*. On the basis of a careful analysis of Yang's statements about his compilation of *Detailed Explanations*, Guo has found that passages in larger type under the headings "Method" (*shu* 術 or *fa* 法), "Notes" (*cao* 草), and "Topic" (*ti* 題) are excerpts from Jia's *Detailed Notes*; passages in smaller type under the heading "Explanation of topic" (*jie ti* 解題), "Parallel category" (*bilei* 比類), and "Detailed explanations" (*xiang jie* 詳解) are by Yang Hui.[22] The determinantal calculation is recorded in both commentaries.

Calculation of a Determinant

The following is a translation from the commentaries by Jia and Yang on the well problem, as preserved in Yang's *Detailed Explanations*. In *Detailed Explanations*, the original text for the well problem, together with the commentary by Liu Hui, is the same, character-for-character, as the version Dai Zhen gives in the *Wuyingdian juzhen ban congshu* edition, translated above in the section "Translation of the Well Problem" (page 112), and so will not be repeated here.[23] Jia's commentary is written in full-size characters and Yang's in smaller characters; for clarity, I have indicated likely authorship by the interpolations "[Jia:]" and "[Yang:]." The actual size of the smaller characters varies from chapter to chapter, and in this chapter, there are three sizes of characters—full-size, slightly smaller, and half-size (see Figure 7.2 on page 128). In the translation below, I have transcribed Jia's commentary in full-size type, and Yang's in smaller type. The half-size characters are presumably the interpolations of a later commentator; I have not included these in my translation.

> [Yang:] Explanation of the problem: Namely *fangcheng* with "major terms" and "minor terms" (*fen mu zi fangcheng*). The ancients' transformation of "five families bor-

[22] There may be exceptions. For example, as Guo notes, the headings with which Yang begins his commentaries, such as "Explanation of the Problem" (*jie ti* 解題), are in full-size characters. And on the other hand, the example below, which Yang presents under the heading "Parallel Categories" (*bilei* 比類), though written in smaller characters, might have been copied from another passage of Jia's *Detailed Notes*, since the example also includes "Notes" (*cao* 草) presumably written by Jia. For Guo's analysis of *Detailed Explanations*, see Guo 1988.

[23] As was the case with the edition of the *Nine Chapters* reconstructed by Dai Zhen, there is no commentary (*zhushi* 注釋) by Li Chunfeng et al. for chapter 8.

row well-ropes to reach the depth [of the well]" into a [mathematical] problem can be called an extraordinary creation.

解題。即分母子方程也。古人變五家借綆逮深為問，可謂佳作。

[Jia:] Method: The numbers of the well-ropes of each family are the "major terms" [*fen mu*] and are multiplied together. [Yang:] Find the "common denominator." [Jia:] The number of well-ropes borrowed is the "minor term" [*fen zi*]. [Yang:] Add the "numerator."[24] [Jia:] First obtain the depth of the well, then place for each family the [number of] original well-ropes, [the number for] those that were borrowed, including the "product" [the depth of the well]. [Yang:] "Product" of the depth of the well. [Jia:] Use the [method for] *fangcheng* and positive and negative [numbers]. [Yang:] The preceding method [*fa*].

術曰：戶綆數為分母相乘。通其分也。借綆數為分子。併內其子也。　先得井深，副列各戶本綆所借及積。井深之積。如方程正負入之。前法。

[Jia:] Notes:[25] Take the numbers of the five well-ropes as the "major terms" and multiply them together. [Yang:] Obtaining 720. [Jia:] As for the number of the borrowed well-ropes, 1 borrowed [well-rope] is the "minor term," add it. [Yang:] Obtaining 721. [Jia:] Take this as the "depth product." Next place for each family the [number of] original well-ropes, [the number for] those [well-ropes] that were borrowed, including the "depth product" [of the well], to find [the solution].

草曰：五綆數為分母相乘。得七百二十。借綆數借一為分子，併之。得七百二十一。為深積。副列各戶本綆所借及深積求。

Detailed Explanations then presents the earliest extant written representation of the problem in the form of an array, as shown below on the left; the translation is presented on the right. Note that Chinese numbers are used instead of counting rods (again, see Figure 7.2 on the following page):

一				二	甲
			三	一	乙
		四	一		丙
	五	一			丁
六	一				戊
七百二十一	七百二十一	七百二十一	七百二十一	七百二十一	深積

1				2	*A*
			3	1	*B*
		4	1		*C*
	5	1			*D*
6	1				*E*
721	721	721	721	721	"Depth product"

[24] For these operations, Jia's commentary uses terms that are usually used for fractions. Yang's subcommentary also uses terminology for fractions, but in a different manner. For example, the phrase *tong fen na zi* 通分內子 usually means to convert a "mixed" fraction into an improper fraction, as explained by Libbrecht: "to multiply the integer by the denominator (*tong fen*) and add the numerator (*na zi*) in order to convert to an improper fraction," for example, $132 + \frac{1}{2} = \frac{264}{2}$ [*tong fen*] $+ \frac{1}{2} = \frac{265}{2}$ [*na zi*] (Libbrecht 1973, 490).

[25] The "notes" (*cao*) present the calculations in sequence.

一綆丙四綆不足如丁一綆丁五綆不足如戊一綆戊六
綆不足如甲一綆如各得所不足一綆皆逮問井深綆長
各幾何
荅曰井深七丈二尺一寸　甲綆長二丈六尺五寸
乙綆長一丈九尺一寸　丙綆長一丈四尺八寸
丁綆長一丈二尺九寸　戊綆長七尺六寸
術曰如方程以正負術入之此率初如方程爲之名各
一逮井其後法得七百二十一實七十六是爲七百二
十一綆而七十六逮井用逮之數以法除實者而戍一
綆逮井之數定逮七百二十一分之七十六是故七百
二十一爲井深七十六爲戊綆之長舉率以言之
詳解九章算法　宜稼堂叢書
解題卽分母子方程也古人變五家借綆逮深爲問可
謂佳作
術曰戶綆數爲分母相乘通其分也借綆數爲分子併
內其子也　先得井深副列各戶本綆所借及積井深
之積如方程正負入之前法
草曰五綆數爲分母相乘得七百二十借綆數借一爲
分子併之得七百二十一爲深積副列各戶本綆所借
及深積求
甲　乙　丙　丁　戊　深積
二　一　七百二十一
三　一　七百二十一

四　一　七百二十一
五　一　七百二十一
一　六　七百二十一
如方程正負入之只求戊行可取諸綆　二乘戊行以
甲行同名減之甲空乙正無入負其一乙戊一十二積
七百二十一　三乘戊行以乙行異名減之乙空丙負
無入正其一丙戊三十六同名加積得二千八百八十
四　四乘戊行以丙行同名減之丙空丁負無入負其
一丁戊一百四十四同名減積得一萬八百一十五
五乘戊行以丁行異名減之丁空同名加戊爲七百二
十一加積得五萬四千七百九十六積爲實戊爲法除
詳解九章算法　宜稼堂叢書
得戊綆七尺六寸遞除丁丙乙甲所借以求四綆合問
比類三人易物甲以朱二兩粉一兩　乙以粉三兩丹
一兩　丙以丹四兩朱一兩皆得椒一斤問各價幾何
荅曰椒二貫五百　朱九百　粉七百　丹四百
草曰以朱二粉三丹四爲分母相乘加內子一　粉丹
朱皆一也　得二十五　前術約綆爲寸今問約錢上
百卽二貫五百文　以三人出物列位如方程正負術
入之
甲　朱二　一　無入　價二貫五百
乙　無入　粉三　丹一　價二貫五百
丙　朱一　無入　丹四　價二貫五百

Fig. 7.2: Photolithographic reproduction of the *Yijiatang congshu* 宜稼堂叢書 edition reprint (1843) of Yang Hui's *Nine Chapters on the Mathematical Arts, with Detailed Explanations.* In this edition, the well problem from the *Nine Chapters* is laid out in an array (using Chinese numbers instead of counting rods) beginning in the left-hand side of the upper left-hand page, and continuing on the right-hand side of the lower right-hand page.

The terminology used here is lost. This problem is described as being of the type *fen mu zi fangcheng* 分母子方程, a term that appears nowhere else in the entire *Complete Collection of the Four Treasuries*. The use of the character *fen* 分 here is unusual—it is most commonly used to mean "fractions." For instance, *fen mu* 分母 and *fen zi* 分子 are most commonly used as "numerator" and "denominator" for fractions. The depth of the well (*jing shen* 井深) is referred to as *shen ji* 深積, and as *jing shen zhi ji* 井深之積, where the term *ji* 積 usually means "product," for example in "square paces" (*ji bu* 積步, an area as the product of two dimensions measured in paces).

We can, however, attempt to reconstruct the meaning of these terms. Representing the array in modern terms as an augmented matrix, we have

$$\begin{bmatrix} k_1 & 1 & 0 & 0 & 0 & b \\ 0 & k_2 & 1 & 0 & 0 & b \\ 0 & 0 & k_3 & 1 & 0 & b \\ 0 & 0 & 0 & k_4 & 1 & b \\ 1 & 0 & 0 & 0 & k_5 & b \end{bmatrix}.$$

The text explicitly identifies the "major terms" (*fen mu*) as 2, 3, 4, 5, and 6, which, when multiplied together, give 720, so these are the diagonal elements $a_{ii} = k_i$ $(1 \le i \le 5)$. The text also identifies the "minor terms" (*fen zi*) as the number of well-ropes borrowed, which are all equal to 1, so these are the superdiagonal elements $a_{i,i+1} = 1$ $(1 \le i \le 4)$ together with $a_{51} = 1$. Finally, the text identifies the "depth product" (*shen ji* 深積) as $k_1k_2k_3k_4k_5 + 1$, the multiplication of elements along the diagonal, to which is added 1 (the product of the $a_{i,i+1}$ and a_{51}, which are all 1). This is precisely the determinant of the coefficient matrix

$$\det A = \begin{vmatrix} k_1 & 1 & 0 & 0 & 0 \\ 0 & k_2 & 1 & 0 & 0 \\ 0 & 0 & k_3 & 1 & 0 \\ 0 & 0 & 0 & k_4 & 1 \\ 1 & 0 & 0 & 0 & k_5 \end{vmatrix} = k_1k_2k_3k_4k_5 + 1.$$

That is, in this commentary on the well problem, the term "depth product" (*shen ji* 深積) is used for what we would now call the determinant.

Further Examples from Yang's Commentary

To further explain this category of problems, Yang Hui presents a second example, possibly copied from elsewhere in Jia's *Detailed Notes*:

25 [Yang:] Parallel Category 比類

[Jia?:] Three persons exchange objects. A takes two *liang*[26] of vermilion and one *liang* of powder; B takes three *liang* of powder and one *liang* of cinnabar; C takes four *liang*

[26] *Liang* is a unit of weight, equal to approximately 15 to 16 grams during this period. On units of measure in early China, see the sources given in footnote 9 on page 114.

of cinnabar and one *liang* of vermilion;[27] each obtains 1 *jin*[28] of pepper.[29] Problem: how much is the value of each? Solution: Pepper is 2 *guan*[30] 500. Vermilion is 900. Rice-
30 powder is 700. Cinnabar is 400.

三人易物。甲以朱二兩，粉一兩， 乙以粉三兩，丹一兩， 丙以丹四兩，朱一兩，皆得椒[31]一斤。問各價幾何。 荅曰：椒二貫五百。 朱九百。 粉七百。丹四百。

[Jia?:] Notes: Take the 2 [, the coefficient for the] vermilion, 3 [, the coefficient for the]
35 rice-powder, and 4 [, the coefficient for the] cinnabar as the "major terms," multiply them with one another, and add the "minor term."[32] [The "minor terms" for] powder, cinnabar, and vermilion are all one. Obtain 25. In the preceding method [for solving the well problem], the well-ropes were all reduced to [the units] *cun*; in this problem, reduce the [units for] coins to hundreds, namely 2 *guan* 500 *wen*. Taking the objects given by the
40 three persons, arrange in position [in an array], and follow the [method for] *fangcheng* and positive and negative [numbers] to solve it.

草曰：以朱二、粉三、丹四為分母，相乘，加內子一。 粉、丹、朱，皆一也。得二十五。 前術，約綆為寸。今問，約錢上百，即二貫五百文。 以三人出物，列位，如方程、正負術入之。

The diagram shown in Figure 7.3 on the facing page is then given. This problem is set up as a system of 3 conditions in 4 unknowns, x_1, x_2, x_3, and y. First, the fourth unknown, y, is found by assigning to it the value of the determinant of the coefficient matrix, $2 \times 3 \times 4 + 1 = 25$ (here multiplied by 100, using *wen* as units).

---

[27] The unknowns in this problem are represented by powders of three different colors, vermilion (*zhu* 朱), white powder (*fen* 粉), and cinnabar (*dan* 丹). What the exact content of the powders might have been is more difficult to determine. Cinnabar, *dan*, is naturally occurring mercury sulfide (HgS), a deep red, translucent crystal; it has been found as crystal ornaments in burial sites dating back to the Shang Dynasty, and was used as a source of mercury and to make bright red ink. Vermilion, *zhu*, is the deep red pigment derived from cinnabar. *Systematic Materia Medica* (本草綱目 *Bencao gangmu*, 1596) suggests in an annotation to the explanation of *yin zhu* 銀朱 (lit., "silver red") that vermilion may have been man-made from mercury sulfide, produced through smelting: "Men of the past held that mercury comes from cinnabar, which, smelted again [presumably with sulfur], forms vermilion, namely this" 昔人謂水銀出於丹砂，鎔化還復為朱者，即此也 (*juan* 9, 19b). Powder, *fen*, refers to any of a variety of white powders: *Explanation of Simple Graphs and Analysis of Compound Characters* (說文解字 *Shuo wen jie zi*, 100 C.E.), states that *fen* was used as facial makeup 粉，傅面者也 (*juan* 7 *shang*, 21b). Both examples are from DKWJ.

[28] One *jin* is 16 *liang*, about 250 grams. Again, see footnote 9 on page 114.

[29] Pepper, which seems out of place here, may have been chosen to indicate the different role played by the final unknown, which is found by calculating the determinant of the matrix of the coefficients of the other unknowns.

[30] *Guan* (lit., "string") is a monetary unit equal to 1000 *qian* (lit., "coins"). Coins in early China were made from metal, and from the Warring States on, had holes in the middle, and thus could be collected together using a string. One thousand coins was then called one *guan*.

[31] Corrected for a misprint for the character, following the CSJC edition.

[32] See footnote 24 on page 127.

丙	乙	甲
朱一	無入	朱二
無入	粉三	一
丹四	丹一	無入
價二貫五百	價二貫五百	價二貫五百

C	B	A
Vermilion 1	Vermilion No entry	Vermilion 2
White No entry	White 3	1
Red 4	Red 1	Red No entry
Price 2 *guan* 500	Price 2 *guan* 500	Price 2 *guan* 500

Fig. 7.3: Problem with 3 conditions in 4 unknowns, from Yang's "Detailed Explanations."

In modern notation, this can be written

$$\begin{bmatrix} 2 & 1 & 0 & 2500 \\ 0 & 3 & 1 & 2500 \\ 1 & 0 & 4 & 2500 \end{bmatrix},$$

which has as its solution,

$$x_1 = 900, \quad x_2 = 700, \quad x_3 = 400.$$

This analysis of Yang Hui's *Detailed Explanations*, then, allows us to draw the following conclusions:

1. The two problems here—the well problem from the *Nine Chapters* together with a second example presented by Yang Hui—were recognized as a *category* of problems, which Yang termed "*fangcheng* with 'major terms' and 'minor terms.'"
2. Although in modern terms these problems are equivalent to indeterminate systems of n equations in $n+1$ unknowns, they were not considered indeterminate at the time. On the contrary, an explicit method was provided for finding the remaining unknown: the problem was transformed into the equivalent of a determinate system of n conditions in n unknowns by a determinantal calculation, which, in modern terms, gives the determinant of the matrix of the coefficients of the first n unknowns. The $n+1^{\text{th}}$ unknown, y, is then set to this result.
3. An explicit formula is given for a determinantal calculation, which is valid for all *fangcheng* of this form.

4. Expressed in modern notation, the formula used for calculating the $n+1^{\text{th}}$ unknown is the following:

$$y = \begin{vmatrix} k_1 & 1 & 0 & \cdots & 0 \\ 0 & k_2 & 1 & \ddots & \vdots \\ \vdots & 0 & \ddots & \ddots & 0 \\ 0 & \vdots & \ddots & k_{n-1} & 1 \\ 1 & 0 & \cdots & 0 & k_n \end{vmatrix} = k_1 k_2 k_3 \ldots k_n + 1, \tag{7.12}$$

   where $n = 3$ or $n = 5$. In the case of the well problem, the term "depth product" was used for what in modern terminology we would call the determinant.
5. The formula given in equation (7.12) is valid in general for all n ($n \geq 3$) if n is odd, and for even n if we change the final + sign to −, that is, $y = k_1 k_2 k_3 \ldots k_n - 1$. Since these problems were meant as exemplars, it seems reasonable to assume that this result was known for at least some values of n other than 3 and 5, such as $n = 4$.
6. If the "minor terms" are values other than 1, as the commentary recognizes they might be, we have the more general form given in augmented matrix (7.2) with $b_1 = \ldots = b_n = y$, and the formula $y = k_1 k_2 k_3 \ldots k_n \pm l_1 l_2 l_3 \ldots l_n$ (+ if n is odd, − if n is even), which again is valid for all n ($n \geq 3$).
7. Finally, it is possible that for problems of this special form, this calculation was recognized as a "determinant" in the modern sense, indicating, that is, that a unique solution exists: if the depth of the well is not zero, there will be a unique solution. More generally, in modern terms, assuming that we are not assigning zero to the $n+1^{\text{th}}$ unknown y, there will always be a unique solution, since the determinant of the coefficient matrix is not zero.

The Earliest Extant Record of a Determinantal Solution

In the preceding section, we saw that Jia Xian's *Detailed Notes*, compiled in about 1025, excerpts from which are preserved in Yang Hui's *Detailed Explanations*, uses a determinantal calculation in the process of solving problems of the form given in the well problem. This is not, however, the same as a determinantal solution to a system of n conditions in n unknowns. This section will present the earliest extant record of what in modern terms we would call a determinantal solution (beyond, that is, solutions to problems with 2 conditions in 2 unknowns).

Fang's Numbers and Measurement

The source is Fang Zhongtong's 方中通 (1634–1698) *Numbers and Measurement, An Amplification* (*Shu du yan* 數度衍), completed in about 1661. Among extant mathematical treatises of the period, it is in many ways unique.[33] Fang was the son of the preeminent scholar Fang Yizhi 方以智 (1611–1671), noted for his interest in Chinese and Western science.[34] Fang Zhongtong thus presumably had access to a considerable number of mathematical treatises. His *Numbers and Measurement* is an apparently indiscriminate compilation from the disparate sources available to him, including, along with traditional Chinese mathematics and Western mathematics introduced by the Jesuits, popular material, such as numerology, and the "Diagram of the Yellow River" (*He tu* 河圖) and "Diagram of the Luo River" (*Luo shu* 洛書), usually not included in mathematical treatises authored by the literati elite. In including this popular material, often considered by the literati elite to be vulgar, Fang's work is similar to the merchant Cheng Dawei's 程大位 (1533–1606) *Comprehensive Source of Mathematical Arts* (*Suanfa tong zong* 算法统宗, 1592). But perhaps because of his status and patronage networks, Fang Zhongtong's work was included in the imperially sponsored *Complete Collection of the Four Treasuries*, whereas Cheng's work later became the object of ridicule and scorn.[35]

Numbers and Measurement preserves a single example of a determinantal solution. Given its contents, it seems clear that this was not Fang's own original work, but rather that he simply copied this example from another source. That this example of a determinantal solution has survived would then appear to be the result of the fortuitous combination of Fang's indiscriminate eclecticism and his status among the elite, which led to the preservation of his work.

Translation of Fang's Determinantal Solution

The determinantal solution that Fang records is the second of two solutions; to understand it, we must first examine relevant portions of the first solution. Fang's statement of the problem and of the solution for the unknowns does not differ in any important respect from previous examples, and so will not be translated here. The wording is similar to that found in the *Guide to Calculation*, suggesting that Fang's source was either the *Guide to Calculation* or the unknown source from which the problem in the *Guide to Calculation* was copied. After the depth of the well is calculated, the array is set as is shown in Figure 7.4 on the next page. The diagram is then similar to that found in Yang's *Detailed Explanations*, Wu's *Complete Compendium of Mathematical Arts*, and the *Guide to Calculation*. However,

[33] Although this is the only surviving record of a determinantal solution that I have found in several years of research on this project, I believe that in the future more will be found.

[34] For an important study of Fang Yizhi, see Peterson 1979. Peterson argues that Fang Yizhi helped redirect the focus of the "investigation of things" (*gewu* 格物) from the moral interpretation of the Cheng-Zhu school to an empirical interpretation of the Qing learning. Fang Yizhi was well acquainted with the works of the Jesuits and their collaborators.

[35] See the critical notes in the *Comprehensive Catalogue of the "Complete Collection of the Four Treasuries"* (*Siku quan shu zongmu* 四庫全書總目), *juan* 107, 28a–b and 49a, respectively.

五	四	三	二	一
甲一	○	○	○	甲二
○	○	○	乙三	乙一
○	○	丙四	丙一	○
○	丁五	丁一	○	○
戊六	戊一	○	○	○
七百二十一	七百二十一	七百二十一	七百二十一	七百二十一

5	4	3	2	1
A 1	0	0	0	*A* 2
0	0	0	*B* 3	*B* 1
0	0	*C* 4	*C* 1	0
0	*D* 5	*D* 1	0	0
E 6	*E* 1	0	0	0
721	721	721	721	721

Fig. 7.4: The arrangement of entries in the well problem in Fang's *Numbers and Measurement*

the determinantal solution in Fang's *Numbers and Measurement* is more general: as we will see, the calculations do not assume that the superdiagonal elements $a_{i,i+1}$ and a_{n1} are identical, nor do they assume that constant terms are identical. So we will write the array as a system with 5 conditions in 5 unknowns as follows,

l_5	0	0	0	k_1
0	0	0	k_2	l_1
0	0	k_3	l_2	0
0	k_4	l_3	0	0
k_5	l_4	0	0	0
b_5	b_4	b_3	b_2	b_1

$$\begin{bmatrix} k_1 & l_1 & 0 & 0 & 0 & b_1 \\ 0 & k_2 & l_2 & 0 & 0 & b_2 \\ 0 & 0 & k_3 & l_3 & 0 & b_3 \\ 0 & 0 & 0 & k_4 & l_4 & b_4 \\ l_5 & 0 & 0 & 0 & k_5 & b_5 \end{bmatrix}.$$

The second solution to the well problem recorded in *Numbers and Measurement* is presented under the following heading:

1 "Method for Calculating Only the Lowermost [Terms] in Two Parts."

2 止推下二段法 (SDY, *juan* 21, 20b)

The exact meaning of the name for this method is unknown; we will return to an analysis of it once we have examined the entire solution.

"Setting the Negatives"

The first step of this method, as recorded by Fang, is to "set the negatives" (*li fu* 立負). The exact meaning of the term "set the negatives," which appears in other mathematical treatises of this period, is not clear. The procedure presented in Tong's *Numbers and Measurement* is as follows:

> Tong[36] says: Following the previous method, calculate the depth of the well. After setting up the array, first take 1, [the entry] for A in the fifth column [l_5], and multiply each column: multiply [it] by [the entry] 1 for B [in the second column, l_2] and set it as -1 [$-l_5 l_2$]; multiply [the previous result] by [the entry] 1 for C [in the third column, l_3] and set it as -1 [$-l_5 l_2 l_3$]; multiply [the previous result] by [the entry] 1 for D [in the third column, l_4] and set it as -1 [$-l_5 l_2 l_3 l_4$]; then halt.
>
> 通曰：如右式推出井深。列定之後，先以五行甲一乘各行：乘乙一立負一，乘丙一立負一，乘丁一立負一，乃止。(SDY, *juan* 21, 20b)

The statement "first take 1, for A in the fifth column, and multiply each column" (lines 4 and 5 on this page), interpreted literally, might seem to instruct us to multiply the entries for B in the second column (l_2), C in the third column (l_3), and D in the fourth column (l_4), each by the entry for A in the fifth column (l_5), giving, respectively, after setting each to be negative, $-l_5 l_2$, $-l_5 l_3$, and $-l_5 l_4$. But this interpretation is incorrect. This passage is a very abbreviated instruction to perform calculations described in more detail in the previous method, which Fang has presented immediately above. The first step of the row reductions in the previous method instructs us to "multiply 1, [the entry for] A in the fifth column [l_5], by [the entry for] B in the first column [l_1], obtaining 1 [$l_1 l_5$], then set -1 [in the entry for B] in the fifth column" 以五行甲一乘一行乙，仍得一，五行即立負一 (SDY, *juan* 21, 19a). In other words, set the entry for B in the fifth column to $-l_1 l_5$. The next step of row reductions includes the instruction, "multiply -1, [the transformed entry for] B in the fifth column [$-l_1 l_5$], by 1, [the entry for] C in the second column [l_2], obtaining 1 [$l_1 l_2 l_5$], and then set -1 [in the entry for C] in the fifth column" 以五行乙負一乘二行丙，仍得一，五行即立負一 (SDY, *juan* 21, 19a). That is, set the entry for C in the fifth column to $-l_1 l_2 l_5$ (not $-l_2 l_5$). The following step of row reductions includes the instruction, "multiply -1, [the current entry for] C in the fifth column [$-l_1 l_2 l_5$], by 1, [the entry for] D in the third column [l_3], obtaining 1, and then set -1 [in the entry for D] in the fifth column" 以五行丙負一乘三行丁，仍得一五行即立負一 (SDY, *juan* 21, 19a–b). That is, set D in the fifth column to $-l_1 l_2 l_3 l_5$ (not $-l_3 l_5$).

Though Fang provides no diagrams, it is easier to follow the calculations if they are set out in an array, which, following the above instructions, would look as follows, with the products $-l_1 l_5 = -1$, $-l_1 l_2 l_5 = -1$, and $-l_1 l_2 l_3 l_5 = -1$, as in Figure 7.5 on the next page. It is important to remember that the entries -1 in column 5 represent the results of a calculation. Fang refers to these entries as "B in the fifth column" (*wu hang yi* 五行乙, or $-l_1 l_5$), "C in the fifth column" (*wu hang bing* 五行丙, or $-l_1 l_2 l_5$), and "D in the fifth column" (*wu hang ding* 五行丁, or $-l_1 l_2 l_3 l_5$). The meaning of "setting the negatives" is somewhat clearer when we

[36] Fang modestly refers to himself by the second character in his given name, Tong.

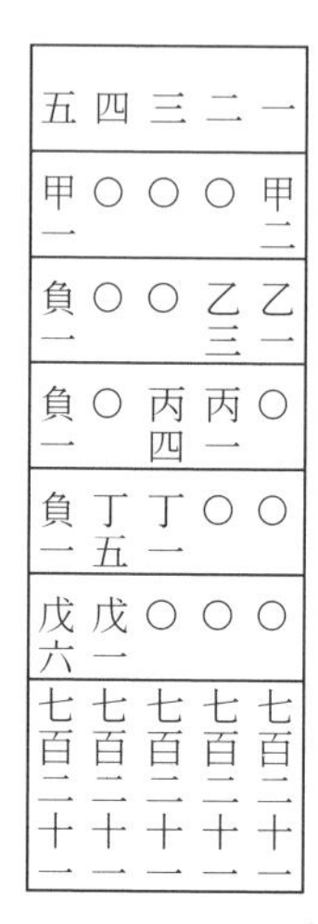

五	四	三	二	一
甲一	○	○	○	甲二
負一	○	○	乙三	乙一
負一	○	丙四	丙一	○
負一	丁五	丁一	○	○
戊六	戊一	○	○	○
七百二十一	七百二十一	七百二十一	七百二十一	七百二十一

5	4	3	2	1
A 1	0	0	0	*A* 2
−1	0	0	*B* 3	*B* 1
−1	0	*C* 4	*C* 1	0
−1	*D* 4	*D* 1	0	0
E 5	*E* 1	0	0	0
721	721	721	721	721

Fig. 7.5: "Setting the negatives" $-l_1 l_5$, $-l_1 l_2 l_5$, and $-l_1 l_2 l_3 l_5$ in Fang's *Numbers and Measurements.*

visualize the result in two dimensions in an array. In modern notation, we have the following:

l_5	0	0	0	k_1
$-l_1 l_5$	0	0	k_2	l_1
$-l_1 l_2 l_5$	0	k_3	l_2	0
$-l_1 l_2 l_3 l_5$	k_4	l_3	0	0
k_5	l_4	0	0	0
b_5	b_4	b_3	b_2	b_1

$$\begin{bmatrix} k_1 & l_1 & 0 & 0 & 0 & b_1 \\ 0 & k_2 & l_2 & 0 & 0 & b_2 \\ 0 & 0 & k_3 & l_3 & 0 & b_3 \\ 0 & 0 & 0 & k_4 & l_4 & b_4 \\ l_5 & -l_1 l_5 & -l_1 l_2 l_5 & -l_1 l_2 l_3 l_5 & k_5 & b_5 \end{bmatrix}.$$

Note that this is no longer an augmented matrix, but rather uses the last column [row] to store the results of the calculations.[37]

Calculating the "Dividend"

After "setting the negatives," the next step is to calculate the first part, here called the "dividend" or numerator:

> Find the lowermost divisor and dividend in two parts. The part for finding the dividend is as follows. Multiply [the constant term in] the fifth column [b_5] by 2, [the entry for] A in the first column [k_1], obtaining 1442 in the lowest position [$k_1 b_5$]; multiply [the constant term in] the first column [b_1] by 1, [the entry for] A in the fifth column [l_5], obtaining 721 [$l_5 b_1$], and, subtracting one from the other, the remainder is 721 [$k_1 b_5 - l_5 b_1$].

[37] See the discussion in "General Forms for n Conditions in n Unknowns" on page 139.

求下法實二段。如求實段：以一行甲二乘五行下，得一十四百四十二，以五行甲一乘一行下，仍得七百二十一，相減，餘七百二十一。

20 Then multiply it [the previous result, $k_1b_5 - l_5b_1$] by 3, [the entry for] B in the second column [k_2], obtaining 2163 [$(k_1b_5 - l_5b_1)k_2$], and multiply [the constant term in] the second column [b_2] by −1, [the value stored in the position for] B in the fifth column [$-l_5l_1$], obtaining 721 [$l_5l_1b_2$], then add, obtaining 2884 [$(k_1b_5 - l_5b_1)k_2 + l_5l_1b_2$].

25 又以二行乙三乘之，得二千一百六十三，以五行乙負一乘二行下，仍得七百二十一，并，得二千八百八十四。

Then multiply it [the previous result, $(k_1b_5 - l_5b_1)k_2 + l_5l_1b_2$] by 4, [the entry for] C in the third column [k_3], obtaining 11536 [$((k_1b_5 - l_5b_1)k_2 + l_5l_1b_2)k_3$], and multiply [the constant term in] the third column [b_3] by
30 −1, [the value stored in the position for] C in the fifth column [$-l_5l_1l_2$], obtaining 721 [$l_5l_1l_2b_3$], and subtract, obtaining 10815 [$((k_1b_5 - l_5b_1)k_2 + l_5l_1b_2)k_3 - l_5l_1l_2b_3$].

又以三行丙四乘之，得一萬一千五百三十六，以五行丙負一乘三行下，仍得七百二十一，相減，餘一萬〇八百一十五。

35 Then multiply it [the previous result, $((k_1b_5 - l_5b_1)k_2 + l_5l_1b_2)k_3 - l_5l_1l_2b_3$] by 5, [the entry for] D in the fourth column [k_4], obtaining 54075 [$(((k_1b_5 - l_5b_1)k_2 + l_5l_1b_2)k_3 - l_5l_1l_2b_3)k_4$], and multiply [the constant term in] the fourth column [b_4] by −1, [the number stored in the position for] D in the fifth column [$-l_5l_1l_2l_3$], obtaining 721 [$l_5l_1l_2l_3b_4$], and add, obtaining
40 54796 [$(((k_1b_5 - l_5b_1)k_2 + l_5l_1b_2)k_3 - l_5l_1l_2b_3)k_4 + l_5l_1l_2l_3b_4$].

又以四行丁五乘之，得五萬四千〇七十五，以五行丁負一乘四行下，仍得七百二十一，并，得五萬四千七百九十六為實也。

(SDY, *juan* 21, 20b–21a)

The "dividend" or numerator computed in this first "part" is then

$$(((k_1b_5 - l_5b_1)k_2 + l_5l_1b_2)k_3 - l_5l_1l_2b_3)k_4 + l_5l_1l_2l_3b_4.$$

This is the same as the numerator found using determinants,

$$\begin{vmatrix} k_1 & l_1 & 0 & 0 & b_1 \\ 0 & k_2 & l_2 & 0 & b_2 \\ 0 & 0 & k_3 & l_3 & b_3 \\ 0 & 0 & 0 & k_4 & b_4 \\ l_5 & 0 & 0 & 0 & b_5 \end{vmatrix} = \begin{aligned} k_1k_2k_3k_4b_5 - k_2k_3k_4l_5b_1 \\ + k_3k_4l_1l_5b_2 - k_4l_1l_2l_5b_3 \\ + l_1l_2l_3l_5b_4. \end{aligned} \tag{7.13}$$

Calculating the "Divisor"

Immediately following the computation of the "dividend," instructions are given for computing the "divisor" or denominator:

The part for finding the divisor is as follows: multiply 2, [the entry for] A in the first column [k_1], by 6, [the entry for] E in the fifth column [k_5], obtaining 12 [$k_1 k_5$]; and again multiply it by 3, [the entry for] B in the second column [k_2], obtaining 36 [$k_1 k_5 k_2$]; and again multiply it by 4, [the entry for] C in the third column [k_3], obtaining 144 [$k_1 k_5 k_2 k_3$]; and again multiply it by 5, [the entry for] D in the fourth column [k_4], obtaining 720 [$k_1 k_5 k_2 k_3 k_4$]. Then multiply −1, [the number stored in the position for] D in the fifth column [$l_5 l_1 l_2 l_3$], by 1, [the entry for] E in the fourth column [l_4], obtaining 1 [$l_5 l_1 l_2 l_3 l_4$], then add, obtaining 721 [$k_1 k_5 k_2 k_3 k_4 + l_5 l_1 l_2 l_3 l_4$].

如求法段，以一行甲二乘五行戊六，得十二；又以二行乙三乘之，得三十六；又以三行丙四乘之，得一百四十四；又以四行丁五乘之，得七百二十。乃以五行丁負一乘四行戊一，仍得一，并得七百二十一為法也。

(SDY, *juan* 21, 21a)

The "divisor" computed in this second "part" is thus

$$k_1 k_5 k_2 k_3 k_4 + l_5 l_1 l_2 l_3 l_4,$$

which, again, is the same as the denominator calculated using Cramer's Rule,

$$\begin{vmatrix} k_1 & l_1 & 0 & 0 & 0 \\ 0 & k_2 & l_2 & 0 & 0 \\ 0 & 0 & k_3 & l_3 & 0 \\ 0 & 0 & 0 & k_4 & l_4 \\ l_5 & 0 & 0 & 0 & k_5 \end{vmatrix} = k_1 k_2 k_3 k_4 k_5 + l_1 l_2 l_3 l_4 l_5. \tag{7.14}$$

Dividing the "dividend" by the "divisor," expressed in modern terminology, gives the following formula for the fifth unknown, the length of E's rope,

$$\frac{(((k_1 b_5 - l_5 b_1) k_2 + l_5 l_1 b_2) k_3 - l_5 l_1 l_2 b_3) k_4 + l_5 l_1 l_2 l_3 b_4}{k_1 k_5 k_2 k_3 k_4 + l_5 l_1 l_2 l_3 l_4}, \tag{7.15}$$

which is the same as the solution given by calculating the determinants,

$$\begin{aligned} x_5 &= \frac{\det B_5}{\det A} \\ &= \frac{k_1 k_2 k_3 k_4 b_5 - k_2 k_3 k_4 l_5 b_1 + k_3 k_4 l_1 l_5 b_2 - k_4 l_1 l_2 l_5 b_3 + l_1 l_2 l_3 l_5 b_4}{k_1 k_2 k_3 k_4 k_5 + l_1 l_2 l_3 l_4 l_5}. \end{aligned}$$

Fang's Terminology

We should now return to the name of the method presented here, *zhi tui xia er duan fa* 止推下二段法, the exact significance of which is never explained. The first term, *zhi* 止, appears only once in this passage, and there it means "cease" or "halt" (line 9 on page 135), apparently referring to further calculations: it seems to mean that after we calculate the depth of the well and set the array, we "set the negatives" (lines 5 to 9 on page 135) but do not proceed to row reductions, or other methods of calculating the solution. However, here, in the name of this method, it seems more likely that *zhi* is used in the sense of "only," since only the "divisor" and "dividend" are calculated, as opposed to elimination by row operations, which transforms many of the entries. The second term, *tui* 推, seems to mean "calculate," as in "calculate the depth of the well" (line 3 on page 135). The third character, *xia* 下, which literally means "lower," is most often used to refer to the constant terms, b_i, since in the array the constant terms are all placed in the lowermost row. It is also sometimes used to refer to the entries for the ith unknown, x_i, since in the array, these entries occupy the lowermost row among the unknowns. One of the key instructions, "find the lowermost divisor and dividend" 求下法實 (line 12 on page 136), explains the use of *xia* here: the "lowermost divisor" refers to the entry for the nth unknown, and the term "dividend" refers to the corresponding constant term in the nth column, once eliminations by row reductions are complete. Expressed in modern terms, the "lowermost divisor" is just the final pivot, a'_{nn}, and the "dividend" is the corresponding transformed n^{th} constant term, b'_n, when the augmented matrix is reduced to upper-diagonal form. The solution for the nth unknown is given by

$$x_n = \frac{b'_n}{a'_{nn}}.$$

Thus *tui xia* would seem to mean "calculating the transformed entries for the lowermost unknown and the constant term." Though the term *duan* often means a row of an array, here it seems to mean "part," as in "two parts" (line 12 on page 136), and the solution is indeed calculated in two parts: the "part for finding the dividend" or denominator (lines 12 and 13 on page 136); and the "part for finding the divisor" or numerator (line 43 on the facing page). Thus it seems that the title means "method for calculating only [the transformed entries for] the lowermost [unknown, a'_{nn}, and the corresponding constant term, b'_n] in two parts."

In the following section we will analyze this method further, and show that it is valid for any system of n conditions in n unknowns.

General Forms for n Conditions in n Unknowns

Chinese mathematical treatises contain numerous examples of *fangcheng* problems, ranging from systems of 3 conditions in 3 unknowns to 9 conditions in 9 unknowns, that, when written in modern terms, are of the form given in equation (7.1). Though this chapter cannot present a translation and analysis of all of these problems, we will demonstrate that the determinantal solution preserved in Fang's *Numbers and Measurements* is a general solution to equation (7.1), valid for any $n \geq 3$.

We begin with the general form for n conditions in n unknowns given in equation (7.2). We can rewrite this system of conditions in the following form:

$$l_{n-1}x_n = b_{n-1} - k_{n-1}x_{n-1}, \tag{7.16}$$
$$l_{n-2}x_{n-1} = b_{n-2} - k_{n-2}x_{n-2}, \tag{7.17}$$
$$l_{n-3}x_{n-2} = b_{n-3} - k_{n-3}x_{n-3}, \tag{7.18}$$
$$l_{n-4}x_{n-3} = b_{n-4} - k_{n-4}x_{n-4}, \tag{7.19}$$
$$\vdots$$
$$l_2x_3 = b_2 - k_2x_2, \tag{7.20}$$
$$l_1x_2 = b_1 - k_1x_1, \tag{7.21}$$
$$l_nx_1 = b_n - k_nx_n. \tag{7.22}$$

We can then solve for x_n in the following manner. Taking equation (7.16), we multiply both sides by l_{n-2},

$$l_{n-2}l_{n-1}x_n = l_{n-2}b_{n-1} - k_{n-1}l_{n-2}x_{n-1}, \tag{7.23}$$

and then, using equation (7.17), substitute $b_{n-2} - k_{n-2}x_{n-2}$ for $l_{n-2}x_{n-1}$, giving

$$\begin{aligned} l_{n-2}l_{n-1}x_n &= l_{n-2}b_{n-1} - k_{n-1}(b_{n-2} - k_{n-2}x_{n-2}) \\ &= l_{n-2}b_{n-1} - k_{n-1}b_{n-2} + k_{n-1}k_{n-2}x_{n-2}. \end{aligned} \tag{7.24}$$

We continue in this manner, until we have

$$\begin{aligned} l_1l_2\cdots l_{n-1}x_n = {} & l_1l_2\cdots l_{n-2}b_{n-1} - l_1l_2\cdots l_{n-3}k_{n-1}b_{n-2} \\ & + l_1\cdots l_{n-4}k_{n-1}k_{n-2}b_{n-3} - \ldots \pm l_1l_2k_{n-1}k_{n-2}\cdots k_4b_3 \\ & \mp l_1k_{n-1}k_{n-2}\cdots k_3b_2 \pm k_{n-1}k_{n-2}\cdots k_2b_1 \\ & \mp k_{n-1}k_{n-2}\cdots k_1x_1, \end{aligned} \tag{7.25}$$

where $\pm$ and $\mp$ are $+$ and $-$ respectively for n even, and $-$ and $+$ respectively for n odd.[38] Finally, we multiply both sides by l_n, and, using equation (7.22), substitute $b_n - k_nx_n$ for l_nx_1,

$$\begin{aligned} l_nl_1l_2\cdots l_{n-1}x_n = {} & l_nl_1l_2\cdots l_{n-2}b_{n-1} - l_nl_1l_2\cdots l_{n-3}k_{n-1}b_{n-2} \\ & + l_nl_1l_2\cdots l_{n-4}k_{n-1}k_{n-2}b_{n-3} - \ldots \\ & \pm l_nl_1l_2k_{n-1}k_{n-2}\cdots k_4b_3 \\ & \mp l_nl_1k_{n-1}k_{n-2}\cdots k_3b_2 \pm l_nk_{n-1}k_{n-2}\cdots k_2b_1 \\ & \mp k_{n-1}k_{n-2}\cdots k_1b_n \pm k_{n-1}k_{n-2}\cdots k_1k_nx_n. \end{aligned} \tag{7.26}$$

[38] Throughout this chapter I will use the convention that $\pm$ and $\mp$ are $+$ and $-$ for n even, and $-$ and $+$ for n odd. That is, for n even, we use the upper sign of $\pm$ and $\mp$, while for n odd we use the lower of $\pm$ and $\mp$. In this case, the $+$ and $-$ signs alternate, beginning with $-$, and so ending with $-$ if n is odd and $+$ if n is even.

Then, solving for x_n, and factoring, gives the result

$$x_n = \frac{\begin{array}{c}((\cdots((((k_1 b_n - l_n b_1)k_2 + l_n l_1 b_2)k_3 - l_n l_1 l_2 b_3)k_4 + l_n l_1 l_2 l_3 b_4)\cdots \\ \cdot k_{n-2} \pm l_n l_1 l_2 \cdots l_{n-3} b_{n-2})k_{n-1} \mp l_n l_1 l_2 \cdots l_{n-2} b_{n-1})\end{array}}{(k_1 k_2 k_3 \cdots k_{n-1} k_n \mp l_n l_1 l_2 l_3 \cdots l_{n-1})}, \qquad (7.27)$$

where again $\pm$ is $+$ for n even and $-$ for n odd.

Thus the determinantal solution given in Fang's *Numbers and Measurement* is perfectly general for systems of n conditions in n unknowns of the form of equation (7.2). But the equations presented above are quite complicated, especially for systems of 9 conditions in 9 unknowns. As the above translation suggests, expressing this problem in words was even more complicated for Chinese mathematical treatises of this period, which did not use symbols such as x_1, x_2, or x_3 for unknowns or a_{ij} for matrix coefficients, but instead positional descriptions such as "2, [the entry for] A in the first column" (line 14 on page 136).

Diagrammatic Reconstruction of the Calculations

In practice, the calculation of the solution given by equation (7.27) is not that complicated on the counting board, as can be seen if we construct diagrams showing how the solution was found. The following diagrams are my own reconstruction—Fang offers no diagrams, and I have yet to find any such diagrams in extant Chinese mathematical treatises. But as the following diagrams will show, it is not difficult to calculate the solutions for *fangcheng* of any size. The calculations can be visualized as a kind of repeated cross multiplication, as follows:

STEP 1. The following diagram illustrates the first step of this calculation. The opposite corners of the array are multiplied together, and subtracted, giving the result $k_1 b_n - l_n b_1$:

$\mathbf{l_n}$	0	$\cdots$	0	0	$\mathbf{k_1}$
$-l_n l_1$	0	$\cdots$	0	k_2	l_1
$-l_n l_1 l_2$	$\vdots$	$\cdot^{\cdot^{\cdot}}$	k_3	l_2	0
$\vdots$	0	$\cdot^{\cdot^{\cdot}}$	$\cdot^{\cdot^{\cdot}}$	0	0
$-l_n l_1 \cdots l_{n-2}$	k_{n-1}	l_{n-2}	$\cdot^{\cdot^{\cdot}}$	$\vdots$	$\vdots$
k_n	l_{n-1}	0	$\cdots$	0	0
$\mathbf{b_n}$	b_{n-1}	$\cdots$	b_3	b_2	$\mathbf{b_1}$

$k_1 b_n - l_n b_1$

STEP 2. In each of the remaining steps, we simply move in one step. We first multiply the results from the previous step, $k_1 b_n - l_n b_1$, by the next entry along the diagonal, k_2, then subtract from the result the product of the next entries at the corners, $-l_n l_1 b_2$:

l_n	0	$\cdots$	0	0	k_1
$\mathbf{-l_n l_1}$	0	$\cdots$	0	$\mathbf{k_2}$	l_1
$-l_n l_1 l_2$	$\vdots$	$\iddots$	k_3	l_2	0
$\vdots$	0	$\iddots$	$\iddots$	0	0
$-l_n l_1 \cdots l_{n-2}$	k_{n-1}	l_{n-2}	$\iddots$	$\vdots$	$\vdots$
k_n	l_{n-1}	0	$\cdots$	0	0
b_n	b_{n-1}	$\cdots$	b_3	$\mathbf{b_2}$	b_1

$(k_1 b_n - l_n b_1)\cdot k_2 + l_n l_1 b_2$

STEP 3. The next step is the same as above, as are all the following steps. We first multiply the results from the previous step, $(k_1 b_n - l_n b_1)k_2 + l_n l_1 b_2$, by the next entry along the diagonal, k_3, then add to the result the product of the next entries at the corners, $-l_n l_1 l_2 b_3$:

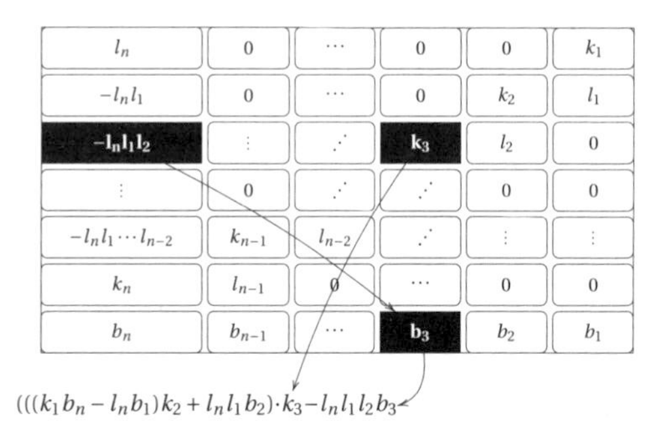

FINAL STEP. We continue in this manner until we reach k_{n-1}, which is the final step. Then we multiply the result from the previous step, $(((k_1 b_n - l_n b_1)k_2 + l_n l_1 b_2)\cdots)\cdot k_{n-2} \pm l_n l_1 \cdots l_{n-3} b_{n-2})$, by k_{n-1}, and then add (or subtract, depending on n) the product of the final corners, $l_n l_1 \cdots l_{n-2} b_{n-1}$:

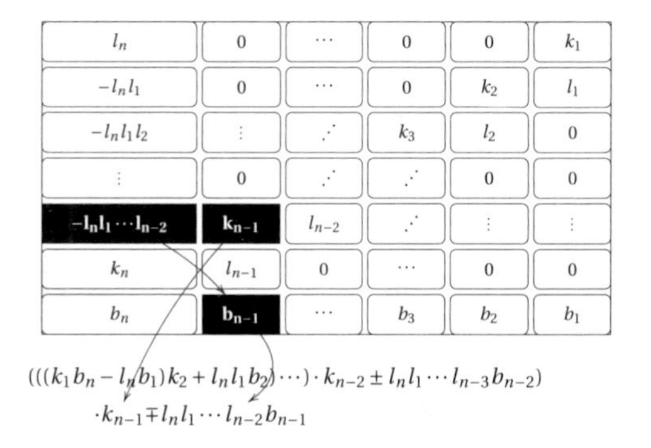

This then gives the result.

Diagrammatic Reconstruction in Modern Terminology

For comparison, it may be helpful to illustrate the calculation using modern notation, even though the calculation is the same (up to rotation). These matrices are unique to Chinese mathematical treatises—I have yet to find any in Western mathematical treatises—and so they have never been analyzed.

STEP 1. Following this method, we begin by "setting the negatives." That is, we set the values $a_{n2} = -l_n l_1$, $a_{n3} = -l_n l_1 l_2$, $a_{n4} = -l_n l_1 l_2 l_3$, and so on, up to $a_{n,n-2} = -l_n l_1 l_2 \cdots l_{n-3}$:

$$\begin{bmatrix} k_1 & l_1 & 0 & \cdots & 0 & b_1 \\ 0 & k_2 & l_2 & \ddots & \vdots & b_2 \\ \vdots & \ddots & \ddots & \ddots & 0 & \vdots \\ 0 & \cdots & 0 & k_{n-1} & l_{n-1} & b_{n-1} \\ l_n & -l_n l_1 & \cdots & -l_n l_1 \cdots l_{n-2} & k_n & b_n \end{bmatrix}$$

Again, it should be noted that the above matrix is no longer an augmented matrix, but rather uses the last row to store the results of the calculations.

STEP 2. Next, we begin the process of multiplying the corners and subtracting:

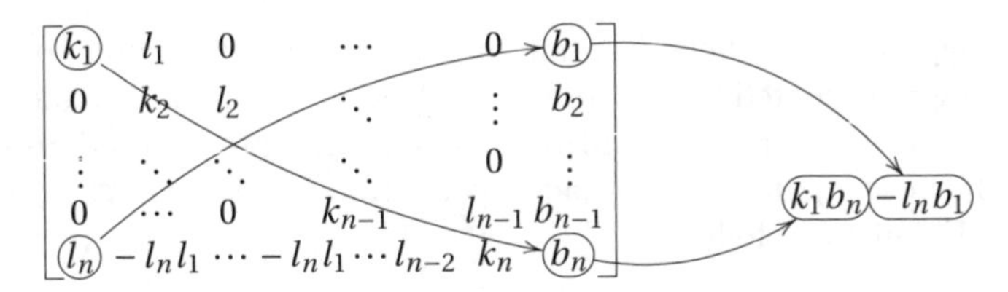

STEP 3. For the following step, we move in, multiply the previous result by k_2, and then subtract the product $-l_n l_1 b_2$:

$$\begin{bmatrix} k_1 & l_1 & 0 & \cdots & 0 & b_1 \\ 0 & k_2 & l_2 & \ddots & \vdots & b_2 \\ \vdots & \ddots & \ddots & \ddots & 0 & \vdots \\ 0 & \cdots & 0 & k_{n-1} & l_{n-1} & b_{n-1} \\ l_n & -l_n l_1 & \cdots & -l_n l_1 \cdots l_{n-2} & k_n & b_n \end{bmatrix} \qquad \begin{matrix} (k_1 b_n - l_n b_1) \\ \cdot k_2 + l_n l_1 b_2 \end{matrix}$$

PENULTIMATE STEP. We continue in this manner, the penultimate step being to multiply the previous result, $(((k_1 b_n - l_n b_1) \cdot k_2 - l_n l_1 b_2) \cdots)$, by k_{n-2}, and add (or subtract, depending on n) the product of the corners, $-l_n l_1 \cdots l_{n-3} b_{n-2}$:

$$\begin{bmatrix} k_1 & l_1 & 0 & \cdots & 0 & b_1 \\ 0 & \ddots & \ddots & \ddots & \vdots & \vdots \\ \vdots & \ddots & \boxed{k_{n-2}} & \ddots & 0 & \boxed{b_{n-2}} \\ 0 & \cdots & 0 & k_{n-1} & l_{n-1} & b_{n-1} \\ l_n & \cdots & \boxed{-l_1 \cdots l_{n-3} l_n} & -l_1 \cdots l_{n-2} l_n & k_n & b_n \end{bmatrix} \quad \begin{array}{l} (((k_1 b_n - l_n b_1) \\ \cdot k_2 + l_n l_1 b_2) \cdots) \\ \boxed{\cdot k_{n-2}} \boxed{\pm l_n l_1 \cdots l_{n-3} b_{n-2}} \end{array}$$

FINAL STEP. The final step in calculating the "dividend" is to multiply the previous result by k_{n-1} and then subtract (or add, depending on n) the product $-l_n l_1 \cdots l_{n-2} b_{n-1}$:

$$\begin{bmatrix} k_1 & l_1 & 0 & \cdots & 0 & b_1 \\ 0 & k_2 & l_2 & \ddots & \vdots & \vdots \\ \vdots & \ddots & \ddots & \ddots & 0 & b_{n-2} \\ 0 & \cdots & 0 & \boxed{k_{n-1}} & l_{n-1} & \boxed{b_{n-1}} \\ l_n & -l_n l_1 & \cdots & \boxed{-l_n l_1 \cdots l_{n-2}} & k_n & b_n \end{bmatrix} \quad \begin{array}{l} ((((k_1 b_n - l_n b_1) \\ \cdot k_2 + l_n l_1 b_2) \cdots) \\ \cdot k_{n-2} \pm l_n l_1 \cdots l_{n-3} b_{n-2}) \\ \boxed{\cdot k_{n-1}} \boxed{\mp l_n l_1 \cdots l_{n-2} b_{n-1}} \end{array}$$

We then calculate the "divisor," divide, and the calculation is complete.

Further Examples

Mei Wending's "On *Fangcheng*" contains numerous further examples of the form of *fangcheng* represented by equation (7.2). The most complex example of this form is the following problem, which gives an array with 9 conditions in 9 unknowns, shown in Figure 7.6 on the facing page. Translated into English, the problem and solution are as follows:

> Suppose that the Moon[39] moves in 3 days from the beginning position of the Tail Lodge to meet Jupiter between the Dipper and Ox [Lodges].[40] After 30 more days, Jupiter moves to the Ox [Lodge]. In these 3 days for the Moon, and 30 days for Jupiter, the sum of their movements is 45 gradations.[41] In making use of the gradations from the Tail to Ox [Lodges], its numbers are approximate. The following is similar.[42]

5

[39] Though the term *taiyin* 太陰 sometimes refers to the Moon, it also sometimes refers to an imaginary astronomical body. The annotations below, however, state that "Luohou, Jidu, and Yueji all have numbers but are without form, for the purpose of clarifying retrograde motion," not mentioning *taiyin*, so it seems that here *taiyin* refers to the Moon. See footnote 43 on the next page.

[40] For translations of the names of the lodges, I have followed Schafer 1977.

[41] Cullen translates the term *du* 度 as "gradation," since 360° = 365.25 *du* (Cullen 1996).

[42] This annotation seems to be an admission that the numbers used for the Lodges are not accurate.

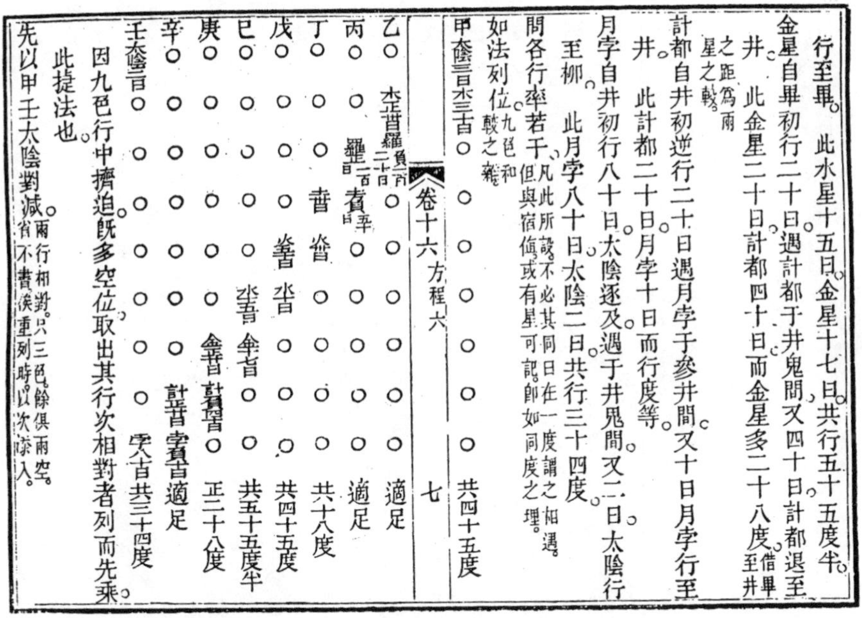

Fig. 7.6: An array of the form given in equation (7.2), representing a problem with 9 conditions in 9 unknowns, from Mei Wending's "On *Fangcheng*."

假如太陰，自尾宿初度行三日遇木星於斗、牛間。又三十日，木星行至牛。此太陰三日，木星三十日共行四十五度。借尾至牛之度，約略其數。後倣此。

10 Jupiter moves in 30 days from the beginning position of the Ox Lodge to meet Luohou[43] between the Ox and Woman Lodges. After 120 more days, Luohou returns to the Ox Lodge. In these 30 days for Jupiter, and 120 days for Luohou, the gradations [traversed] are equal. Luohou, Jidu, and Yueji all have numbers but are without form, for the purpose of clarifying retrograde motion.

---

[43] I have left the terms "Luohou," "Jidu," and "Yueji" untranslated because they do not correspond to actual astronomical bodies. In *Investigation of Harmonics and Calendars, Ancient and Contemporary* (*Gu jin lü li kao* 古今律歷考), the Eleven Luminaries (*shiyi yao* 十一曜) are described as follows: "In addition to the Seven Governors there are also the Four Hidden Luminaries, Ziqi, Yuebei, Luohou, and Jidu. Adepts of astronomy use them to prognosticate fate, and call them the Four Remainders[?]. Together with the Seven Governors, they form the Eleven Luminaries" 七政之外，又有四隱曜：紫氣、月孛、羅睺、計都。星家以之占命，謂之四餘。共七政為十一曜，是也。(*juan* 64, 7b). The Seven Governors are the following: Sun (*ri* 日), Moon (*yue* 月), Jupiter (*mu xing* 木星), Mars (*huo xing* 火星), Saturn (*tu xing* 土星), Mercury (*shui xing* 水星), and Venus (*jin xing* 金星).

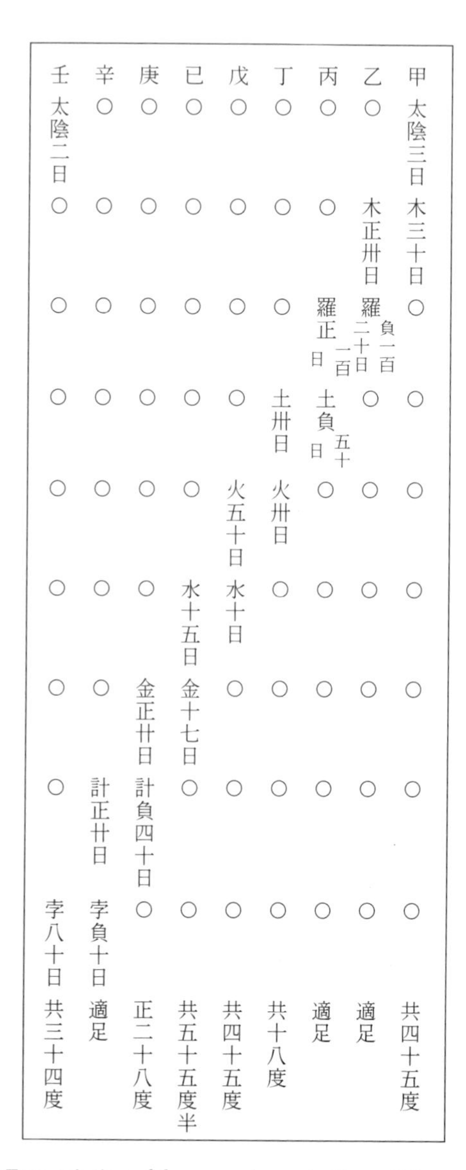

Fig. 7.7: Transcription of the array shown in Figure 7.6 on page 145.

15 木星自牛初行三十日，與羅睺[44]遇於牛、女間。又一百二十日，羅睺退至牛。此木星行三十日，羅睺一百二十日而度等。羅睺、計都、月孛，有數無形，借顯[45]逆行之用。

[44] Substituted for a variant throughout.

[45] Substituted for a variant.

I	H	G	F	E	D	C	B	A
Moon 2 Days	0	0	0	0	0	0	0	Moon 3 Days
0	0	0	0	0	0	0	Jupiter +30 Days	Jupiter 30 Days
0	0	0	0	0	0	Luo[hou] +100 Days	Luo[hou] −120 Days	0
0	0	0	0	0	Saturn 30 Days	Saturn −50 Days	0	0
0	0	0	0	Mars 50 Days	Mars 30 Days	0	0	0
0	0	0	Mercury 15 Days	Mercury 10 Days	0	0	0	0
0	0	Venus +20 Days	Venus 17 Days	0	0	0	0	0
0	Ji[du] +20 Days	Ji[du] −40 Days	0	0	0	0	0	0
[Yue]ji 80 Days	[Yue]ji −10 Days	0	0	0	0	0	0	0
Sum 34 Gradations	Equal	+28 Gradations	Sum $55+\frac{1}{2}$ Gradations	Sum 45 Gradations	Sum 18 Gradations	Equal	Equal	Sum 45 Gradations

Fig. 7.8: Translation into English of the array in Figures 7.6 and 7.7.

In 100 days Luohou goes in retrograde motion from the beginning position of the Ox [Lodge] to meet Saturn between the Winnower and Dipper [Lodges]. After 50 more days Saturn moves to the Ox [Lodge]. In these 100 days for Luohou, and 50 days for Saturn, the gradations [traversed] are equal.

羅睺自牛初退行一百日，遇土星於箕、斗間。又五十日，土星行至牛。此羅睺一百日，土星五十日行度等。

Saturn moves in 30 days from the beginning position of the Ox [Lodge] to catch Mars, meeting [it] between the Ox and Woman [Lodges]. After 30 more days, Mars moves to the Barrens [Lodge]. In these 30 days for Saturn, and these 30 days for Mars, the sum of their movements is 18 gradations.

土星自牛初行三十日，火星逐及，遇於牛、女間。又三十日，火星行至虛。此土星三十日，（水）〔火〕[46]星三十日，而共行十八度。

[46] Emended following the ZKJD edition, *juan* 16, 6b.

Mars moves in 50 days from the beginning position of the Barren [Lodge] to catch Mercury, meeting [it] between the Rooftop and House [Lodges]. After 10 more days, Mercury moves to the Straddler [Lodge]. In these 10 days for Mars, and 10 days for Mercury, the sum of their movements is 45 gradations.

35 火星自虛初行五十日，水星逐及，遇於危、室間。又十日，水星行至奎。此火星行五十日，水星十日，共行四十五度。

Mercury moves in 15 days from the beginning position of the Straddler [Lodge] to catch Venus, meeting [it] between the Mane and Net [Lodges]. After 17 more days, Venus moves to the Net [Lodge]. In these 15 days for 40 Mercury, and these 17 days for Venus, the sum of their movements is 55 and $\frac{1}{2}$ gradations.

水星自奎初行十五日，逐及金星，遇於昴、畢間。又十七日，金星行至畢。此水星十五日，金星十七日，共行五十五度半。

Venus moves in 20 days from the beginning position of the Net [Lodge] to 45 meet Jidu between the Well and Ghost [Lodges]. After 40 more days, Venus moves to the Well [Lodge]. In these 20 days for Venus, and 40 days for Jidu, Jidu moves 28 more gradations.

金星自畢初行二十日，遇計都於井、鬼間。又四十日，計都退至井。此金星二十日，計都四十日，而金星多二十八（日）〔度〕[47]。借畢至 50 井之距為兩星之較。

In 20 days Jidu goes in retrograde motion from the beginning position of the Triaster [Lodge] to meet Yueji between the Triaster and Well [Lodges]. After 10 more days, Yueji moves to the Well [Lodge]. In these 20 days for Jidu, and 10 days for Yueji, their movements are equal.

55 計都自井初逆行二十日，遇月孛於參、井間。又十日，月孛行至井。此計都二十日，月孛十日，而行度等。

Yueji moves in 80 days from the beginning position of the Well [Lodge] to catch the Moon, meeting [it] between the Well and Ghost [Lodges]. After 2 more days, the Moon moves to the Willow [Lodge]. In these 80 days for 60 Yueji, and 2 days for the Moon, the sum of their movements is 34 gradations.

月孛自井初行八十日，太陰逐及，遇於井、鬼間。又二日，太陰行至柳。此月孛八十日，太陰二日，共行三十四度。

Problem: How much is the rate of movement for each?

[47] Emended following the ZKJD edition, *juan* 16, 7a.

Problem Written in Modern Terminology

The array is then arranged as shown in Figures 7.6, 7.7, and 7.8. In modern notation, we write this array in the following manner:

$$\begin{bmatrix} k_1 & l_1 & 0 & 0 & 0 & 0 & 0 & 0 & 0 & b_1 \\ 0 & k_2 & l_2 & 0 & 0 & 0 & 0 & 0 & 0 & b_2 \\ 0 & 0 & k_3 & l_3 & 0 & 0 & 0 & 0 & 0 & b_3 \\ 0 & 0 & 0 & k_4 & l_4 & 0 & 0 & 0 & 0 & b_4 \\ 0 & 0 & 0 & 0 & k_5 & l_5 & 0 & 0 & 0 & b_5 \\ 0 & 0 & 0 & 0 & 0 & k_6 & l_6 & 0 & 0 & b_6 \\ 0 & 0 & 0 & 0 & 0 & 0 & k_7 & l_7 & 0 & b_7 \\ 0 & 0 & 0 & 0 & 0 & 0 & 0 & k_8 & l_8 & b_8 \\ l_9 & 0 & 0 & 0 & 0 & 0 & 0 & 0 & k_9 & b_9 \end{bmatrix}.$$

Then, if we follow the method given in Fang's *Numbers and Measurement* (as generalized in equation 7.2), we have

$$\begin{bmatrix} k_1 & l_1 & 0 & 0 & 0 & 0 & 0 & 0 & 0 & b_1 \\ 0 & k_2 & l_2 & 0 & 0 & 0 & 0 & 0 & 0 & b_2 \\ 0 & 0 & k_3 & l_3 & 0 & 0 & 0 & 0 & 0 & b_3 \\ 0 & 0 & 0 & k_4 & l_4 & 0 & 0 & 0 & 0 & b_4 \\ 0 & 0 & 0 & 0 & k_5 & l_5 & 0 & 0 & 0 & b_5 \\ 0 & 0 & 0 & 0 & 0 & k_6 & l_6 & 0 & 0 & b_6 \\ 0 & 0 & 0 & 0 & 0 & 0 & k_7 & l_7 & 0 & b_7 \\ 0 & 0 & 0 & 0 & 0 & 0 & 0 & k_8 & l_8 & b_8 \\ l_9 & -l_1 l_9 & -l_1 l_2 l_9 & \begin{matrix} -l_1 l_2 \\ \cdot l_3 l_9 \end{matrix} & \begin{matrix} -l_1 l_2 l_3 \\ \cdot l_4 l_9 \end{matrix} & \begin{matrix} -l_1 l_2 l_3 \\ \cdot l_4 l_5 l_9 \end{matrix} & \begin{matrix} -l_1 l_2 l_3 l_4 \\ \cdot l_5 l_6 l_9 \end{matrix} & \begin{matrix} -l_1 l_2 l_3 \\ \cdot l_4 l_5 l_6 \\ \cdot l_7 l_9 \end{matrix} & k_9 & b_9 \end{bmatrix}.$$

The solution is easily calculated, giving

$$x_9 = \frac{\begin{aligned} &((((((((k_1 b_9 - l_9 b_1) k_2 + l_9 l_1 b_2) k_3 - l_9 l_1 l_2 b_3) k_4 + l_9 l_1 l_2 l_3 b_4) \\ &\quad \cdot k_5 - l_9 l_1 l_2 l_3 l_4 b_5) k_6 + l_9 l_1 l_2 l_3 l_4 l_5 b_6) \\ &\quad \cdot k_7 - l_9 l_1 l_2 l_3 l_4 l_5 l_6 b_7) k_8 + l_9 l_1 l_2 l_3 l_4 l_5 l_6 l_7 b_8) \end{aligned}}{(k_1 k_2 k_3 k_4 k_5 k_6 k_7 k_8 k_9 + l_9 l_1 l_2 l_3 l_4 l_5 l_6 l_7 l_8)}.$$

Conclusions

On the basis of the analysis presented in this chapter, we are now in a position to offer important revisions to the early history of the development of determinants and the history of Chinese mathematics. The results here challenge the received view that determinants seem to have been discovered only in the seventeenth century, by Leibniz and Seki Takakazu, with little precedent or history of development.

This chapter has established two important results:

1. The first is a *terminus ante quem* of about 1025 for the use of determinantal calculations to find the $n+1^{\text{th}}$ unknown y for problems in a category exemplified by the well problem with n conditions in $n+1$ unknowns, by setting
$$y = k_1 k_2 k_3 \ldots k_n + 1,$$
where $n = 3$ or $n = 5$.
2. The second result is a *terminus ante quem* of 1661 for the more general determinantal solution to problems of the form in equation (7.2), given in modern terminology by the formula
$$x_n = \frac{\begin{array}{c}((((((k_1 b_n - l_n b_1) k_2 + l_n l_1 b_2) k_3 - l_n l_1 l_2 b_3) k_4 + l_n l_1 l_2 l_3 b_4) \cdots \\ \cdot k_{n-2} \pm l_n l_1 l_2 \cdots l_{n-3} b_{n-2}) k_{n-1} \mp l_n l_1 l_2 \cdots l_{n-2} b_{n-1})\end{array}}{(k_1 k_2 k_3 \cdots k_{n-1} k_n \mp l_n l_1 l_2 l_3 \cdots l_{n-1})},$$
where $\pm$ is + for n even and − for n odd.

The early history of the development of determinants must include the discovery of formulas for the solution of systems of linear equations for specific cases—formulas that necessarily preceded any generalization into a theory of determinants.

Chapter 8
Evidence of Early Determinantal Solutions

The preceding chapter presented written records demonstrating a *terminus ante quem* of about 1025 for determinantal calculations and a *terminus ante quem* of 1661 for determinantal solutions for the "well problem." This chapter will examine evidence that determinantal methods may have been known as early as the writing of the *Nine Chapters*, by focusing on the well problem and four similar problems in chapter 8 of the *Nine Chapters*. The chapter will be divided into the following sections:

1. The first, "The Classification of Problems," analyzes the likely purpose for including individual problems in the *Nine Chapters*.
2. The second section, "Five Problems from the *Nine Chapters*," then examines five problems from chapter 8 of the *Nine Chapters* that have determinantal solutions. From the point of view of the *fangcheng* procedure, as we will see, there was no reason to include these problems in chapter 8.

The Classification of Problems

One of the difficulties in analyzing the problems in "*Fangcheng*," chapter 8 of the *Nine Chapters*, is classification. Each of these problems is an exemplar of a wider class of problems; the question, then, is what each problem is supposed to exemplify.

Modern Classifications

These problems are all equivalent to systems of linear equations, and reasonably enough, it has seemed natural to modern historians of early Chinese mathematics to classify the problems in chapter 8 by their order, which ranges from 2 to 5, in the following manner:[1]

1. 2 conditions in 2 unknowns (problems 2, 4, 5, 6, 7, 9, 10, and 11);
2. 3 conditions in 3 unknowns (problems 1, 3, 8, 12, 15, and 16);
3. 4 conditions in 4 unknowns (problems 14 and 17);
4. 5 conditions in 5 unknowns (problem 18);
5. 5 conditions in 6 unknowns (problem 13).

This manner of classifying the problems has more to do with modern perceptions of progress (explicitly or implicitly represented by greater numbers of conditions and unknowns in the systems to be solved) than with the technical concerns of the literati writing in the historical period.

Inspection of the conspectus of all of the problems in chapter 8, presented in Table 6.1 (on pages 87–88), suggests that classifying the problems by the number of conditions and unknowns is in fact not particularly appropriate. First, this is not the approach to classification used by chapter 8—that is, the problems are not presented there in the order of the numbers of conditions and unknowns. Indeed, the *fangcheng* procedure—elimination and back substitution—can be applied equally well to systems with any number of conditions and unknowns; and in fact it is so easy to extend the *fangcheng* procedure to higher orders that chapter 8 provides only a single exemplar to be used in solving all 18 problems.

Classification of Fangcheng *Problems by Their Pedagogic Function*

Instead, we should ask what particular technique or difficulty—in historical context—each problem is meant to exemplify. Unfortunately, this is often difficult—the *Nine Chapters* provides little in the way of explanations. As a starting point, it seems reasonable that we might expect that related problems would be grouped together, and within each group each problem might be selected to exemplify a different point. The main pedagogic features of the problems in chapter 8 might be summarized as follows:[2]

---

[1] I have not seen any exceptions to this approach to the classification of the problems in chapter 8.

[2] I have also noted whether the problem is formulated using the terminology of "excess and deficit" problems; in the footnotes I have noted other possibly interesting features of the problem.

PROBLEM 1. This problem demonstrates the *fangcheng* procedure—a general solution to problems with n conditions in n unknowns, using elimination and back substitution. It is formulated as equalities.[3]

$$\begin{bmatrix} 3 & 2 & 1 & 39 \\ 2 & 3 & 1 & 34 \\ 1 & 2 & 3 & 26 \end{bmatrix}.$$

PROBLEM 2. The second problem is presented using the terms "increase" and "decrease," suggesting that it may be an example of how the *fangcheng* procedure can be used to solve "excess-deficit" problems such as those in chapter 7, "Excess and Deficit," of the *Nine Chapters.*

$$\begin{bmatrix} 8 & 2 & 9 \\ 2 & 7 & 11 \end{bmatrix}.$$

PROBLEM 3. The next four problems, problems 3 to 6, seem to demonstrate how to solve problems with negative numbers. In problem 3, negative numbers emerge in the process of elimination, and immediately following it the procedure for positive and negative numbers is introduced. It is formulated using "deficits." It is also of the form of equation (7.2) on page 112.

$$\begin{bmatrix} 2 & 1 & 0 & 1 \\ 0 & 3 & 1 & 1 \\ 1 & 0 & 4 & 1 \end{bmatrix}.$$

PROBLEM 4. It seems reasonable to assume that the following three problems, 4–6, are exemplars meant to demonstrate the solution to different kinds of problems formulated with negative numbers in the coefficients and constant terms. In problem 4, two negative numbers appear as coefficients of the unknowns, both in the second unknown. It is formulated as equalities.[4]

$$\begin{bmatrix} 5 & -7 & 11 \\ 7 & -5 & 25 \end{bmatrix}.$$

PROBLEM 5. In problem 5, again there are two negative numbers as coefficients of the unknowns, but this time one for the first unknown, and one for the second. It is formulated as equalities.

$$\begin{bmatrix} 6 & -10 & 18 \\ -5 & 15 & 5 \end{bmatrix}.$$

[3] Also note that each row of the matrix A of the coefficients of the unknowns is a permutation of the first row; it is not clear what, if any, significance that may have.

[4] Also, the rows of coefficients of unknowns are a type of permutation.

PROBLEM 6. In problem 6, negative numbers are used in the coefficients and also in the constant terms b_i. It is formulated as equalities.

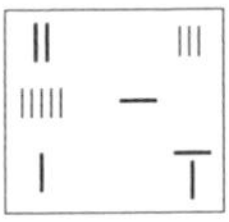

$$\begin{bmatrix} 3 & -10 & -6 \\ -2 & 5 & -1 \end{bmatrix}.$$

PROBLEM 7. It is not clear what the purpose of problem 7 might be. It is formulated as equalities; the rows of coefficients of unknowns are a simple permutation.

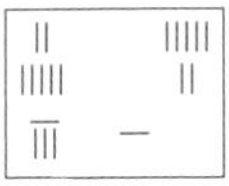

$$\begin{bmatrix} 5 & 2 & 10 \\ 2 & 5 & 8 \end{bmatrix}.$$

PROBLEM 8. Problem 8 demonstrates the solution to a problem in which the constants b_i are equal to a positive number, zero, and a negative number, respectively; positive and negative numbers also appear as coefficients of the unknowns a_{ij}. It is formulated as "excess-deficit."

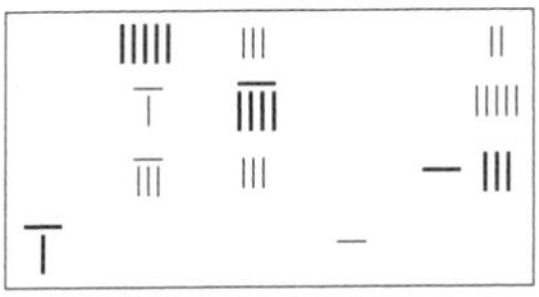

$$\begin{bmatrix} 2 & 5 & -13 & 1000 \\ 3 & -9 & 3 & 0 \\ -5 & 6 & 8 & -600 \end{bmatrix}.$$

PROBLEM 9. It is not clear what the purpose of problem 9 might be. It is possible that it may have been solved by a special method: its constant terms are the same, and it has one for an entry in each row.

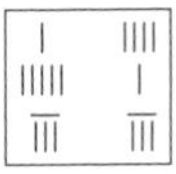

$$\begin{bmatrix} 4 & 1 & 8 \\ 1 & 5 & 8 \end{bmatrix}.$$

PROBLEM 10. Problems 10 and 11 demonstrate how to solve problems with fractions. Problem 10 has proper fractions, which are transformed into integers at the outset. It is formulated as "excess-deficit."

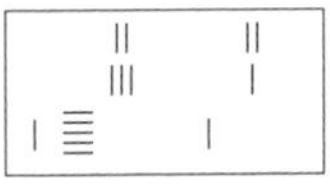

$$\begin{bmatrix} 1 & \frac{1}{2} & 50 \\ \frac{2}{3} & 1 & 50 \end{bmatrix}.$$

PROBLEM 11. Problem 11 has "mixed" fractions, which, again, are transformed into integers at the outset. It is formulated as "excess-deficit."

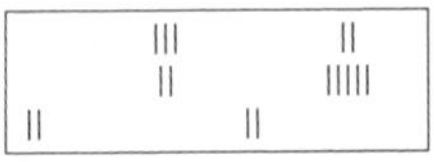

$$\begin{bmatrix} 1 & 2+\frac{1}{2} & 10000 \\ 2-\frac{1}{2} & 1 & 10000 \end{bmatrix}.$$

PROBLEM 12. Problems 12 through 15 are all variants of the form of augmented matrix in equation (7.2)—these are the problems with determinantal solutions, which will be analyzed in more detail in this chapter. Problem 12 is formulated as "excess-deficit."

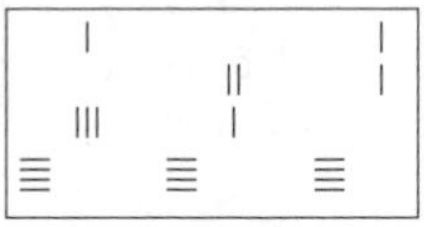

$$\begin{bmatrix} 1 & 1 & 0 & 40 \\ 0 & 2 & 1 & 40 \\ 1 & 0 & 3 & 40 \end{bmatrix}.$$

PROBLEM 13. The well problem is an exemplar for the form of augmented matrix in equation (7.2). It is formulated as "excess-deficit."

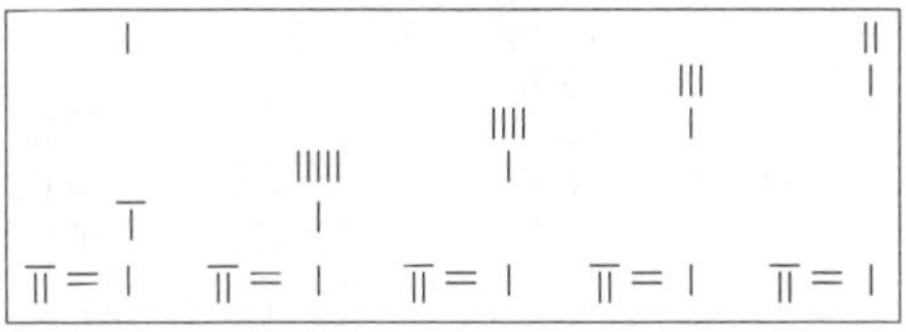

$$\begin{bmatrix} 2 & 1 & 0 & 0 & 0 & 721 \\ 0 & 3 & 1 & 0 & 0 & 721 \\ 0 & 0 & 4 & 1 & 0 & 721 \\ 0 & 0 & 0 & 5 & 1 & 721 \\ 1 & 0 & 0 & 0 & 6 & 721 \end{bmatrix}.$$

PROBLEM 14. Problem 14 is a more complex variant of the form of augmented matrix in equation (7.2). It is formulated as "excess-deficit."

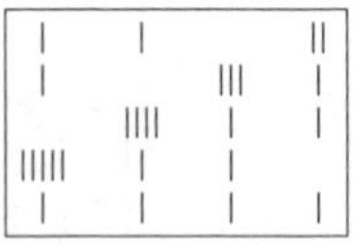

$$\begin{bmatrix} 2 & 1 & 1 & 0 & 1 \\ 0 & 3 & 1 & 1 & 1 \\ 1 & 0 & 4 & 1 & 1 \\ 1 & 1 & 0 & 5 & 1 \end{bmatrix}.$$

PROBLEM 15. Problem 15 is another variant of the form of augmented matrix in equation (7.2). It is formulated as "excess-deficit."

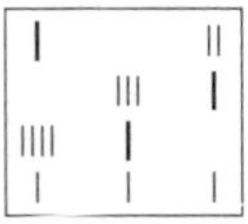

$$\begin{bmatrix} 2 & -1 & 0 & 1 \\ 0 & 3 & -1 & 1 \\ -1 & 0 & 4 & 1 \end{bmatrix}.$$

PROBLEM 16. Again, unfortunately, it is not clear what the purpose of problem 16 might be. In this problem, each of the rows is a permutation of the first row.[5] In problem 16, the constants b_i are not all equal. But if we set them to be all equal, $b_i \equiv b$, then the problem has a particularly simple solution; problems 1, 4, and 7 also share this particular solution. So these problems may be additional examples of a form for which an alternative solution was known.[6]

[5] Though technically, the matrix A of the coefficients of the unknowns is, in modern terminology, a circulant matrix (that is, each row is a cyclic permutation of the entries in the first row, in this case shifted by one place), this is not particularly relevant, since the special properties of circulant matrices involve eigenvalues, a theory not known at the time. However, it is also a Toeplitz matrix (entries along the diagonals are all the same), which may be relevant, because solutions of Toeplitz matrices can be more easily computed.

[6] More specifically, if the rows of the matrix of the coefficients of the unknowns A are all permutations of the entries in the first row, if $\det A \neq 0$, and if the constants are all equal, $b_i \equiv b$, and if the sum of the entries in the first row is not zero, then the solution is simply $x_j = b/(a_{11} + a_{12} + \cdots + a_{1n})$ for $1 \leq j \leq n$.

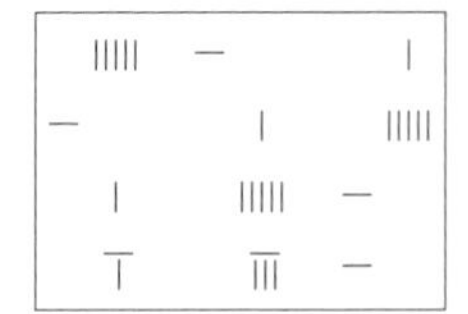

$$\begin{bmatrix} 1 & 5 & 10 & 10 \\ 10 & 1 & 5 & 8 \\ 5 & 10 & 1 & 6 \end{bmatrix}.$$

PROBLEM 17. Problems 17 and 18 are examples demonstrating the computational power of the *fangcheng* procedure to solve complex problems. In the case of problem 17, it is 4 conditions in 4 unknowns. It is formulated as equalities.

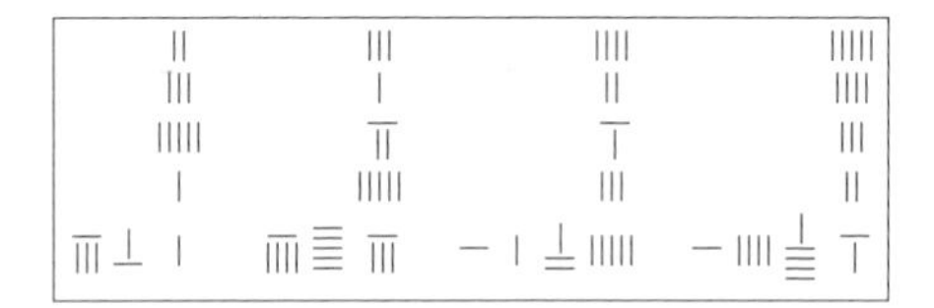

$$\begin{bmatrix} 5 & 4 & 3 & 2 & 1496 \\ 4 & 2 & 6 & 3 & 1175 \\ 3 & 1 & 7 & 5 & 958 \\ 2 & 3 & 5 & 1 & 861 \end{bmatrix}.$$

PROBLEM 18. Much like problem 17, problem 18 is quite complex, and has 5 conditions in 5 unknowns. It is formulated as equalities.

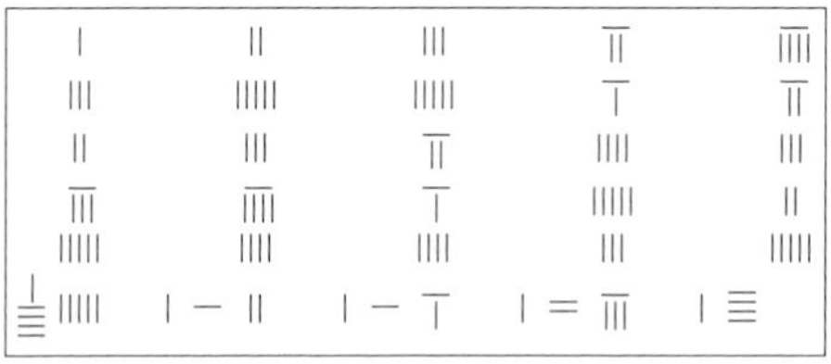

$$\begin{bmatrix} 9 & 7 & 3 & 2 & 5 & 140 \\ 7 & 6 & 4 & 5 & 3 & 128 \\ 3 & 5 & 7 & 6 & 4 & 116 \\ 2 & 5 & 3 & 9 & 4 & 112 \\ 1 & 3 & 2 & 8 & 5 & 95 \end{bmatrix}.$$

Five Problems from the *Nine Chapters*

Although it has been recognized that many of the 18 problems in the chapter "*Fangcheng*" are formulated using the terminology of "excess and deficit" problems, analyses of this chapter, following the *fangcheng* procedure and the earliest extant commentary by Liu Hui, have examined only solutions to these problems that use elimination and back substitution. Though it has been noticed that the excess-deficit problems in chapter 7 of the *Nine Chapters* are in ways equivalent to simultaneous linear equations with 2 conditions in 2 unknowns, no one, evidently, has explored whether methods of solution similar to those for the "excess and deficit" problems might exist for these problems.

As noted above, among the 18 problems in chapter 8, "*Fangcheng*," we can identify problems 3, 12, 13, 14, and 15 as a special class of problems that represent variants of the form of augmented matrix in equation (7.2). We will examine the specific properties of these *fangcheng* problems in the following sections. Just by inspecting the form of these *fangcheng* problems on the counting board, we can first note that these problems (again, 3, 12, 13, 14, and 15) share the following features (explained here using modern mathematical terminology):

1. Each of these problems has zeros as coefficients in the equations; these are the only problems in this chapter having zeros as coefficients.[7] More specifically, in each of these problems: there is at least one zero coefficient in each of the linear equations constituting the problem; all of the linear equations have the same number of zeros; and the placement of the zeros changes in each equation in a cyclical fashion, so that all of the unknowns have zero coefficients an equal number of times. In problem 13, an example of 5 equations in 6 unknowns, most of the coefficients are zero.
2. When these equations are laid out on the counting board, their forms exhibit distinctive patterns, as can be seen above.
3. In each of these problems, the constant terms for all the linear equations are the same. (In addition to these five problems, problems 9, 10, and 11 also have the property that the constant terms are the same.)
4. Each of the these problems is phrased as a deficit-equality problem. That is, among the eighteen problems in this chapter, in nine of them (problems 1, 2, 4, 5, 6, 7, 16, 17, and 18) the conditions are phrased as equalities, as we saw in problem 1 above: "Two bundles of superior paddy, three bundles of medium paddy, and one bundle of inferior paddy yield thirty-four *dou* of grain." In contrast, the problems we are examining here (3, 12, 13, 14, and 15), together with four other problems (8, 9, 10, and 11), are all formulated using the language of excess-deficit problems, that is, explicitly using terms such as "excess" (ying 盈) and "deficit" (*bu zu* 不足). More specifically, the problems we are examining here are phrased as a particular form of excess-deficit problem—deficit-equality problems—and the formulation is similar: $k_1x_1, k_2x_2, \cdots k_nx_n$ are all insufficient (*bu zu* 不足); k_1x_1 added to x_2, k_2x_2 added to x_3, ... k_nx_n added to x_1 are all sufficient. (It should be noted that the phrasing of these problems as deficit-equality problems is, however, completely superfluous. The problems are in fact solved by using the equalities; the phrasing as deficit plays no role in the problems whatsoever.)

But the similarities hardly end there. As we have seen, when displayed on the counting board, the arrangements of counting rods in these five *fangcheng* problems are highly symmetric. And more important, the solutions computed by using determinants turn out to be elegant permutations. In the following subsections, I will analyze each of these problems individually, to show that each has a determinantal solution that is considerably simpler than elimination, and that from the standpoint of the *fangcheng* procedure there is no pedagogical purpose for the inclusion of these problems in chapter 8 of the *Nine Chapters*.

[7] Zeros do sometimes appear in other problems in this chapter, but only as a constant term; none of these five problems has a zero as a constant term.

Problem 13

Problem 13, the well problem, was translated in the previous chapter, and solved following the *fangcheng* procedure. Here we will show that a simpler determinantal solution exists.

Determinantal Solutions

As we saw in the previous chapter, the well problem, written in modern notation in equation (7.1), was solved by first assigning the value of the determinant of the coefficient matrix to the $n+1^{\text{th}}$ unknown, $y = \det A$. This then transformed the well problem into a system of 5 equations in 5 unknowns. If we write this as an augmented matrix, we have

$$\begin{bmatrix} k_1 & 1 & 0 & 0 & 0 & b \\ 0 & k_2 & 1 & 0 & 0 & b \\ 0 & 0 & k_3 & 1 & 0 & b \\ 0 & 0 & 0 & k_4 & 1 & b \\ 1 & 0 & 0 & 0 & k_5 & b \end{bmatrix}, \tag{8.1}$$

where we will not assume, at the outset, that $\det A = b$. We can calculate the solution by applying Cramer's Rule, which states that

$$x_j = \frac{\det B_j}{\det A},$$

for each j $(1 \le j \le 5)$. The determinant of the matrix A of coefficients of the unknowns is given by

$$\det A = \begin{vmatrix} k_1 & 1 & 0 & 0 & 0 \\ 0 & k_2 & 1 & 0 & 0 \\ 0 & 0 & k_3 & 1 & 0 \\ 0 & 0 & 0 & k_4 & 1 \\ 1 & 0 & 0 & 0 & k_5 \end{vmatrix} = k_1 k_2 k_3 k_4 k_5 + 1. \tag{8.2}$$

Solving for B_j, we obtain

$$\det B_1 = \begin{vmatrix} b & 1 & 0 & 0 & 0 \\ b & k_2 & 1 & 0 & 0 \\ b & 0 & k_3 & 1 & 0 \\ b & 0 & 0 & k_4 & 1 \\ b & 0 & 0 & 0 & k_5 \end{vmatrix} = bk_2 k_3 k_4 k_5 - bk_3 k_4 k_5 + bk_4 k_5 - bk_5 + b,$$

$$\det B_2 = \begin{vmatrix} k_1 & b & 0 & 0 & 0 \\ 0 & b & 1 & 0 & 0 \\ 0 & b & k_3 & 1 & 0 \\ 0 & b & 0 & k_4 & 1 \\ 1 & b & 0 & 0 & k_5 \end{vmatrix} = bk_3k_4k_5k_1 - bk_4k_5k_1 + bk_5k_1 - bk_1 + b,$$

$$\det B_3 = \begin{vmatrix} k_1 & 1 & b & 0 & 0 \\ 0 & k_2 & b & 0 & 0 \\ 0 & 0 & b & 1 & 0 \\ 0 & 0 & b & k_4 & 1 \\ 1 & 0 & b & 0 & k_5 \end{vmatrix} = bk_4k_5k_1k_2 - bk_5k_1k_2 + bk_1k_2 - bk_2 + b,$$

$$\det B_4 = \begin{vmatrix} k_1 & 1 & 0 & b & 0 \\ 0 & k_2 & 1 & b & 0 \\ 0 & 0 & k_3 & b & 0 \\ 0 & 0 & 0 & b & 1 \\ 1 & 0 & 0 & b & k_5 \end{vmatrix} = bk_5k_1k_2k_3 - bk_1k_2k_3 + bk_2k_3 - bk_3 + b,$$

$$\det B_5 = \begin{vmatrix} k_1 & 1 & 0 & 0 & b \\ 0 & k_2 & 1 & 0 & b \\ 0 & 0 & k_3 & 1 & b \\ 0 & 0 & 0 & k_4 & b \\ 1 & 0 & 0 & 0 & b \end{vmatrix} = bk_1k_2k_3k_4 - bk_2k_3k_4 + bk_3k_4 - bk_4 + b.$$

Combining the numerators and denominator, we obtain

$$x_1 = \frac{b(k_2k_3k_4k_5 - k_3k_4k_5 + k_4k_5 - k_5 + 1)}{k_1k_2k_3k_4k_5 + 1}, \tag{8.3}$$

$$x_2 = \frac{b(k_3k_4k_5k_1 - k_4k_5k_1 + k_5k_1 - k_1 + 1)}{k_1k_2k_3k_4k_5 + 1}, \tag{8.4}$$

$$x_3 = \frac{b(k_4k_5k_1k_2 - k_5k_1k_2 + k_1k_2 - k_2 + 1)}{k_1k_2k_3k_4k_5 + 1}, \tag{8.5}$$

$$x_4 = \frac{b(k_5k_1k_2k_3 - k_1k_2k_3 + k_2k_3 - k_3 + 1)}{k_1k_2k_3k_4k_5 + 1}, \tag{8.6}$$

$$x_5 = \frac{b(k_1k_2k_3k_4 - k_2k_3k_4 + k_3k_4 - k_4 + 1)}{k_1k_2k_3k_4k_5 + 1}. \tag{8.7}$$

Now if, as stipulated in the well problem, we set $b = \det A = k_1k_2k_3k_4k_5 + 1$, the solution can be simplified to the following:

$$x_1 = k_2k_3k_4k_5 - k_3k_4k_5 + k_4k_5 - k_5 + 1, \tag{8.8}$$

$$x_2 = k_3k_4k_5k_1 - k_4k_5k_1 + k_5k_1 - k_1 + 1, \tag{8.9}$$

$$x_3 = k_4k_5k_1k_2 - k_5k_1k_2 + k_1k_2 - k_2 + 1, \tag{8.10}$$

$$x_4 = k_5k_1k_2k_3 - k_1k_2k_3 + k_2k_3 - k_3 + 1, \tag{8.11}$$

$$x_5 = k_1k_2k_3k_4 - k_2k_3k_4 + k_3k_4 - k_4 + 1. \tag{8.12}$$

This solution, both simple and elegant, is in itself intriguing—the solutions for all the x_j are identical, up to a cycling of the indices i of the k_i. However, we do not have any surviving written sources that record the use of equations (8.8) through (8.12). So what evidence do we have, beyond the simple, elegant solution given in equations (8.8) through (8.12), to suggest that these problems could have been solved using algebraic methods that are equivalent to determinants? The remainder of this section will present further evidence.

First, there is a very simple method for computing the solution to equations (8.8) through (8.12), much simpler than elimination. Rearranging terms,

$$x_1 = (((k_2 - 1)k_3 + 1)k_4 - 1)k_5 + 1, \tag{8.13}$$

$$x_2 = (((k_3 - 1)k_4 + 1)k_5 - 1)k_1 + 1, \tag{8.14}$$

$$x_3 = (((k_4 - 1)k_5 + 1)k_1 - 1)k_2 + 1, \tag{8.15}$$

$$x_4 = (((k_5 - 1)k_1 + 1)k_2 - 1)k_3 + 1, \tag{8.16}$$

$$x_5 = (((k_1 - 1)k_2 + 1)k_3 - 1)k_4 + 1. \tag{8.17}$$

The solutions are thus very easy to calculate, so easy that with a little practice the solutions can be read off the array and easily calculated in the head, without any manipulation of the counting rods.[8] The most important evidence, however, is probably the ease with which such calculations can be made on the counting board.

Determinantal Solutions on the Counting Board

The following diagrams show how this could have been done (again, I must note that I have not found written sources that record these calculations; these diagrams are a reconstruction to show how easily the calculations could have been directly computed from the placement of counting rods as described in the *Nine Chapters*).

1ST UNKNOWN. The first diagram illustrates the solution for the first unknown, the length of *A*'s rope (the calculations at each step have been added to the arrows that indicate the order):

(*i*) $3 - 1 = 2$

(*ii*) $2 \times 4 + 1 = 9$

(*iii*) $9 \times 5 - 1 = 44$

(*iv*) $44 \times 6 + 1 = 265$

---

[8] In a personal communication, John Crossley has pointed out to me that these methods appear to be similar to those used to find n^{th} roots. I will address this important insight in future research.

We proceed along the diagonal, beginning with the entry immediately following the entry for A, from which we subtract 1 [step (*i*)], and multiply by the following entry and add 1 to the result [step (*ii*)], multiply by the following entry and subtract 1 from the result [step (*iii*)], and multiply by the following entry and add 1 to the result [step (*iv*)], yielding the answer, 265. Calculation of the other unknowns uses precisely this same formula, in each case beginning with the entry on the diagonal immediately following, and ending with the entry immediately preceding.

2ND UNKNOWN. The next diagram illustrates the solution for the second unknown, the length of B's rope:

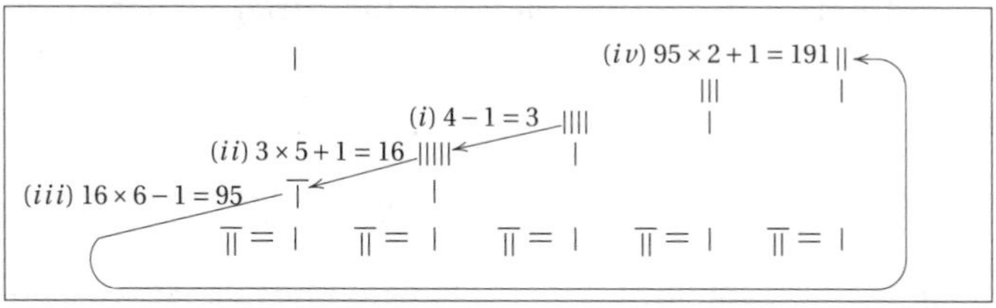

(*iv*) $95 \times 2 + 1 = 191$
(*i*) $4 - 1 = 3$
(*ii*) $3 \times 5 + 1 = 16$
(*iii*) $16 \times 6 - 1 = 95$

We again proceed along the diagonal, starting with the entry immediately following the entry for B, from which we subtract 1 [step (*i*)], multiply by the following entry and add 1 to the result [step (*ii*)], multiply by the following entry and subtract 1 from the result [step (*iii*)], and multiply by the following entry and add 1 to the result [step (*iv*)], yielding the answer, 191.

3RD UNKNOWN. The third diagram illustrates the solution for the third unknown, the length of C's rope:

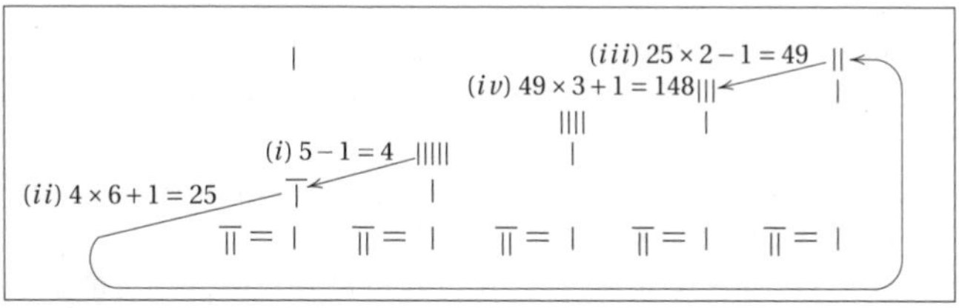

(*iii*) $25 \times 2 - 1 = 49$
(*iv*) $49 \times 3 + 1 = 148$
(*i*) $5 - 1 = 4$
(*ii*) $4 \times 6 + 1 = 25$

The formula is again the same, and the result is 148.

4TH UNKNOWN. The next diagram illustrates the solution for the fourth unknown, the length of D's rope, giving the result 129:

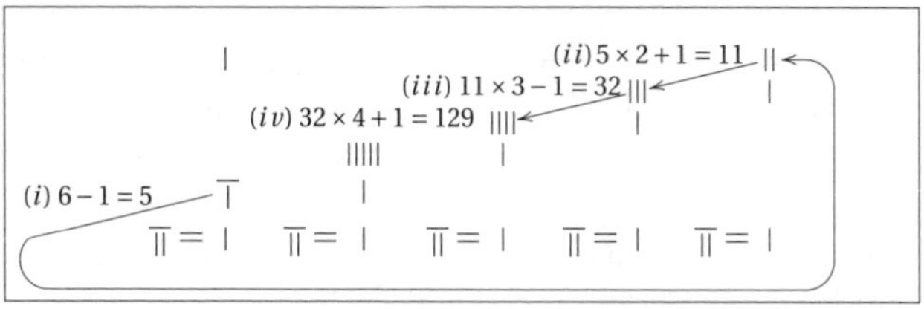

(*ii*) $5 \times 2 + 1 = 11$
(*iii*) $11 \times 3 - 1 = 32$
(*iv*) $32 \times 4 + 1 = 129$
(*i*) $6 - 1 = 5$

5TH UNKNOWN. The final diagram illustrates the calculation for the fifth unknown, the length of E's rope, giving the result 76:

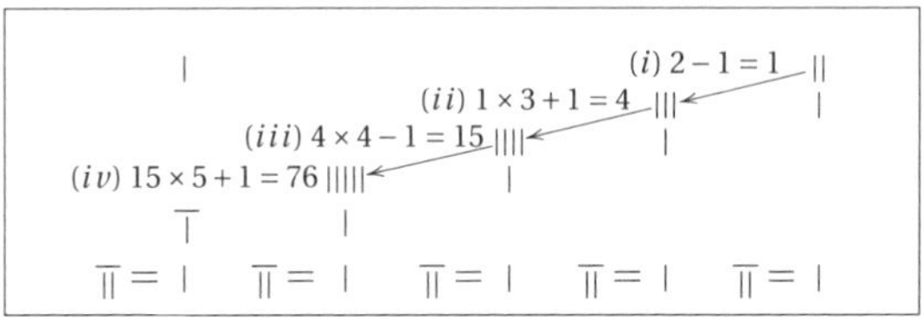

$(i)\ 2-1=1$

$(ii)\ 1\times 3+1=4$

$(iii)\ 4\times 4-1=15$

$(iv)\ 15\times 5+1=76$

It seems reasonable to speculate that adepts of this method might have argued for the superiority of their approach by simply reading the solution off of the counting board and calculating the solution in their head, perhaps as evidence of special powers.

Conclusion—Evidence for Determinantal Solutions to Problem 13

In sum, there is considerable evidence that determinantal-style solutions were known by the time the *Nine Chapters* was compiled in about the 1st century C.E. These solutions had the following features:

1. One simple rule is all that is required: from the current value subtract (alternately, add) 1, then multiply by the following entry on the diagonal.
2. This one rule is then repeatedly applied: to find the value of one of the unknowns, begin with the entry immediately following; taking the entries in order along the diagonal, alternately subtract or add one and multiply by the next entry, until reaching the entry immediately preceding the one representing the unknown.
3. This one simple procedure is used to solve for all the unknowns.
4. This procedure is valid for any problem of the form given above in equation (7.1), if $b = \det A$.
5. This procedure is valid for any system of n linear equations in n unknowns of the form in equation (7.1).
6. The only calculations required are multiplication, addition, and subtraction; division and fractions are not needed.[9]
7. The calculations involve only integers.
8. Negative numbers do not appear in the calculations (assuming, as in problems 3, 12, 13, 14, and 15, that $k_i \geq 1$).
9. Zero does not appear in the calculations if $k_i \geq 2$ for all $(1 \leq i \leq n)$; the only exception is problem 12, where $k_1 = 1$.
10. In contrast to the relative simplicity of these calculations, the solution given in the *fangcheng* procedure is considerably more complex.

9 This assumes, as in the well problem, that the constant terms are all equal to the determinant of the coefficient matrix, $b = \det A$. If $b \neq \det A$, as in some of the problems analyzed later in this chapter, then division is required, but only at the final step.

These features of this solution to the well problem will, in general, remain valid for the solution to similar problems from the *Nine Chapters*, which we will analyze in the remainder of this section.

Problem 3

The first problem in chapter 8 that is of the form given in equation (7.1) is problem 3. Unlike the other four problems of this form in chapter 8, problem 3 serves a clear pedagogical purpose: the appearance of a negative number in the process of elimination following the *fangcheng* procedure provides the motivation for explaining the procedure for positive and negative numbers. It should also be noted that problem 3 is not grouped together with the other four problems of this form, problems 12 to 15. Problem 3 is translated in chapter 5 of this book, and the calculations for elimination and back substitution are shown there (see pages 81–84). Negative numbers, however, appear in the process of elimination in many *fangcheng* problems; the symmetric placement of zeros in problem 3 allows for determinantal solutions.

Determinantal Solutions in Modern Mathematical Terminology

As an exemplar of more general cases, we can write problem 3 in the following manner:

$$\begin{bmatrix} k_1 & 1 & 0 & b \\ 0 & k_2 & 1 & b \\ 1 & 0 & k_3 & b \end{bmatrix}. \tag{8.18}$$

This is of the same form as equation 7.2 on page 112, but with 3 conditions in 3 unknowns. It is easy to calculate that the solutions are as follows:

$$x_1 = \frac{b(1 - k_3 + k_2 k_3)}{k_1 k_2 k_3 + 1},$$
$$x_2 = \frac{b(1 - k_1 + k_3 k_1)}{k_1 k_2 k_3 + 1},$$
$$x_3 = \frac{b(1 - k_2 + k_1 k_2)}{k_1 k_2 k_3 + 1}.$$

In problem 3 we have $b = 1$, so if we write $d = \det A = k_1 k_2 k_3 + 1$, we have the solutions

$$x_1 = \big((k_2 - 1)k_3 + 1\big) \div d, \tag{8.19}$$
$$x_2 = \big((k_3 - 1)k_1 + 1\big) \div d, \tag{8.20}$$
$$x_3 = \big((k_1 - 1)k_2 + 1\big) \div d, \tag{8.21}$$

which are similar to equations (8.13) through (8.17) on page 160, with the exception that $b \neq \det A$. It is also important, as noted above (item 8 on page 162), that solving this problem by this method does not require negative numbers, as opposed to the *fangcheng* procedure, which does.

Determinantal Solutions on the Counting Board

Like problem 13, the well problem, determinantal solutions to problem 3 can easily be calculated once the problem is placed on the counting board.

1ST UNKNOWN. To solve for the first unknown, we take the second entry in the second column, then subtract one, giving $3 - 1 = 2$. We then multiply the third entry in the third column by the result and add one, giving $2 \times 4 + 1 = 9$:

(*i*) $3 - 1 = 2$
(*ii*) $2 \times 4 + 1 = 9$

2ND UNKNOWN. To find the second unknown, we begin with the third entry in the third column and subtract one, giving $4 - 1 = 3$. We then multiply the first entry in the first column by the result and add one, yielding $3 \times 2 + 1 = 7$:

(*ii*) $3 \times 2 + 1 = 7$
(*i*) $4 - 1 = 3$

3RD UNKNOWN. To find the third unknown, we begin with the first entry in the first column and subtract one, giving $2 - 1 = 1$. We multiply the second entry in the second column by the result, and add one, giving $1 \times 3 + 1 = 4$:

(*i*) $2 - 1 = 1$
(*ii*) $1 \times 3 + 1 = 4$

FINAL STEP. Finally, we must divide each of the results above by d, giving the values for the three unknowns,

$$x_1 = 9 \div 25 = \frac{9}{25}, \quad x_2 = 7 \div 25 = \frac{7}{25}, \quad x_3 = 4 \div 25 = \frac{4}{25}.$$

Problem 12

Problem 12 is very similar to problem 3, except for the constant terms. It is the first in a series of four problems with determinantal solutions. From the point of view of elimination and back substitution following the *fangcheng* procedure, there would seem to be no particular reason to include this problem as an exemplar in chapter 8 of the *Nine Chapters*. But from the point of view of determinants, as we will see, problem 12 represents an exemplar of a different determinantal solution.

Translation of Problem 12

> Given one superior horse, two common horses, and three inferior horses, each [group of horses] is able to carry 40 *dan* [1 *dan* ≈ 30 kilograms][10] to the base of an incline, but none [of the groups] are able to ascend. The superior horse together with one common horse, the [group of two] common horses
> 5 together with one inferior horse, and the [group of three] inferior horses together with one superior horse, are all able to ascend. Problem: How much weight do the superior horse, common horse, and inferior horse each have the strength to pull?
>
> 今有武馬一匹，中馬二匹，下馬三匹，皆載四十石至阪，皆不能上。
> 10 武馬借中馬一匹，中馬借下馬一匹，下馬借武馬一匹，乃皆上。問武、中、下馬一匹各力引幾何？
>
> Solution: One superior horse has the strength to pull 22 *dan* and $\frac{6}{7}$ *dan*; one common horse has the strength to pull 17 *dan* and $\frac{1}{7}$ *dan*; one inferior horse has the power to pull 5 and $\frac{5}{7}$ *dan*.
>
> 15 答曰：武馬一匹力引二十二石七分石之六，中馬一匹力引一十七石七分石之一，下馬一匹力引五石七分石之五。
>
> Method: Use the *fangcheng* procedure; place what is taken in each case; and apply the method of positive and negative [numbers].
>
> 術曰：如方程，各置所借，以正負術入之。

Elimination Following the Fangcheng *Procedure*

Elimination following the *fangcheng* procedure is as follows:

STEP 1. The initial placement of the counting rods for problem 12 is as follows:

[10] On Chinese units of measure, again see note 9 on page 114.

$$\begin{bmatrix} 1 & 1 & 0 & 40 \\ 0 & 2 & 1 & 40 \\ 1 & 0 & 3 & 40 \end{bmatrix}.$$

STEP 2. Next, we subtract the first column [row] from the third column [row], giving $(1,0,3,40) - (1,1,0,40) = (0,-1,3,0)$,

$$\begin{bmatrix} 1 & 1 & 0 & 40 \\ 0 & 2 & 1 & 40 \\ 0 & -1 & 3 & 0 \end{bmatrix}.$$

STEP 3. Then we multiply the third column [row] by 2, and add the second column [row], giving $2 \cdot (0,-1,3,0) + (0,2,1,40) = (0,0,7,40)$,

$$\begin{bmatrix} 1 & 1 & 0 & 40 \\ 0 & 2 & 1 & 40 \\ 0 & 0 & 7 & 40 \end{bmatrix}.$$

Back Substitution Following the Fangcheng *Procedure*

Now we solve using back substitution following the *fangcheng* procedure.

STEP 4. First we "cross multiply" the final entry in the second column by the third entry in the final column, $40 \times 7 = 280$, and the third entry in the second column by the final entry in the final column, $1 \times 40 = 40$, giving

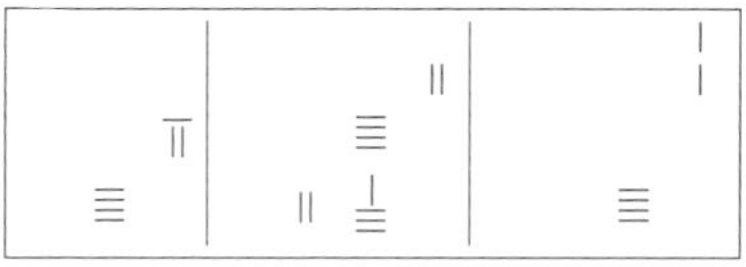

$$\begin{bmatrix} 1 & 1 & 0 & 40 \\ 0 & 2 & 40 & 280 \\ 0 & 0 & 7 & 40 \end{bmatrix}.$$

STEP 5. We then subtract the third entry in the second column from the final entry in the second column, $280 - 40 = 240$, and then divide the result by the second entry in the second column, giving the result $240 \div 2 = 120$,

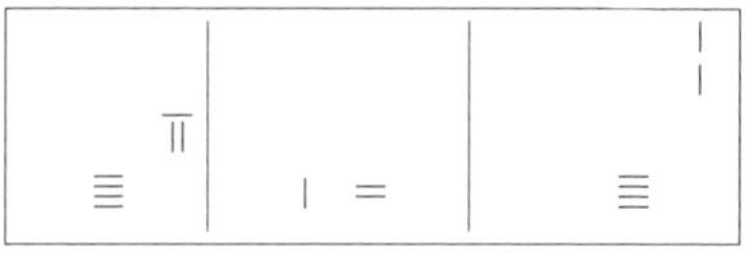

$$\begin{bmatrix} 1 & 1 & 0 & 40 \\ 0 & 0 & 0 & 120 \\ 0 & 0 & 7 & 40 \end{bmatrix}.$$

STEP 6. The next step is to simplify entries in the first column. We "cross multiply" the final entry in the first column by the third entry in the final column, yielding $7 \times 40 = 280$, and the second entry in the first column by the final entry in the second column, $120 \times 1 = 120$,

$$\begin{bmatrix} 1 & 120 & 0 & 280 \\ 0 & 0 & 0 & 120 \\ 0 & 0 & 7 & 40 \end{bmatrix}.$$

STEP 7. We then subtract the second entry in the first column from the final entry in the first column, $280 - 120 = 160$, and divide the result by the first entry in the first column, $160 \div 1 = 160$,

$$\begin{bmatrix} 0 & 0 & 0 & 160 \\ 0 & 0 & 0 & 120 \\ 0 & 0 & 7 & 40 \end{bmatrix}.$$

STEP 8. The final step is then to divide the final entries in all of the columns by the third entry in the final column, 7, giving the solutions to the unknowns as fractions,

$$\begin{bmatrix} 0 & 0 & 0 & 160/7 \\ 0 & 0 & 0 & 120/7 \\ 0 & 0 & 0 & 40/7 \end{bmatrix}.$$

As can be seen from the above calculations of the elimination and back substitution following the *fangcheng* procedure, there is nothing new in problem 12 that would recommend its inclusion in chapter 8 of the *Nine Chapters*. More specifically, using the *fangcheng* procedure, the solution to problem 12 is identical to the solution to problem 3: at each step of the calculation, only the magnitudes of the entries differ, not the signs of the entries or the positions of the entries that are zero.

Determinantal Solutions to Problem 12

To solve this by determinants, the general form of problem 12 is as follows:

$$\begin{bmatrix} k_1 & 1 & 0 & b \\ 0 & k_2 & 1 & b \\ 1 & 0 & k_3 & b \end{bmatrix}.$$

If we write $d = \det A = k_1 k_2 k_3 + 1$, it is easy to check that the general solution is given by

$$x_1 = \big(\big((k_2 - 1)k_3 + 1\big) \cdot b\big) \div d, \tag{8.22}$$

$$x_2 = \big(\big((k_3 - 1)k_1 + 1\big) \cdot b\big) \div d, \tag{8.23}$$

$$x_3 = \big(\big((k_1 - 1)k_2 + 1\big) \cdot b\big) \div d. \tag{8.24}$$

The point of this problem would seem to be that the constant $b = 40$ is not equal to 1 (as in problem 3), nor equal to det A (as in the well problem), which necessitates first multiplying the result by b and then dividing that by d.

Determinantal Solutions on the Counting Board

The computation on the counting board is similar to that in problem 3; the main difference here is in the further calculations necessary when we are finished, namely multiplying the result by b and dividing by d.

1ST UNKNOWN. To find the first unknown, we subtract one from the second entry in the second column, $2 - 1 = 1$. Then we multiply the third entry in the third column by the result and add one, giving $1 \times 3 + 1 = 4$,

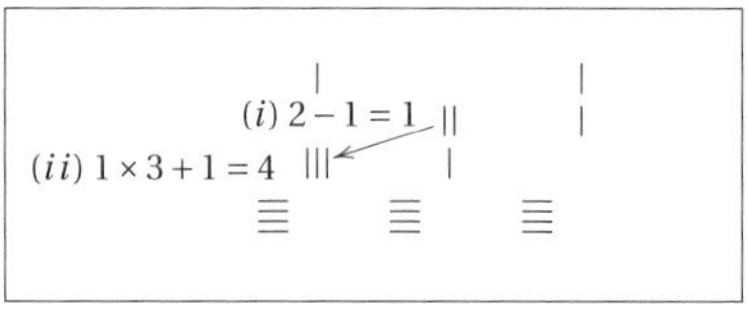

(*i*) $2 - 1 = 1$
(*ii*) $1 \times 3 + 1 = 4$

2ND UNKNOWN. To find the second unknown, we subtract one from the third entry in the third column, which yields $3 - 1 = 2$. We then multiply the first entry in the first column by the result and add one, giving the value $2 \times 1 + 1 = 3$,

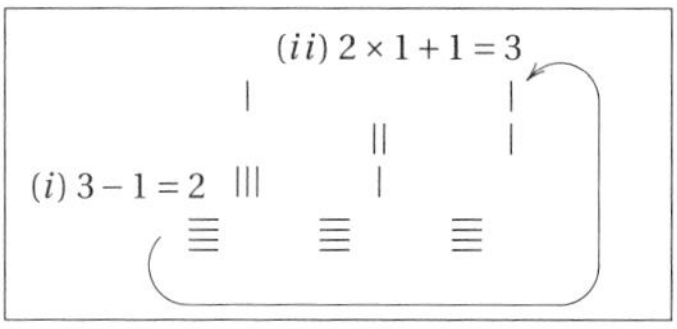

(*ii*) $2 \times 1 + 1 = 3$
(*i*) $3 - 1 = 2$

3RD UNKNOWN. To find the third unknown, we subtract one from the first entry in the first column, giving $1 - 1 = 0$. We then multiply the second entry in the second column by the result and add one, giving the value $0 \times 2 + 1 = 1$,

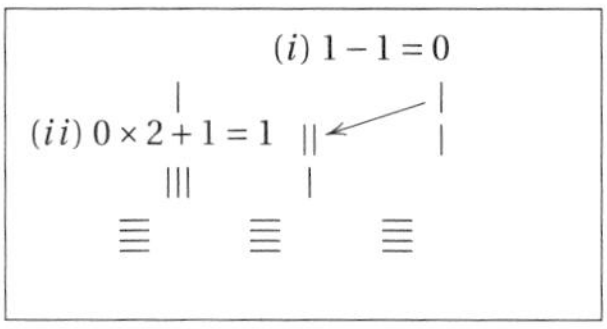

(*i*) $1 - 1 = 0$
(*ii*) $0 \times 2 + 1 = 1$

FINAL STEP. Finally, we must multiply each of the results above by b and divide by d, giving the values for the three unknowns,

$$x_1 = (4 \times 40) \div 7 = \frac{160}{7}, \quad x_2 = (3 \times 40) \div 7 = \frac{120}{7}, \quad x_3 = (1 \times 40) \div 7 = \frac{40}{7}.$$

In sum, as noted above, from the point of view of explaining the *fangcheng* procedure, problem 12 is simply superfluous. However, if this is solved by determinants, problem 12 does exemplify important technical points. First, as noted above, the constant terms for problem 12 are neither 1 nor det A, and thus, in contrast with the solutions for problems 3 (equations 8.19 to 8.21 on page 163) and 13 (equations 8.13 to 8.17 on page 160), problem 12 requires the final multiplication of the results by b and division by d. Problem 12 substitutes the sequence of 1, 2, and 3 on the diagonal for 2, 3, and 4 in problem 3 (equation 8.18), and the calculation for the third unknown is then $(1-1)\times 2+1=1$, and so is the only example of 0 appearing in these calculations (the exception noted in item 9 on page 162).

Problem 14

Problem 14 is an example of a more complicated variant of the form of *fangcheng* problem given by equation (7.1):

Translation of Problem 14

Given [a field of] white rice [measuring] 2 [square] paces,[11] [a field of] green rice [measuring] 3 [square] paces, [a field of] yellow rice [measuring] 4 [square] paces, [a field of] black rice [measuring] 5 [square] paces, the grain from each [field] does not yield 1 *dou.*[12] When [the yield of] one [square] pace of the green [rice field] and the yellow [rice field] is added to [the total yield of the field of] the white [rice], when [the yield of] one [square] pace of the yellow [rice field] and the black [rice field] is added to [the total yield of the field of] the green [rice], when [the yield of] one [square] pace of the black [rice field] and the white [rice field] is added to [the total yield of the field of] the yellow [rice], when [the yield of] one [square] pace of the white [rice field] and the green [rice field] is added to [the total yield of the field of] the black [rice field], the [total amount of] grain fills one *dou*. Problem: what is [the yield of] one [square] pace of each [of the fields of] white, green, yellow, and black rice?

今有白禾二步。青禾三步。黄禾四步。黑禾五步。禾實各不滿斗。白取青、黄。青取黄、黑。黄取黑、白。黑取白、青。各一步。而實滿斗。問白、青、黄、黑禾實一步各幾何？

---

[11] One pace (*bu* 步) is two steps, measured from the heel of one foot to the heel of the same foot. More precisely, whereas during the Zhou Dynasty one pace was 8 *chi* 尺, following the Qin Dynasty the length of a pace was 6 *chi*, or approximately $6\times 23.1\,cm = 1.386$ m.

[12] Again, 1 *dou* is approximately equal to 2 dry liters—see note 45 on page 81.

Solution: One [square] pace of white rice yields $\frac{33}{111}$ [= $\frac{11}{37}$][13] *dou* of grain, one [square] pace of green rice yields $\frac{28}{111}$ *dou* of grain, one [square] pace of yellow rice yields $\frac{17}{111}$ *dou* of grain, and one [square] pace of black rice yields $\frac{10}{111}$ *dou* of grain.

答曰：白禾一步實一百一十一分斗之三十三。青禾一步實一百一十一分斗之二十八。黃禾一步實一百一十一分斗之一十七。黑禾一步實一百一十一分斗之一十。

Method: Use the *fangcheng* procedure, place what is taken in each case, and use the procedure for positive and negative [numbers].

術曰：如方程，各置所取，以正負術入之。

Elimination Following the Fangcheng *Procedure*

Elimination using the *fangcheng* procedure is as follows:

STEP 1. First we place the counting rods as follows:

$$\begin{bmatrix} 2 & 1 & 1 & 0 & 1 \\ 0 & 3 & 1 & 1 & 1 \\ 1 & 0 & 4 & 1 & 1 \\ 1 & 1 & 0 & 5 & 1 \end{bmatrix}.$$

STEP 2. We then proceed by eliminating entries, starting with the first column [row], proceeding from the second column [row], and going left [down]. The first entry in the second column [row] is already 0, so we begin with the third column [row]. We multiply it by 2 and subtract the first column [row], yielding $2 \cdot (1,0,4,1,1) - (2,1,1,0,1) = (0,-1,7,2,1)$,

$$\begin{bmatrix} 2 & 1 & 1 & 0 & 1 \\ 0 & 3 & 1 & 1 & 1 \\ 0 & -1 & 7 & 2 & 1 \\ 1 & 1 & 0 & 5 & 1 \end{bmatrix}.$$

STEP 3. Then we eliminate the first entry of the final column [row]. We multiply the final column [row] by 2 and subtract the first column [row], giving $2 \cdot (1,1,0,5,1) - (2,1,1,0,1) = (0,1,-1,10,1)$,

$$\begin{bmatrix} 2 & 1 & 1 & 0 & 1 \\ 0 & 3 & 1 & 1 & 1 \\ 0 & -1 & 7 & 2 & 1 \\ 0 & 1 & -1 & 10 & 1 \end{bmatrix}.$$

[13] This fraction is not reduced in the *Nine Chapters*.

STEP 4. Next we eliminate entries in the third column [row]. We multiply the third column [row] by 3 and add the second column [row], yielding the result $3\cdot(0,-1,7,2,1)+(0,3,1,1,1,1)=(0,0,22,7,4)$,

$$\begin{bmatrix} 2 & 1 & 1 & 0 & 1 \\ 0 & 3 & 1 & 1 & 1 \\ 0 & 0 & 22 & 7 & 4 \\ 0 & 1 & -1 & 10 & 1 \end{bmatrix}.$$

STEP 5. Multiply the fourth column [row] by 3 and subtract the second column [row], giving $3\cdot(0,1,-1,10,1)-(0,3,1,1,1)=(0,0,-4,29,2)$,

$$\begin{bmatrix} 2 & 1 & 1 & 0 & 1 \\ 0 & 3 & 1 & 1 & 1 \\ 0 & 0 & 22 & 7 & 4 \\ 0 & 0 & -4 & 29 & 2 \end{bmatrix}.$$

STEP 6. We multiply the fourth column [row] by 22, and add four times the third column [row], $22\cdot(0,0,-4,29,2)+4\cdot(0,0,22,7,4)=(0,0,0,666,60)$,

$$\begin{bmatrix} 2 & 1 & 1 & 0 & 1 \\ 0 & 3 & 1 & 1 & 1 \\ 0 & 0 & 22 & 7 & 4 \\ 0 & 0 & 0 & 666 & 60 \end{bmatrix}.$$

Back Substitution Following the Fangcheng *Procedure*

Although back substitution is somewhat more complicated in this example than in problems 3 or 12, it is considerably less complicated here than it is in other problems included in chapter 8 of the *Nine Chapters*. So the reason for the inclusion of problem 14 in chapter 8 of the *Nine Chapters* is not because of this back substitution.

STEP 7. We begin by "cross multiplying" the final entry in the third column by the fourth entry in the final column, $4\times 666=2664$, and the fourth entry in the third column by the final entry in the fourth column, $7\times 60=420$:

$$\begin{bmatrix} 2 & 1 & 1 & 0 & 1 \\ 0 & 3 & 1 & 1 & 1 \\ 0 & 0 & 22 & 420 & 2664 \\ 0 & 0 & 0 & 666 & 60 \end{bmatrix}.$$

STEP 8. We then subtract the fourth entry in the third column from the final entry in the third column, $2664-420=2244$, and divide the result by the third entry in the third column, yielding $2244\div 22=102$:

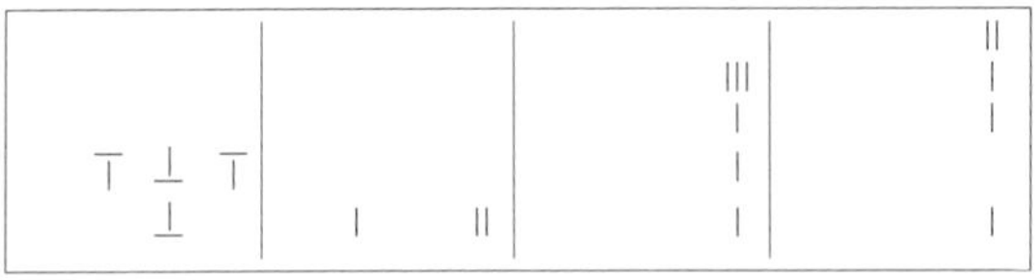

$$\begin{bmatrix} 2 & 1 & 1 & 0 & 1 \\ 0 & 3 & 1 & 1 & 1 \\ 0 & 0 & 0 & 0 & 102 \\ 0 & 0 & 0 & 666 & 60 \end{bmatrix}.$$

STEP 9. The next step is to operate on the second column. We "cross multiply" the final entry of the second column by the fourth entry in the final column, $1 \times 666 = 666$, and the fourth entry in the second column by the final entry in the final column, $1 \times 60 = 60$, together with the third entry in the second column by the final entry in the third column, $1 \times 102 = 102$:

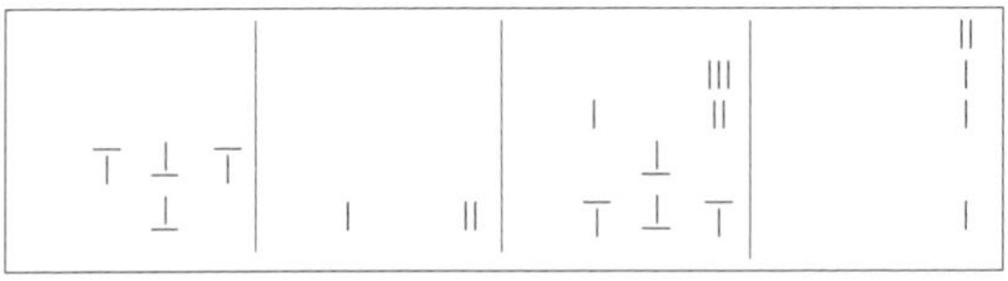

$$\begin{bmatrix} 2 & 1 & 1 & 0 & 1 \\ 0 & 3 & 102 & 60 & 666 \\ 0 & 0 & 0 & 0 & 102 \\ 0 & 0 & 0 & 666 & 60 \end{bmatrix}.$$

STEP 10. We subtract from the final entry in the second column the third and fourth entries in the second column, $666 - 102 - 60 = 504$. We then divide this result by the second entry in the second column, yielding $504 \div 3 = 168$,

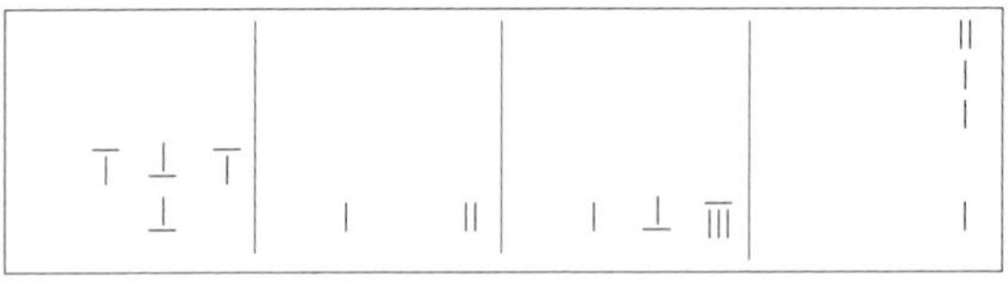

$$\begin{bmatrix} 2 & 1 & 1 & 0 & 1 \\ 0 & 0 & 0 & 0 & 168 \\ 0 & 0 & 0 & 0 & 102 \\ 0 & 0 & 0 & 666 & 60 \end{bmatrix}.$$

STEP 11. In the next step, we simplify the first column [row]. We "cross multiply" the final entry in the first column by the fourth entry in the final column, $1 \times 666 = 666$, the third entry in the first column by the final entry in the third column, $1 \times 102 = 102$, and the second entry in the first column by the final entry in the second column, $1 \times 168 = 168$,

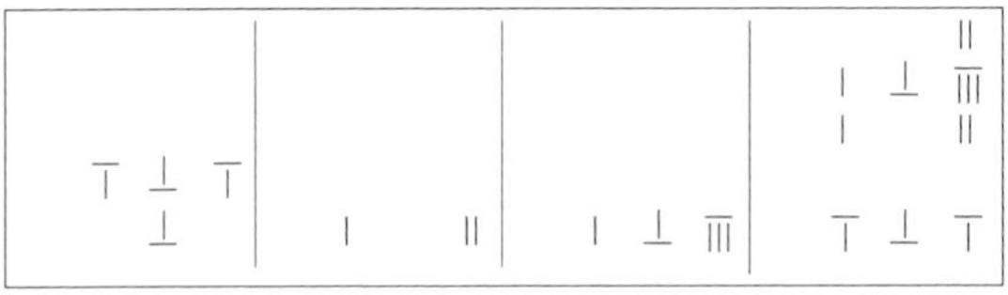

$$\begin{bmatrix} 2 & 168 & 102 & 0 & 666 \\ 0 & 0 & 0 & 0 & 168 \\ 0 & 0 & 0 & 0 & 102 \\ 0 & 0 & 0 & 666 & 60 \end{bmatrix}.$$

STEP 12. We then subtract from the final entry in the first column the second and third entries in the first column, $666 - 102 - 168 = 396$. We then divide the result by the first entry in the first column, yielding $396 \div 2 = 198$:

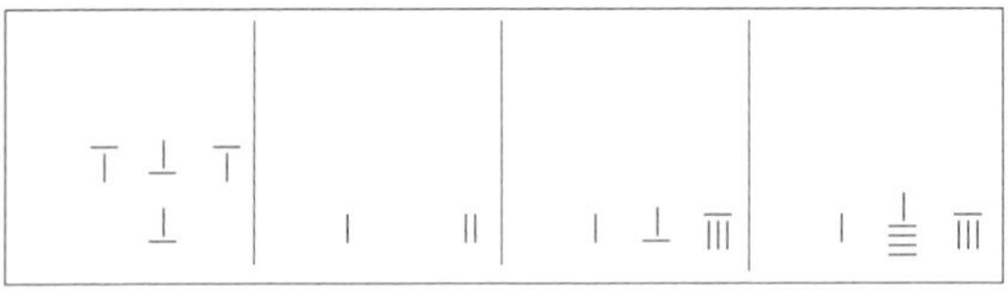

$$\begin{bmatrix} 0 & 0 & 0 & 0 & 198 \\ 0 & 0 & 0 & 0 & 168 \\ 0 & 0 & 0 & 0 & 102 \\ 0 & 0 & 0 & 666 & 60 \end{bmatrix}.$$

STEP 13. We then divide the final entry in each column by the fourth entry in the final column, 666,

$$\begin{bmatrix} 0 & 0 & 0 & 0 & 198/666 \\ 0 & 0 & 0 & 0 & 168/666 \\ 0 & 0 & 0 & 0 & 102/666 \\ 0 & 0 & 0 & 0 & 60/666 \end{bmatrix}.$$

STEP 14. Finally, we reduce the fractions,[14] giving the following values for the four unknowns:

$$\begin{bmatrix} 0 & 0 & 0 & 0 & 33/111 \\ 0 & 0 & 0 & 0 & 28/111 \\ 0 & 0 & 0 & 0 & 17/111 \\ 0 & 0 & 0 & 0 & 10/111 \end{bmatrix}.$$

Determinantal Solutions to Problem 14

We can generalize this as follows, as a variant of the *fangcheng* problem of the form given in equation (7.1) with 4 equations in 4 unknowns, identical except that instead of being zero, the coefficients $a_{13} = a_{24} = a_{31} = a_{42} = 1$,

$$\begin{bmatrix} k_1 & 1 & 1 & 0 & b \\ 0 & k_2 & 1 & 1 & b \\ 1 & 0 & k_3 & 1 & b \\ 1 & 1 & 0 & k_4 & b \end{bmatrix}. \tag{8.25}$$

Again, it is easy to check that the solution is given by the following:

$$x_1 = \frac{b(k_2 + k_4 - k_2k_4 - k_3k_4 + k_2k_3k_4)}{k_1 + k_2 + k_3 + k_4 - k_1k_3 - k_2k_4 + k_1k_2k_3k_4},$$
$$x_2 = \frac{b(k_1 + k_3 - k_1k_3 - k_1k_4 + k_1k_3k_4)}{k_1 + k_2 + k_3 + k_4 - k_1k_3 - k_2k_4 + k_1k_2k_3k_4},$$
$$x_3 = \frac{b(k_2 + k_4 - k_1k_2 - k_2k_4 + k_1k_2k_4)}{k_1 + k_2 + k_3 + k_4 - k_1k_3 - k_2k_4 + k_1k_2k_3k_4},$$
$$x_4 = \frac{b(k_1 + k_3 - k_1k_3 - k_2k_3 + k_1k_2k_3)}{k_1 + k_2 + k_3 + k_4 - k_1k_3 - k_2k_4 + k_1k_2k_3k_4},$$

where the denominator is the determinant of the coefficient matrix,

$$\det A = k_1 + k_2 + k_3 + k_4 - k_1k_3 - k_2k_4 + k_1k_2k_3k_4.$$

[14] Again, note that in the original *Nine Chapters*, $\frac{198}{666}$ is reduced to $\frac{33}{111}$ and not to $\frac{11}{37}$.

Surprisingly, although the numerator appears complicated, it is again easy to compute, and still would be easy to compute in the head, without manipulating the counting rods. That is, as was the case in the well problem, we can rearrange this solution into a much simpler form,

$$x_1 = ((k_4(k_3 - 1)(k_2 - 1) + k_2) \cdot b) \div d, \tag{8.26}$$

$$x_2 = ((k_1(k_4 - 1)(k_3 - 1) + k_3) \cdot b) \div d, \tag{8.27}$$

$$x_3 = ((k_2(k_1 - 1)(k_4 - 1) + k_4) \cdot b) \div d, \tag{8.28}$$

$$x_4 = ((k_3(k_2 - 1)(k_1 - 1) + k_1) \cdot b) \div d, \tag{8.29}$$

writing for the determinant of the coefficient matrix $\det A = d$. Though this solution is more complicated than that for the well problem, the steps are similar to those for the well problem, and the solution shares many of its features.

Determinantal Solutions on the Counting Board

The following diagrams illustrate how this might be calculated on a counting board.

1ST UNKNOWN. To calculate the value of the first unknown, we begin with the entry immediately following, proceeding this time in a clockwise direction. We take that entry [step (*i*)], multiply it by the result of 1 subtracted from the following entry [step (*ii*)], multiply that by the result of 1 subtracted from the following entry and to that result add that entry [step (*iii*)], giving 33. Finally, since $b = 1$, we divide by $d = 111$ to give the first unknown, $\frac{33}{111} = \frac{11}{37}$,

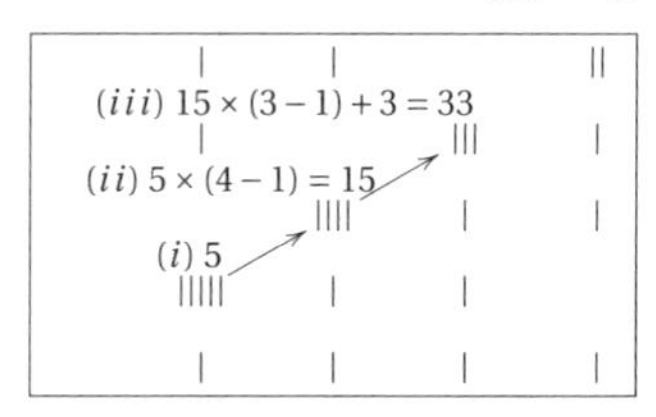

2ND UNKNOWN. To calculate the value of the second unknown, we begin with the entry for the first unknown, and using exactly the same procedure, we arrive at 28. Then we divide by $d = 111$ to give the second unknown, $\frac{28}{111}$.

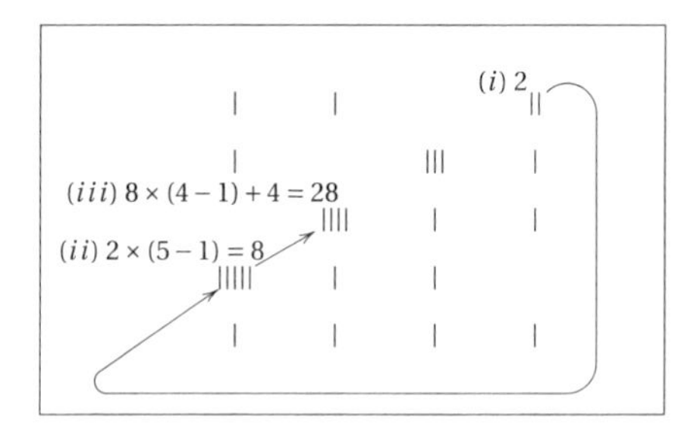

3RD UNKNOWN. To calculate the value for the third unknown, again we follow the same procedure, resulting in 17, and dividing by $d = 111$, we have $\frac{17}{111}$.

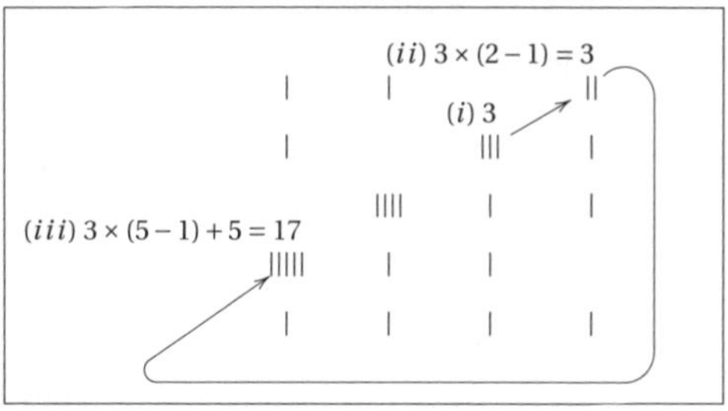

(i) 3

(ii) $3 \times (2-1) = 3$

(iii) $3 \times (5-1) + 5 = 17$

4TH UNKNOWN. The calculation for the fourth unknown again follows the same procedure, yielding 10, then divide by 111, giving $\frac{10}{111}$.

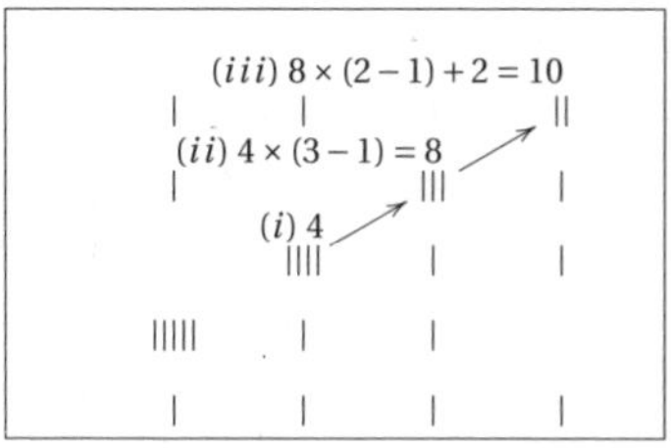

(i) 4

(ii) $4 \times (3-1) = 8$

(iii) $8 \times (2-1) + 2 = 10$

Summary

Problem 14 is the most significant variant of the basic form of the *fangcheng* problem represented by equation (7.1) that we have been exploring, and so we should compare its features with those listed for the well problem (items 1 to 10 on page 162):

1. Though the procedure for calculation in problem 14 is not as simple as that in the well problem, it is the same for all the unknowns.
2. Though this procedure is valid for any *fangcheng* problem of the form represented in equation (8.25), this result is not valid in general for n equations in n unknowns, and not even for 5 equations in 5 unknowns.
3. The calculations again involve only positive integers—zero and negative numbers cannot emerge (assuming $k_i \geq 2$).
4. Though the calculations here are more difficult than those for the well problem, they are still considerably simpler than solving the problem by elimination and back substitution in the *fangcheng* procedure.

Problem 15

Problem 15 is a second important variant of the form of augmented matrix given in equation (7.1). Again, we will see that from the point of view of the *fangcheng* procedure, the inclusion of this problem in chapter 8 of the *Nine Chapters* is superfluous. But from the point of view of determinants, problem 15 is an exemplar for another class of determinantal solutions.

Translation of Problem 15

Given 2 bundles of [grade] A paddy, 3 bundles of [grade] B paddy, and 4 bundles of [grade] C paddy, each weighs more than 1 *dan*.[15] 2 bundles of [grade] A paddy exceed the weight [of 1 *dan*] by [an amount equal to the weight of] 1 bundle of [grade] B paddy; 3 bundles of [grade] B paddy exceed the weight [of 1 *dan*] by [an amount equal to the weight of] 1 bundle of [grade] C paddy; 4 bundles of [grade] C paddy exceed the weight [of 1 *dan*] by [an amount equal to the weight of] 1 bundle of [grade] A paddy. Problem: How much does 1 bundle each of [grades] A, B, and C paddy weigh?

今有甲禾二秉、乙禾三秉、丙禾四秉，重皆過於石。甲二重如乙一，乙三重如丙一，丙四重如甲一。問甲、乙、丙禾一秉各重幾何。

Solution: One bundle of [grade] A paddy weighs $\frac{17}{23}$ *dan*, one bundle of [grade] B paddy weighs $\frac{11}{23}$ *dan*, and one bundle of [grade] C paddy weighs $\frac{10}{23}$ *dan*.

荅曰：甲禾一秉重二十三分石之一十七，乙禾一秉重二十三分石之一十一，丙禾一秉重二十三分石之一十。

Method: Use the *fangcheng* procedure; set [the number corresponding to] the weight by which the object exceeds 1 *dan* to be negative.

術曰：如方程，置重過於石之物為負。

Elimination Following the Fangcheng *Procedure*

STEP 1. The initial placement of counting rods is shown in the following diagram:

<table>
<tr><td>|</td><td></td><td>||</td></tr>
<tr><td></td><td>|||</td><td>|</td></tr>
<tr><td>||||</td><td>|</td><td></td></tr>
<tr><td>|</td><td>|</td><td>|</td></tr>
</table>

$$\begin{bmatrix} 2 & -1 & 0 & 1 \\ 0 & 3 & -1 & 1 \\ -1 & 0 & 4 & 1 \end{bmatrix}.$$

STEP 2. First we multiply the third column [row] by 2, and add the first column [row], $2 \cdot (-1, 0, 4, 1) + (2, -1, 0, 1) = (0, -1, 8, 3)$,

[15] Again, 1 *dan* ≈ 30 kilograms.

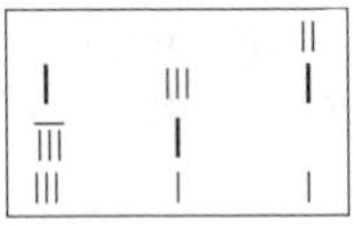

$$\begin{bmatrix} 2 & -1 & 0 & 1 \\ 0 & 3 & -1 & 1 \\ 0 & -1 & 8 & 3 \end{bmatrix}.$$

STEP 3. Then we multiply the third column [row] by 3, and add the second column [row], $3 \cdot (0, -1, 8, 3) + (0, 3, -1, 1) = (0, 0, 23, 10)$,

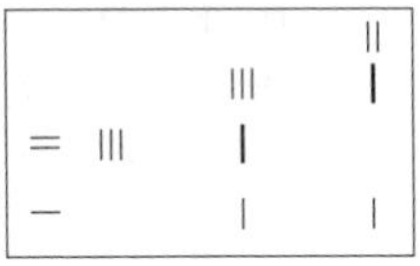

$$\begin{bmatrix} 2 & -1 & 0 & 1 \\ 0 & 3 & -1 & 1 \\ 0 & 0 & 23 & 10 \end{bmatrix}.$$

Back Substitution Following the Fangcheng *Procedure*

Again, in the process of elimination following the *fangcheng* procedure we found no pedagogical reason for including problem 15. We will find little instructive in the calculations for back substitution: though it is true that we must deal with negative numbers in the process of back substitution, this is not the only problem for which this is true.

STEP 4. We begin back substitution by operating on the second column. We "cross multiply" the final entry in the second column by the third entry in the final column, yielding $23 \times 1 = 23$. We then "cross multiply" the third entry in the second column by the final entry in the final column, $1 \times -10 = -10$,

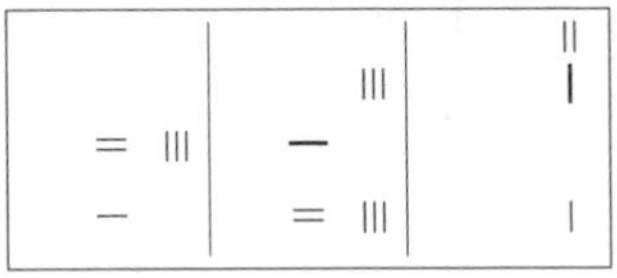

$$\begin{bmatrix} 2 & -1 & 0 & 1 \\ 0 & 3 & -10 & 23 \\ 0 & 0 & 23 & 10 \end{bmatrix}.$$

STEP 5. Next we subtract the third entry in the second column from the final entry in the second column, $23 - (-10) = 33$. We then divide the result by the second entry in the second column, $33 \div 3 = 11$:

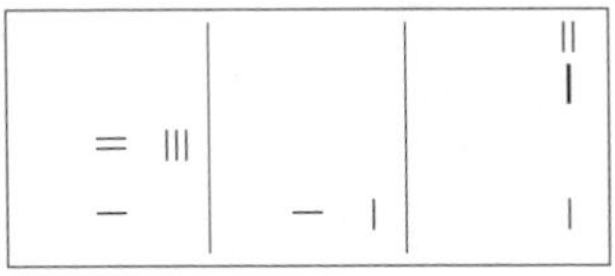

$$\begin{bmatrix} 2 & -1 & 0 & 1 \\ 0 & 0 & 0 & 11 \\ 0 & 0 & 23 & 10 \end{bmatrix}.$$

STEP 6. Now we operate on the first column. We multiply the final entry in the first column by the third entry in the final column, $23 \times 1 = 23$, and the second entry in the first column by the final entry in the second column, $-1 \times 11 = -11$:

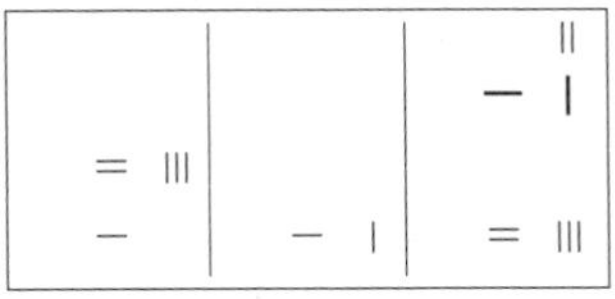

$$\begin{bmatrix} 2 & -11 & 0 & 23 \\ 0 & 0 & 0 & 11 \\ 0 & 0 & 23 & 10 \end{bmatrix}.$$

STEP 7. Then we subtract from the final entry in the first column the second entry in the first column, $23-(-11)=34$, and then divide the result by the first entry in the first column, $34 \div 2 = 17$,

$$\begin{bmatrix} 0 & 0 & 0 & 17 \\ 0 & 0 & 0 & 11 \\ 0 & 0 & 23 & 10 \end{bmatrix}.$$

STEP 8. The last step is then to divide the final entries in each column by the third entry in the final column, 23, giving the three unknowns:

$$\begin{bmatrix} 0 & 0 & 0 & 17/23 \\ 0 & 0 & 0 & 11/23 \\ 0 & 0 & 0 & 10/23 \end{bmatrix}.$$

Determinantal Solutions to Problem 15

The class of problems for which this is an exemplar is another variant of the form represented in equation (7.1), with the alteration that the superdiagonal coefficients $a_{i,i+1} = -1$, and $a_{n1} = -1$, instead of $+1$,

$$\begin{bmatrix} k_1 & -1 & 0 & b \\ 0 & k_2 & -1 & b \\ -1 & 0 & k_3 & b \end{bmatrix}. \tag{8.30}$$

If we write $d = \det A = k_1 k_2 k_3 - 1$, since $b_i = 1$, the solution for equation (8.30) is

$$x_1 = \frac{1}{d}(k_2 k_3 + k_3 + 1),$$
$$x_2 = \frac{1}{d}(k_1 k_3 + k_1 + 1),$$
$$x_3 = \frac{1}{d}(k_1 k_2 + k_2 + 1),$$

or alternatively,

$$x_1 = \big((k_2+1)k_3+1\big) \div d, \tag{8.31}$$
$$x_2 = \big((k_3+1)k_1+1\big) \div d, \tag{8.32}$$
$$x_3 = \big((k_1+1)k_2+1\big) \div d. \tag{8.33}$$

Determinantal Solutions on the Counting Board

The solution to this variant form is particularly simple, and satisfies all the features listed for the well problem, items 1 to 10 on page 162. Again, this is quite easy to solve on the counting board.

1ST UNKNOWN. To find the first unknown,

(*i*) $3+1=4$
(*ii*) $4\times4+1=17$

2ND UNKNOWN. To find the second unknown,

(*ii*) $5\times2+1=11$
(*i*) $4+1=5$

3RD UNKNOWN. To find the third unknown,

(*i*) $2+1=3$
(*ii*) $3\times3+1=10$

Conclusions

In conclusion, there can be little doubt that these five problems—problems 3, 12, 13, 14, and 15—from chapter 8 of the *Nine Chapters* were known to have solutions by methods other than the *fangcheng* procedure, most likely by determinantal methods similar to the entry-by-entry cross multiplication in "excess and deficit" problems.

Chapter 9
Conclusions

The central thesis of this book is that it was the visualization in two dimensions of problems with n conditions in n unknowns that led to the discovery of solutions—solutions similar to those used today in modern linear algebra. This book attempts to reconstruct these mathematical practices, performed on the counting board so long ago, from narrative descriptions preserved in extant treatises on the mathematical arts of imperial China. In the preceding chapters of this book, I presented this material in a manner more expository than chronological, presenting first the necessary background material (chapters 1 to 3), then analyzing the actual practices for which we have explicit records (chapters 4 to 7), and finishing with further reconstructions based on those recorded practices (chapter 8). In this concluding chapter, below, the first section summarizes the early history of these developments chronologically, the second outlines areas for further research, the third discusses methodological issues, and the fourth section explores the significance of these findings.

The Early History of Linear Algebra

This book demonstrates that the early history of what today is called linear algebra differs considerably from what we might have expected to find. Extant records of mathematical arts in imperial China show that solutions to problems with n conditions in n unknowns, using what we would now call augmented matrices, Gaussian elimination, and determinantal-style calculations, date back to at least the first century C.E. The central purpose of this book has been to reconstruct these mathematical practices, and to demonstrate that they are considerably more varied and sophisticated than has previously been understood. Here I shall summarize the results of this book chronologically, with a view toward contributing to a revised history of these early developments.

The **Book of Computation**, *c. 186* B.C.E.

The earliest extant source from which we can reconstruct mathematical practices in China is the *Book of Computation.* As we saw in chapter 3, it is a funerary good written on bamboo strips, which has important implications for the reconstruction of early mathematical practices in China. On the positive side, the text comes to us directly without intervening alterations, editing, or revisions; the date 186 B.C.E. can be assigned with some certainty as the *terminus ante quem* for the practices recorded therein. There are also, however, limitations: there is no information on the authors or compilers; there is no preface, introduction, or postface; little effort was made to organize the problems; there is no commentary or subcommentary; some bamboo strips are missing, and some may be out of place; and there are numerous mistakes and scribal errors in the statements of the problems and solutions. Despite these shortcomings, the *Book of Computation* preserves sufficient evidence from which we can reliably reconstruct ancient mathematical practices, practices that may date from as much as several centuries prior to its compilation.

The visualization—in two dimensions, on the counting board—of problems with 2 conditions in 2 unknowns is documented in the *Book of Computation,* as we saw in chapter 4. This is the earliest known record of such solutions. The problem titled "Dividing Coins," for example, explicitly instructs us to "cross multiply" (*hu cheng* 互乘) individual entries on the counting board, and add them together to find the solution. The solution is in some ways analogous to adding fractions. This problem explicitly employs the terms "excess and deficit," suggesting continuity with practices recorded in later treatises on the mathematical arts, where solutions to "excess and deficit" problems are given a fuller explanation. Solutions preserved in the *Book of Computation* are, however, limited to problems with 2 conditions in 2 unknowns.

The **Nine Chapters on the Mathematical Arts**, *c. 100* C.E.

The most important source for reconstructing mathematical practices in early imperial China is the *Nine Chapters on the Mathematical Arts,* which is examined in chapter 3 of this book. It is the earliest transmitted text focused on mathematics. The provenance of the material preserved in extant editions is, however, not entirely clear. There is little reliable information on the authorship, composition, or dates of the material (most of the information we have about the text is from the preface by Liu Hui, which provides a history of the *Nine Chapters,* a history likely written the for the purpose of obtaining imperial patronage). The text has been reconstructed from fragments from various editions. Despite all this, as with the *Book of Computation,* the material preserved in extant editions of the *Nine Chapters* is sufficient to reliably reconstruct mathematical practices in early imperial China. The original text can be tentatively dated at about 100 C.E., and the earliest extant commentary, attributed to Liu Hui, is generally accepted to date from 263 C.E. The practices recorded in the original text and commentary may date from several centuries before their compilation.

From the material preserved in the *Nine Chapters* we can reconstruct the following mathematical practices, visualized in two dimensions on the counting board, as analyzed in detail in chapters 4 to 8 of this book: (1) solutions of problems with 2 conditions in 2 unknowns, explained more fully than in the *Book of Computation*; (2) the display in two dimensions of problems with n conditions in n unknowns; (3) the elimination of entries by the cross multiplication of columns by individual entries; (4) a sophisticated approach to back substitution that in most cases avoids computations with fractions; (5) determinantal-style calculations for special problems with n conditions in $(n+1)$ unknowns, used to assign a value to the $(n+1)^{\text{th}}$ unknown; and (6) determinantal-style solutions to special problems with n conditions in n unknowns. I shall summarize these practices individually below.

Visualization in Two Dimensions of Problems with 2 Conditions in 2 Unknowns

The *Nine Chapters* affords us further information by which to reconstruct the mathematical practices for solving problems with 2 conditions in 2 unknowns by visualizing them in two dimensions on the counting board. As we saw in chapter 4 of this book, in the *Nine Chapters*, "excess and deficit" serves as the name of a category of problems, as the title of a chapter in the *Nine Chapters*, and as the name of the main procedure for finding their solution. This procedure explicitly describes displaying the entries on the counting board in two dimensions, and "cross multiplying" (*wei cheng* 維乘) the entries. These methods are presented in more detail in the *Nine Chapters* than found in the *Book of Computation*.

The "Excess and Deficit" chapter of the *Nine Chapters* appears to have been compiled earlier than the *Fangcheng* chapter, as suggested by the limitations of the "excess and deficit" problems it presents: (1) There are no problems with more than 2 conditions in 2 unknowns. (2) Negative numbers are not used. (3) As a result, there is no general method—different methods are presented for positive and negative numbers, such as the "double excess" and "double deficit" procedures.

Although "excess and deficit" problems were often solved by visualizing them in two dimensions on the counting board, this visualization is not a prerequisite for solving these problems, which are relatively simple compared to the later *fangcheng* problems. Other methods of solution are recorded in the *Book of Computation* and the "Excess and Deficit" chapter of the *Nine Chapters*, and some of these problems may have been solved in the head by esoteric procedures. It is in the extension to problems with n conditions in n unknowns that this visualization in two dimensions on the counting board takes on a more crucial role.

Visualization in Two Dimensions of Problems with n Conditions in n Unknowns

Solutions to problems with n conditions in n unknowns ($n \geq 3$)—the focus of chapters 5 to 8 of this book—begin with their display as entries in a two-dimensional array on the counting board. Problems thus displayed were termed *fangcheng*. The term *fang* means rectangular, and the term *cheng* apparently means "measures," a set of numbers (in this case arranged in a column) that are to remain in balance or proportion: columns can be multiplied by a number, entry by entry; and one column can be subtracted from another, entry by entry. The term *fangcheng* serves as the name of a category of problems, as the name of a procedure provided for their solution, and as the title of chapter 8 of the *Nine Chapters*. *Fangcheng*, displayed as arrays of numbers, are equivalent to what in modern linear algebra are called augmented matrices. The only difference between the two is the writing system—the symbols used (counting rods or Chinese characters instead of Arabic numerals) and the ordering of the writing systems (Chinese writing proceeds from top to bottom then right to left, in contrast to European languages, which proceed from left to right and then top to bottom). This two-dimensional visualization of these problems led to a multiplicity of solutions.

In the *Nine Chapters*, immediately following the first *fangcheng* problem, a general solution for problems with n conditions in n unknowns, termed the *fangcheng* procedure, is recorded. All eighteen problems there are followed by the instruction that they are to be solved by this procedure, and this procedure succeeds for all of the 18 problems recorded in the chapter. It is, in fact, valid for all problems with n conditions in n unknowns for which there is a unique solution. The procedure can be analyzed as comprising two parts, which in modern linear algebra are termed elimination and back substitution.

Elimination of Entries on the Counting Board

It is the visualization in two dimensions of problems of n conditions in n unknowns that permits the expedient elimination of entries, as we saw in chapter 5. We eliminate the first entry of the second column by cross multiplication (*bian cheng* 徧乘) and subtraction, and continue eliminating entries until we reach what we would now call triangular form. This first part of the *fangcheng* procedure is just what we now call Gaussian elimination (more precisely, one variant of it).

The *fangcheng* procedure recorded in the *Nine Chapters* is abbreviated, so much so that the description is unintelligible without a prior understanding of the mathematical practices being described. The problems presented demonstrate how exemplary problems are solved, rather than explaining all that can go wrong, and the procedure fails to address many circumstances that would have been routinely encountered in practice (for example, there is no discussion of the fact that some problems will fail to have a unique solution, and others will fail to have any solution at all). The record of the procedure is, however, adequate to reconstruct at least the main outlines of the practice.

Back Substitution on the Counting Board

The method for back substitution in the *fangcheng* procedure recorded in the original text of the *Nine Chapters* is quite sophisticated, one of the central findings of this book. The key features for this method, as we saw in chapters 5 and 6 above, are the following: (1) it is calculated on the counting board by a series of cross multiplications; (2) it is valid in general; and (3) it avoids calculations with fractions.

This method has heretofore not been fully explained or analyzed. It is described only briefly in the original text of the *Nine Chapters*; it differs from the methods presented by Liu Hui in his commentary; and subsequent commentators often did not follow this method. Nor has this method been fully explained in modern studies of Chinese mathematics, which have often explained it using approaches familiar from modern linear algebra.

The most obvious approach to back substitution, familiar today from modern linear algebra, begins with the augmented matrix reduced by elimination to triangular form. A simple division then gives the nth unknown, which is then substituted back to find the the $(n-1)^{\text{th}}$ unknown, and the remaining unknowns are thus found successively. This is the most intuitive approach: the results found at the end of each step are the values of the unknowns. This explains why this approach is commonly used, not just in modern linear algebra, but also in extant Chinese treatises on the mathematical arts.

But in the original text of the *fangcheng* procedure, the approach to back substitution differs from the intuitive approach, and visualization is the key to understanding it. Visualized on the counting board, following simple patterns, all that is required is addition, subtraction, multiplication, and division of integers. This is arguably simpler than other approaches, but translated into narrative form in classical Chinese, or into modern mathematical terminology, it is difficult to understand.

This approach to back substitution is valid for any problem that has been successfully reduced to what we now call triangular form, as we saw in chapter 6. Extant Chinese texts offer little explanation of this approach, and modern studies of Chinese mathematics have not analyzed its validity. To confirm its validity, it was necessary to translate it into modern mathematical terminology. In modern terminology, it was equivalent to modern back substitution, but with the division of the unknowns by the final pivot postponed until the last step.

This approach to back substitution avoids calculations with fractions, as was shown in chapter 6 by further analysis of the *fangcheng* procedure using modern mathematics. This mathematical analysis yielded the following two results:

1. By using modern mathematical terminology to compute the results at each stage of the *fangcheng* procedure, we found that in most cases, fractions

would be avoided. More precisely, we found that as long as each pivot was used at least once to cross multiply a higher column [row], fractions would not emerge until the last step.

2. By creating at random augmented matrices similar to those found in the *Nine Chapters*, we found that fractions emerge in only a small percentage of cases (on the same order as the percentage of problems that fail to have unique solutions).

On the counting board, on which the two dimensions were already used to display n conditions in n unknowns, computations with fractions would result in considerable complications. The *fangcheng* procedure avoided fractions.

In sum, then, this method for back substitution recorded in the original text of the *Nine Chapters* evinces considerable sophistication in these mathematical practices from early imperial China.

Positive and Negative Numbers

The procedure for positive and negative numbers is a second important procedure introduced in "*Fangcheng*," chapter 8 of the *Nine Chapters*. The *Book of Computation* does not use negative numbers; and "Excess and Deficit," chapter 7 of the *Nine Chapters*, does not mention negative numbers. The use of negative numbers is an important additional feature in the solution of *fangcheng* problems.

Determinantal-Style Calculations

Determinantal-style calculations (calculations, that is, made in a manner we might now call determinantal, but which do not provide solutions to all the unknowns) were likely known by the time of the compilation of the *Nine Chapters*. This is the second key finding offered by this book, presented in chapter 7. More specifically, evidence from the *Nine Chapters* suggests that determinantal calculations, used to find the the $(n+1)^{\text{th}}$ unknown for a special class of problems with n equations in $n+1$ unknowns, were known at the time of its compilation. The "well problem," problem 13 of the "*Fangcheng*" chapter, is an exemplar for these problems. It is formulated with 5 conditions in 6 unknowns, which does not determine a unique solution for all 6 unknowns. A value for the 6th unknown, the depth of the well, can be assigned by cross multiplying the individual entries and adding the results, which yields what we would now call the determinant of the coefficient matrix. The original text of the *Nine Chapters* assigns this value to the 6th unknown without any explanation of how it is found. A commentary on the well problem from 1261 C.E. records this calculation, which suggests that the value of the depth of the well was likely calculated in this manner by the 1st century C.E.

Determinantal-Style Solutions

Solutions to a special class of *fangcheng* problems with n equations in n unknowns, found by determinantal-style methods, were also likely known by the time of the compilation of the *Nine Chapters*, the third key finding presented in this book. Although there are no extant records describing these methods, the *Nine Chapters* contains considerable evidence of such solutions, as we saw in chapters 7 and 8 of this book. This evidence may be summarized as follows:

1. The visual form of these problems when displayed on the counting board is quite distinctive: (i) When these problems are laid out on the counting board, their forms exhibit visually distinctive patterns (each condition has at least one zero coefficient, all of the conditions have the same number of zeros, and the placement of the zeros changes in a cyclical fashion). (ii) In the *Nine Chapters*, these problems, and only these problems, have zeros as coefficients in the conditions. (iii) In these problems, and only these problems, the constant terms for the conditions are all the same. (iv) These problems are quite unusual—among thousands of matrices in modern works on linear algebra, there are only one or two examples of similar problems.
2. Alternative solutions with the following properties were shown to exist: (i) For each of these problems, the value of an unknown can be found by the application of a single simple rule, following patterns easily visualized on the counting board, consisting of multiplying or adding individual entries to intermediate results, repeatedly applied. (ii) For each problem, the corresponding rule produces solutions for all the unknowns. (iii) For each problem, the corresponding rule is valid for any problem with n conditions in n unknowns of the form for which the problem serves as an exemplar.[1] (iv) These calculations, which are similar to the cross multiplication of entries in "excess and deficit" problems, are relatively simple. (v) In contrast, solutions given by the *fangcheng* procedure are considerably more complex.
3. Five of the eighteen problems (problems 3, 12, 13, 14, and 15) from "*Fangcheng*," chapter 8 of the *Nine Chapters*, are all variants of one form, and share these properties.
4. The eighteen *fangcheng* problems recorded in the *Nine Chapters* presumably serve as exemplars, and there is no reason for these problems to have been included if they were solved only by the *fangcheng* procedure.
5. Evidence preserved in later texts further supports this finding: (i) In later texts, these problems continue to represent a substantial percentage of the *fangcheng* problems recorded. (ii) In later texts, the problems recorded are not just copied from early texts—variations continue to develop, suggesting a continuing practice, evolving over time. (iii) Later texts contain references to puzzling terminology and methods, suggesting alternative solutions. (iv) In

[1] The exception is the form for which problem 14 serves as an exemplar, which is valid only for 4 conditions in 4 unknowns.

particular, Mei Wending's "On *Fangcheng*" (c. 1674) records numerous examples of this form, including one problem with 9 conditions in 9 unknowns. Mei criticizes alternative methods used to solve these problems, methods he never delineates.

6. Fang Zhongtong's *Numbers and Measurement*, from about 1661, preserves an explicit record of one such determinantal solution.

The reconstructions I presented in chapter 8 of this book seem to me to offer the most plausible approach to solving these problems, but other solutions may also exist.

Yang Hui's Nine Chapters on the Mathematical Arts, with Detailed Explanations, *c. 1261*

The earliest extant record of a determinantal-style calculation comes from Yang Hui's *Nine Chapters on the Mathematical Arts, with Detailed Explanations*, which we can date from about 1261 C.E.; this material likely dates from *Jia Xian's Nine Chapters*, c. 1050, which is no longer extant.

Detailed Explanations records a category of problems termed "*fangcheng* with 'major terms' and 'minor terms,' " which are problems with n conditions in $n+1$ unknowns. In modern linear algebra, such problems are considered to be indeterminate, but *Detailed Explanations* demonstrates that these problems were not considered indeterminate at the time. An auxiliary procedure is provided: the $(n+1)^{\text{th}}$ unknown is assigned by multiplying together individual entries and then adding the results. In modern terms, this is just the determinant of the matrix of coefficients. The assignment of this value to the $(n+1)^{\text{th}}$ unknown transforms the problem into one with n conditions in n unknowns, which does have a unique solution (assuming the value assigned to the $(n+1)^{\text{th}}$ unknown is not zero). The solution to the remaining n unknowns can then found by the *fangcheng* procedure.

It is also possible that, for problems of this special form, the calculation of the $(n+1)^{\text{th}}$ unknown was recognized as a "determinant" in the modern sense. That is, as long as the value assigned to the $(n+1)^{\text{th}}$ unknown is not zero, a unique solution exists.

Fang Zhongtong's Numbers and Measurement, an Amplification, *c. 1661*

The earliest extant record of a determinantal-style solution is preserved in *Numbers and Measurement, an Amplification*, completed in about 1661. *Numbers and Measurement* is an eclectic compilation from disparate sources, including materials from popular treatises. It records a determinantal solution for a generalized form of the well problem (represented, in modern terms, by the augmented matrix 7.2 on page 112).

The record of this determinantal solution is preserved in narrative form in classical Chinese. Translated into classical Chinese, this method is almost impenetrable, and the diagrams provided do not explain the calculations in any detail. Entries on the counting board are referred to by their column and the unknown they represent (for example, "A in the fifth column"). The nth unknown is found by a two-part calculation. In modern terms, the result is the same as that given by Cramer's Rule: the first part yields what we would now call the determinant of B_n;[2] the second part yields the determinant of the coefficient matrix A. The calculations are described in detail, but in narrative form, which would be incomprehensible without prior knowledge of the practice. This narrative record is, however, sufficient to reconstruct the mathematical practice being described.

Visualization of the problem on the counting board is again the key to this solution. The solution is found by following simple patterns of cross multiplication on the counting board, using nothing more than the addition, subtraction, multiplication, and division of integers.

This method provides a general solution for all the unknowns, and for all problems of this form with n conditions in n unknowns ($n \geq 3$), as was shown in chapter 7 of this book. That is, although the solution is given only for the fifth unknown for the well problem, this is an exemplar for a more general method. Writing the formula in modern mathematical terms has permitted us to check its validity, but the resulting calculations, written in modern mathematical notation, are also quite complicated.

Questions for Further Research

The results presented in this book represent an initial inquiry into what we would now call linear algebra in imperial China. Areas that deserve additional study include further research into *fangcheng* practices, the social and historical context of these practices, and questions about their origins, dissemination, and transmission.

Further Analysis of **Fangcheng** *Practices*

One important area for future research is further analysis of *fangcheng* practices, including study of the following: (1) The *fangcheng* procedures presented by Liu Hui, author of the earliest extant commentary on the *Nine Chapters.* (2) *Fangcheng* problems from late imperial China, that is, from the fourteenth to seven-

[2] B_n is the matrix obtained by replacing the nth column of the coefficient matrix A with the solution vector $b = (b_i)$. See equation 2.10 on page 20.

teenth centuries, along with further work to locate rare editions of mathematical compilations from this period. (3) Mei Wending's "On *Fangcheng*" (c. 1674), evidently the most comprehensive extant compilation of *fangcheng* problems.

The Social and Historical Context

This study has focused primarily on reconstructing mathematical practices from extant treatises on mathematics. Another important area for future research is the search for further information on the social, historical, and political context of the practitioners. There may be records of these practitioners beyond the occasional (and often negative) statements preserved in treatises on the mathematical arts. An additional area of inquiry is the applications, if any, to which *fangcheng* methods were applied.

Origins, Dissemination, and Transmission

The fortuitous preservation of written records of these practices by literati-officials in imperial China does not, of course, prove that these practices originated there, though this seems to be the most likely hypothesis. These practices could have spread easily, since they do not require literacy, and they are not tied specifically to Chinese language or culture. It also seems possible that these practices could have been disseminated by merchants or travelers to other areas—perhaps to Japan or Korea, or along the Silk Routes. It is thus possible that these practices were more widespread than just within imperial China. Of particular interest is the possible transmission of these techniques to Japan and to Europe in the seventeenth century, and in particular the possibility that early Chinese work on determinants is the source for later work on determinants, the discovery of which is usually credited to Leibniz and Seki Takakazu.

Methodological Issues

Here I address broader methodological issues, including the importance of interdisciplinary research, and ways in which this research challenges some of the conventional assumptions about the history of mathematics.

An Interdisciplinary Approach

This inquiry into the history of mathematics in imperial China required that I pursue a highly interdisciplinary approach, combining approaches from recent work in the cultural history of science with more traditional approaches drawn from sinology and technical histories of the exact sciences. More specifically, this inquiry into *fangcheng* practices in imperial China combined the following elements: (1) compiling a comprehensive list of treatises on the mathematical arts recorded in Chinese bibliographies from throughout the imperial period; (2) collection of extant rare editions; (3) critical evaluation of extant editions; (4) textual analysis of variations in extant editions; (5) textual reconstructions, with the aid of modern critical editions, digitized corpora, and mathematical analysis; (6) translations of passages, commentaries, and philological notes into English, together with detailed explanatory footnotes; (7) close readings of prefaces to mathematical treatises in the context of authorship, credit, and patronage in the imperial court; (8) critical historical analysis of the production, legitimation, and preservation of mathematical treatises in historical context; (9) the reconstruction of mathematical practices from narrative descriptions preserved in extant texts; (10) mathematical analysis of *fangcheng* procedures and calculations using modern mathematics and computational tools; and (11) visual reconstructions of *fangcheng* using diagrams of calculations as performed on the counting board. The findings presented here required the integration of more technical approaches with those of cultural studies.

Mathematics and Texts

One implication of this study is that we should distinguish between mathematical practices, on the one hand, and written records of those practices translated into narrative form, on the other hand. Extant treatises on the mathematical arts from imperial China are written records of these practices, but rarely attempt to fully explain these practices in writing. Perhaps most significant is that there are no diagrams in the received versions of the earliest edition of the *Nine Chapters*, and there are relatively few diagrams in most later treatises. As a result, written records of these practices are often difficult to understand. The translation of these procedures into narrative form presumably reflects a preference of the literati for the beauty, difficulty, and prestige of classical Chinese writing.

We should also ask whether modern research on the history of Chinese mathematics might consider critically reflecting on its central focus—inherited perhaps from our literati predecessors—on the recovery of ancient classics and their meanings. To focus too narrowly on texts, without reconstructing the mathematical practices, may even leave the impression that Chinese mathematics is comparatively primitive, in contrast to mathematics found elsewhere in the world.

Authorship and Credit

Another implication of this study is that we might do well to be skeptical about assigning credit for these discoveries to the literati who sought to claim that credit for themselves, that is, those literati who, in their pursuit of sinecure, patronage, and office, presented compilations of mathematical arts to the imperial court. It is more likely that these arts were developed by numerous anonymous and illiterate adepts over many centuries. The literati who recorded *fangcheng* practices seem to have understood the basics, but existing records suggest that they did not in general understand the more esoteric calculations used in special cases.

Elite and Popular Cultures

We should also reconsider the assumption that only in relatively "higher" civilizations, and only within relatively elite subcultures, could such complex mathematics be invented and develop. Despite their mathematical complexity and sophistication, these arts required little more than a set of counting rods, along with instruction in the practice. The evidence presented here suggests that *fangcheng* was not a practice of the elite, and as noted above, the literati who recorded these practices seem not to have completely understood them, and in fact often wrote about its practitioners in a derisive manner. Furthermore, as noted above, we have no clear evidence that these practices originated in China, only that they were recorded and preserved in Chinese compilations. Such practices may easily have circulated across political and linguistic boundaries.

Significance and Implications

I conclude now with some brief reflections on the significance and broader implications of this study for the history of mathematics, the history of Chinese mathematics, and the world history of science.

Visualization and the History of Mathematics

One of the most surprising results of the investigation presented here is the importance of visualization. The essential feature in the solution of linear algebra problems is, I have argued, visualization. For it is the visualization of problems in two dimensions as an array of numbers on a counting board and the "cross-multiplication" of entries there that led to general solutions of systems of linear equations not found in Greek or early European mathematics. And ultimately, I have argued, this visualization is the key to reconstructing *fangcheng* practices: the translation of *fangcheng* calculations into classical Chinese prose, modern English prose, or modern mathematical terminology renders these practices almost incomprehensible. In contrast, as my reconstructions show, *fangcheng* procedures are easily visualized on the counting board, and in practice the calculations are not difficult.

The History of Chinese Mathematics

It has often been assumed that Chinese mathematics, and science in general, was in a state of decline in late imperial China. Much of this view is based on the evaluations by the Jesuits and their Chinese collaborators in the seventeenth century. For example, as noted in the preface, Xu Guangqi, the most prominent of the Chinese officials who collaborated with the Jesuits, asserts that Western mathematics is in every way superior to Chinese mathematics. This book provides considerable evidence opposing such views. One implication of this study, then, is that we should look more broadly at the global circulation of mathematical practices, rather than focusing narrowly on the introduction of Western science and mathematics into China.

The World History of Science

This book seeks to contribute to the history of science and mathematics conceived of as world history, suggesting that unexpected findings may await us if we allow historical evidence to revise our assumptions and preconceptions.

As we have seen, *fangcheng* problems are essentially equivalent to the solution of systems of n equations in n unknowns in modern linear algebra. The *fangcheng* procedure is in essence what we now call Gaussian elimination. Extant Chinese mathematical treatises also record determinantal calculations and solutions for special forms of *fangcheng* problems. Throughout the imperial period in China, *fangcheng* was one of the most significant branches of the mathematical arts. There are no known records of similar solutions anywhere else in the world before Leibniz's work on elimination and determinants beginning in about 1678. These procedures were transmitted to Japan and possibly Europe, where they may have served as precursors of modern matrices, Gaussian elimination, and determinants.

The essentials of the methods used today, then, in modern linear algebra were not first discovered by Leibniz or by Gauss. The essentials of these methods—augmented matrices, elimination, and determinantal-style calculations—were known at least by the first century C.E. in China, and were practiced by anonymous and likely illiterate adepts throughout the imperial period. These practices are considerably more sophisticated than has previously been understood. It was the visualization in two dimensions of problems with n conditions in n unknowns that was the key to their solution. Thus visualized, simple patterns made the efficient calculation of solutions practicable.

Appendix A
Examples of Similar Problems

Generally speaking, outside the Chinese sources analyzed in this book, there is comparatively little in the way of records of work on what we would now call systems of simultaneous linear equations before about 1678.[1] There are some examples of systems of simultaneous linear equations in ancient Babylonia and Egypt, but no problems of the form of the "well problem."[2] From the Greek tradition, several problems in Book I of Diophantus's *Arithmetica*[3] exhibit some interesting similarities with those we have analyzed. There do not seem to be comparable problems from medieval or Renaissance Europe,[4] until the work of Leibniz, which dates from 1678.

This Appendix presents a brief overview of linear algebra problems that are in some ways similar to those found in Chinese treatises. The first section of the Appendix will examine records of solutions to systems of linear equations preserved in Book I of Diophantus's *Arithmetica.* Although these problems share some intriguing similarities with problems in Chinese treatises, they were solved by substitution, rather than by elimination or determinantal methods. The second section will examine modern mathematical literature for similar problems. We will find that the form of augmented matrices examined in chapters 7 and 8 of this book is quite distinctive and unusual.

---

[1] For general historical overviews, see Smith 1923; Neugebauer 1951; Tropfke [1902–1903] 1980; van der Waerden [1950] 1954. See also Libbrecht 1973, 214–66 on indeterminate analysis outside China, and 267–93 in China. For a general bibliography of primary and secondary mathematical works, see Dauben and Lewis 2000; and May 1973.

[2] On Babylonian mathematics, see Høyrup 2002, and Neugebauer, Sachs, and Götze 1945. On Egyptian mathematics, see Clagett 1999, and Imhausen 2003.

[3] Studies of Diophantus's works, their reception, and Diophantine equations include Heath 1885; Sesiano 1977, 1982; Diophantus [c. 250] 2002; Christianidis 1994; and "Linear Diophantine Equations and Congruences," chap. 2 of Dickson [1919] 1966.

[4] Catalogues of early European mathematical treatises include Smith [1847–1939] 1970; and Van Egmond 1980.

Examples from Diophantus's *Arithmetica*

Though the "Greeks did not achieve a general solution method, which would be applicable to all systems of linear equations" (Tropfke [1902–1903] 1980, 391), the method in *Arithmetica* of reducing systems of n equations in m unknowns to a single equation in a single unknown is applied to a surprising range of problems. Analyzing how this method is used will provide interesting points of comparison with the problems we have analyzed from the *Nine Chapters.*

From a modern perspective, the problems that lead to the most interesting comparisons with those we have analyzed are problems 1–25 from Book I of Diophantus's *Arithmetica,*[5] problems we would call determinate and indeterminate systems of n linear equations in m unknowns. I will focus on the methods used to solve the problems.

Conspectus of Selected Problems

In the following table, I will analyze these problems, from a modern, abstract, and formal perspective, by writing problems 1–25 from Book I of *Arithmetica* in the form of augmented matrices, even though these problems were not originally written in this form. The table is based on the translations of these problems presented in two secondary historical studies: the translations and analysis in modern mathematical terminology in Thomas Heath's *Diophantos of Alexandria: A Study in the History of Greek Algebra* (Heath 1885, 129–246); and Jacques Sesiano's translation, in his "Conspectus of the Problems of the *Arithmetica,*" into modern notation as systems of several n equations in m unknowns (Sesiano 1982, 461–83). For each problem, I have taken the summary given in Heath, condensed it as much as possible, and added interpolations that explain the problem in modern mathematical notation. Then, drawing upon Sesiano's translation of these problems into systems of n equations in m unknowns, I write the problem in the form of an augmented matrix, altering his notation where appropriate. Augmented matrices prove to be the most concise way to express these problems, and will facilitate comparisons with the problems from chapter 8 of the *Nine Chapters.*

[5] For a translation of *Arithmetica* into Latin by Paul Tannery, with the Greek original, see Diophantus [c. 250] 1893–1895; for a translation into German by Gustav Wertheim, see Diophantus [c. 250] 1890; for a translation into French by Paul Ver Eecke, see Diophantus [c. 250] 1959. Important historical studies include Heath 1885; and Sesiano 1982.

Table A.1: Conspectus of selected problems from Book 1 of the *Arithmetica*.

Number	*Problem*	*Matrix Notation*
1	"Divide a given number $[b_1]$ into two $[x_1 + x_2 = b_1]$ having a given difference $[b_2 = x_1 - x_2]$."	$\begin{bmatrix} 1 & 1 & b_1 \\ 1 & -1 & b_2 \end{bmatrix}$
2	"Divide a given number into two $[x_1 + x_2 = b_1]$ having a given ratio $[x_1 = r x_2]$."	$\begin{bmatrix} 1 & 1 & b_1 \\ 1 & -r & 0 \end{bmatrix}$
3	"Divide a given number into two $[x_1 + x_2 = b_1]$ having a given ratio and difference $[x_1 = r x_2 + b_2]$."	$\begin{bmatrix} 1 & 1 & b_1 \\ 1 & -r & b_2 \end{bmatrix}$
4	"Find two numbers in a given ratio $[x_1 = r x_2]$, their difference also being given $[x_1 - x_2 = b_2]$."	$\begin{bmatrix} 1 & -r & 0 \\ 1 & -1 & b_2 \end{bmatrix}$
5	"Divide a given number into two $[x_1 + x_2 = b_1]$, such that the sum of given fractions of each is a given number $[\frac{1}{k_1} x_1 + \frac{1}{k_2} x_2 = b_2]$."	$\begin{bmatrix} 1 & 1 & b_1 \\ \frac{1}{k_1} & \frac{1}{k_2} & b_2 \end{bmatrix}$
6	"Divide a given number into two parts $[x_1 + x_2 = b_1]$, such that a given fraction of one exceeds a given fraction of the other by a given difference $[\frac{1}{k_1} x_1 = \frac{1}{k_2} x_2 + b_2]$."	$\begin{bmatrix} 1 & 1 & b_1 \\ \frac{1}{k_1} & -\frac{1}{k_2} & b_2 \end{bmatrix}$
12	"Twice divide a given number into two parts $[x_1 + y_1 = b, x_2 + y_2 = b]$, such that the first parts have a given ratio $[x_1 = r_1 x_2]$, and the second parts have a given ratio $[y_1 = r_2 y_2]$."	$\begin{bmatrix} 1 & 1 & 0 & 0 & b \\ 0 & 0 & 1 & 1 & b \\ 1 & 0 & -r_1 & 0 & 0 \\ 0 & 1 & 0 & -r_2 & 0 \end{bmatrix}$

Continued on next page

Number	*Problem*	*Matrix Notation*
13	"Thrice divide a given number into two parts [$x_1 + y_1 = x_2 + y_2 = x_3 + y_3 = b$], such that one of the first parts and one of the second, the other of the second and one of the third, the other of the third and the remaining of the first, are all in given ratios [$x_1 = r_1 y_2$, $x_2 = r_2 y_3$, $x_3 = r_3 y_1$]."	$\begin{bmatrix} 1 & 1 & 0 & 0 & 0 & 0 & b \\ 0 & 0 & 1 & 1 & 0 & 0 & b \\ 0 & 0 & 0 & 0 & 1 & 1 & b \\ 1 & 0 & 0 & -r_1 & 0 & 0 & 0 \\ 0 & 0 & 1 & 0 & 0 & -r_2 & 0 \\ 0 & -r_3 & 0 & 0 & 1 & 0 & 0 \end{bmatrix}$
15	"Find two numbers such that each, after receiving a given number from the other, forms a given ratio to the remainder of the other [$x_1 + k_1 = r_1(x_2 - k_1)$, $x_2 + k_2 = r_2(x_1 - k_2)$]."	$\begin{bmatrix} 1 & -r_1 & -k_1(r_1+1) \\ -r_2 & 1 & -k_2(r_2+1) \end{bmatrix}$
16	"Find three numbers such that the sums of each pair are given [$x_2 + x_3 = b_1$, $x_1 + x_3 = b_2$, $x_1 + x_2 = b_3$]."	$\begin{bmatrix} 0 & 1 & 1 & b_1 \\ 1 & 0 & 1 & b_2 \\ 1 & 1 & 0 & b_3 \end{bmatrix}$
17	"Find four numbers such that the sums of all sets of three are given [$x_2 + x_3 + x_4 = b_1$, $x_1 + x_3 + x_4 = b_2$, $x_1 + x_2 + x_4 = b_3$, $x_1 + x_2 + x_3 = b_4$]."	$\begin{bmatrix} 0 & 1 & 1 & 1 & b_1 \\ 1 & 0 & 1 & 1 & b_2 \\ 1 & 1 & 0 & 1 & b_3 \\ 1 & 1 & 1 & 0 & b_4 \end{bmatrix}$
18	"Find three numbers such that the sum of any two exceeds the third by a given number [$x_2 + x_3 = x_1 + b_1$, $x_1 + x_3 = x_2 + b_2$, $x_1 + x_2 = x_3 + b_3$]."	$\begin{bmatrix} -1 & 1 & 1 & b_1 \\ 1 & -1 & 1 & b_2 \\ 1 & 1 & -1 & b_3 \end{bmatrix}$
19	"Find four numbers such that the sum of any three exceeds the fourth by a given number [$x_2 + x_3 + x_4 = x_1 + b_1$, $x_1 + x_3 + x_4 = x_2 + b_2$, $x_1 + x_2 + x_4 = x_3 + b_3$, $x_1 + x_2 + x_3 = x_4 + b_4$]."	$\begin{bmatrix} -1 & 1 & 1 & 1 & b_1 \\ 1 & -1 & 1 & 1 & b_2 \\ 1 & 1 & -1 & 1 & b_3 \\ 1 & 1 & 1 & -1 & b_4 \end{bmatrix}$

Continued on next page

Number	*Problem*	*Matrix Notation*
20	"Divide a given number into three $[x_1 + x_2 + x_3 = b]$, such that for each extreme $[x_1, x_3]$ the sum with the middle $[x_2]$ has a given ratio to the remaining extreme $[x_1 + x_2 = r_1 x_3$, $x_3 + x_2 = r_2 x_1]$."	$\begin{bmatrix} 1 & 1 & 1 & b \\ 1 & 1 & -r_1 & 0 \\ -r_2 & 1 & 1 & 0 \end{bmatrix}$
21	"Find three numbers such that the greatest exceeds the middle by a given fraction of the least $[x_3 = x_2 + \frac{1}{k_1} x_1]$; the middle exceeds the least by a given fraction of the greatest $[x_2 = x_1 + \frac{1}{k_2} x_3]$; and the least exceeds a given fraction of the greatest by a given number $[x_1 = \frac{1}{k_3} x_3 + b]$."	$\begin{bmatrix} -\frac{1}{k_1} & -1 & 1 & 0 \\ -1 & 1 & -\frac{1}{k_2} & 0 \\ 1 & 0 & -\frac{1}{k_3} & b \end{bmatrix}$
22	"Find three numbers $[x_1, x_2, x_3]$, such that if each gives to the following a given fraction of itself, the results will all be equal $[x_1 - \frac{1}{k_1} x_1 + \frac{1}{k_3} x_3 = x_2 - \frac{1}{k_2} x_2 + \frac{1}{k_1} x_1 = x_3 - \frac{1}{k_3} x_3 + \frac{1}{k_2} x_2]$."	$\begin{bmatrix} 1-\frac{1}{k_1} & 0 & \frac{1}{k_3} & y \\ \frac{1}{k_1} & 1-\frac{1}{k_2} & 0 & y \\ 0 & \frac{1}{k_2} & 1-\frac{1}{k_3} & y \end{bmatrix}$
23	"Find four numbers such that if each gives to the following a given fraction of itself, the results will all be equal $[x_1 - \frac{1}{k_1} x_1 + \frac{1}{k_4} x_4 = x_2 - \frac{1}{k_2} x_2 + \frac{1}{k_1} x_1 = x_3 - \frac{1}{k_3} x_3 + \frac{1}{k_2} x_2 = x_4 - \frac{1}{k_4} x_4 + \frac{1}{k_3} x_3]$."	$\begin{bmatrix} 1-\frac{1}{k_1} & 0 & 0 & \frac{1}{k_4} & y \\ \frac{1}{k_1} & 1-\frac{1}{k_2} & 0 & 0 & y \\ 0 & \frac{1}{k_2} & 1-\frac{1}{k_3} & 0 & y \\ 0 & 0 & \frac{1}{k_3} & 1-\frac{1}{k_4} & y \end{bmatrix}$
24	"Find three numbers such that if each receives a given fraction of the sum of the other two, the results are equal $[x_1 + \frac{1}{k_1}(x_2 + x_3) = x_2 + \frac{1}{k_2}(x_1 + x_3) = x_3 + \frac{1}{k_3}(x_1 + x_2)]$."	$\begin{bmatrix} 1 & \frac{1}{k_1} & \frac{1}{k_1} & y \\ \frac{1}{k_2} & 1 & \frac{1}{k_2} & y \\ \frac{1}{k_3} & \frac{1}{k_3} & 1 & y \end{bmatrix}$
25	"Find four numbers such that if each receives a given fraction of the sum of the other three, the results are equal $[x_1 + \frac{1}{k_1}(x_2 + x_3 + x_4) = x_2 + \frac{1}{k_2}(x_1 + x_3 + x_4) = x_3 + \frac{1}{k_3}(x_1 + x_2 + x_4) = x_4 + \frac{1}{k_4}(x_1 + x_2 + x_3)]$."	$\begin{bmatrix} 1 & \frac{1}{k_1} & \frac{1}{k_1} & \frac{1}{k_1} & y \\ \frac{1}{k_2} & 1 & \frac{1}{k_2} & \frac{1}{k_2} & y \\ \frac{1}{k_3} & \frac{1}{k_3} & 1 & \frac{1}{k_3} & y \\ \frac{1}{k_4} & \frac{1}{k_4} & \frac{1}{k_4} & 1 & y \end{bmatrix}$

From the above table we can make the following initial observations:

1. In problems 1–6, just as in problems 1–11 in chapter 8 of the *Nine Chapters*, the basic form to be solved is presented in the first problem, and then variants are given in the subsequent problems. Here, these variants include the use of a ratio instead of a difference (problem 2), a combination of a ratio and a difference (problems 3 and 4), and reciprocals (problems 5 and 6).
2. The simple problems are then followed by much more complex problems.
3. Several of the more complex problems are presented in pairs—12 and 13, 16 and 17, 18 and 19, 22 and 23, and 24 and 25. In these pairs, the problem is first solved for n unknowns, and then for $n+1$ unknowns.

Solutions to Selected Problems

We will now examine in detail the methods used to solve several of these problems—the pairs 12 and 13, 16 and 17, 18 and 19, and 22 and 23. In each case, in the analysis I will proceed as follows:

1. For the sake of brevity, I will examine only the second problem of the pair, and in summary form. I have again used Heath's "Appendix" (Heath 1885, 163–244) for the statement of the problem, the given conditions, the method of solution, and the solution itself; as above, I have silently modernized, corrected, and further abbreviated Heath's summaries where appropriate.[6]
2. Next, I will present a general solution for the problem, which will itself consist of two parts. First, although the solutions to these problems are calculated in *Arithmetica* only after specific numerical assignments are made, considerably simplifying the calculations, here, in order to see why, and under what conditions, these solutions are valid, we will find general solutions for the unknowns in terms of the given constants k_i, r_i, and b_i $(1 \leq i \leq n)$. Second, since most of these pairs of problems are solved first for 3 unknowns, and then for 4 unknowns, we will want to determine whether the solutions are true in general for n unknowns. Though there is no indication that any such proofs were given, if the solution is true in general for n, and since examples for 3 and 4 unknowns are given, it seems reasonable to assume that other cases, such as 5, 6, or possibly more unknowns had also been solved.

[6] Again, I am interested here only in the method used to solve the problems in *Arithmetica*, and so have not individually noted my changes to Heath 1885.

Problems 12 and 13

Problem 12 is formulated as 4 equations in 4 unknowns (2 pairs); problem 13 is 6 equations in 6 unknowns (3 pairs). Problem 13, further condensing Heath's summary, is as follows:

> Thrice divide a given number [b] into two parts [$x_1 + y_1 = x_2 + y_2 = x_3 + y_3 = b$], such that one of the first parts and one of the second parts, the other of the second parts and one of the third parts, the other of the third parts and the remaining one of the first parts, are, respectively, in given ratios [that is, $x_1 = r_1 y_2$, $x_2 = r_2 y_3$, and $x_3 = r_3 y_1$].

The problem is not solved in general terms, but by first assigning specific values:

> Given the number 100 [$b = 100$], the ratio of the greater of the first parts to the lesser of the second is 3:1 [$r_1 = 3$]; the ratio of the greater of the second to the lesser of the third is 2:1 [$r_2 = 2$]; the ratio of the greater of the third to the lesser of the first is 4:1 [$r_3 = 4$].

The solution is then found by transforming the problem from one that is equivalent to 6 equations in 6 unknowns (3 pairs) to one single equation written in terms of a single unknown, x:

> Let x be the smaller of the third parts [$x = y_3$]. Therefore, the greater of the second is $2x$ [that is, $x_2 = r_2 y_3$], the lesser of the second is $100 - 2x$ [$y_2 = b - x_2$], the greater of the first is $300 - 6x$ [$x_1 = r_1 y_2$], the lesser of the first is $6x - 200$ [$y_1 = b - x_1$], and therefore the greater of the third is $24x - 800$ [$x_3 = r_3 y_1$]. Therefore, $25x - 800 = 100$ [$x_3 + y_3 = b$], and $x = 36$.[7]

This problem can easily be generalized for an arbitrary number of unknowns. That is, divide a given number b into n pairs,

$$x_1 + y_1 = x_2 + y_2 = x_3 + y_3 = \ldots = x_{n-1} + y_{n-1} = x_n + y_n = b$$

such that the given ratios are as follows:

$$x_1 = r_1 y_2, \quad x_2 = r_2 y_3, \quad x_3 = r_3 y_4, \quad \ldots, \quad x_{n-1} = r_{n-1} y_n, \quad x_n = r_n y_1.$$

The method used to solve this problem is to transform it from the equivalent of $2n$ equations in $2n$ unknowns to a single equation in a single unknown by successive substitutions. We begin by setting the single unknown to be the lesser of the two parts in the last pair, $x = y_n$. Then we use the given equations to find, in

[7] The statement, given conditions, and solution are adapted from Heath 1885, 165–66; see also Tropfke [1902–1903] 1980, 394.

succession, expressions for x_{n-1} and then y_{n-1}, x_{n-2} and then y_{n-2}, x_1 and then y_1, and finally x_n, each in terms of x, b, and r_i:

$$\begin{aligned}
x_{n-1} &= r_{n-1}y_n = r_{n-1}x,\\
y_{n-1} &= b - x_{n-1} = b - r_{n-1}x,\\
x_{n-2} &= r_{n-2}y_{n-1} = r_{n-2}(b - r_{n-1}x),\\
y_{n-2} &= b - x_{n-2} = b - r_{n-2}(b - r_{n-1}x),\\
x_{n-3} &= r_{n-3}y_{n-2} = r_{n-3}(b - r_{n-2}(b - r_{n-1}x)),\\
y_{n-3} &= b - x_{n-2} = b - r_{n-3}(b - r_{n-2}(b - r_{n-1}x)),\\
&\vdots\\
x_1 &= r_1 y_2 = r_1(b - r_2(b - r_3(\ldots(b - r_{n-3}(b - r_{n-2}(b - r_{n-1}x)))\ldots))),\\
y_1 &= b - x_1 = b - r_1(b - r_2(b - r_3(\ldots(b - r_{n-3}(b - r_{n-2}(b - r_{n-1}x)))\ldots))),\\
x_n &= r_n y_1 = r_n(b - r_1(b - r_2(\ldots(b - r_{n-3}(b - r_{n-2}(b - r_{n-1}x)))\ldots))).
\end{aligned}$$

Then, using the equation

$$b = x_n + y_n,$$

we can write a single equation for x strictly in terms of b and r_i $(1 \le i \le n)$:

$$b = r_n(b - r_1(b - r_2(\ldots(b - r_{n-3}(b - r_{n-2}(b - r_{n-1}x)))\ldots))) + x.$$

In *Arithmetica*, because numerical values are assigned at the outset, the right-hand side is simply in the form $mx + p$, where m and p are integers, and the equation $b = mx + p$ is therefore easy to solve for x. But to solve it in more general terms, we expand the expression for x_n,

$$x_n = br_n - br_1r_n + br_1r_2r_n - \ldots \pm br_1r_2\cdots r_{n-2}r_n \mp r_1r_2\cdots r_nx,$$

where $\pm$ and $\mp$ are $+$ and $-$ for n even, and $-$ and $+$ for n odd. This gives

$$b = br_n - br_1r_n + br_1r_2r_n - \ldots \pm br_1r_2\cdots r_{n-2}r_n \mp r_1r_2\cdots r_nx + x.$$

Then, solving for x, we have

$$x = \frac{b(r_1r_2\cdots r_{n-2}r_n - r_1r_2\cdots r_{n-3}r_n + \ldots \pm r_1r_2r_n \mp r_1r_n \pm r_n \mp 1)}{r_1r_2\cdots r_n \mp 1}.$$

Having found $x = y_n$, the remaining unknowns can easily be calculated. There is thus a unique solution if and only if $r_1r_2\cdots r_n \mp 1 \neq 0$. If in addition $r_1r_2\cdots r_n \mp 1$ divides b, then the unknowns will all be integers (note that in problem 12, $2\times 3 - 1 = 5$ divides 100, and in problem 13, $3\times 2\times 4 + 1 = 25$ divides 100). This condition is sufficient (but not necessary) for all the unknowns to be integers.

This generalized case of dividing a given number b into n pairs, yielding the equivalent of $2n$ equations in $2n$ unknowns, corresponds to the following associated augmented matrix:

$$\left[\begin{array}{ccccccccc|c} 1 & 1 & & & & & & & & b \\ & & 1 & 1 & & & & & & b \\ & & & & \ddots & \ddots & & & & \vdots \\ & & & & & & 1 & 1 & & b \\ 1 & & & -r_1 & & & & & & \\ & \ddots & & & \ddots & & & & & \\ & & 1 & & & & -r_{n-1} & & & \\ & -r_n & & & & 1 & & & & \end{array}\right],$$

where zero entries have been omitted, since the nonzero entries are not exactly on diagonals.

Problems 16 and 17

Problem 16 is 3 equations in 3 unknowns, and problem 17 is 4 equations in 4 unknowns. Problem 17 states:

> Find four numbers such that the sums of all sets of three are given [$x_2 + x_3 + x_4 = b_1$, $x_1 + x_3 + x_4 = b_2$, $x_1 + x_2 + x_4 = b_3$, $x_1 + x_2 + x_3 = b_4$].

The four sums are then assigned specific numerical values:

> Given are the sums of the threes, 22, 24, 27, and 20 [$b_1 = 22$, $b_2 = 24$, $b_3 = 27$, $b_4 = 20$].

The problem is then reduced to a single equation in a single unknown, x:

> Let x be the sum of the four unknowns [$x = x_1 + x_2 + x_3 + x_4$], then the four unknowns are $x - 22$, $x - 24$, $x - 27$, and $x - 20$ [$x_1 = x - b_1, x_2 = x - b_2, x_3 = x - b_3, x_4 = x - b_4$]. Therefore, $4x - 93 = x$ [$x_1 + x_2 + x_3 + x_4 = x$], and $x = 31$.[8]

Again, this problem can be generalized to n, in which case we have n equations in n unknowns. In modern notation,

$$\begin{aligned} x_2 + x_3 + \ldots + x_n &= b_1, \\ x_1 + x_3 + \ldots + x_n &= b_2, \\ &\vdots \\ x_1 + x_2 + \ldots + x_{n-1} &= b_n. \end{aligned}$$

[8] Statement, conditions, and solution adapted from Heath 1885, 166–67.

We transform this to a single equation in a single unknown x by setting x to the sum of the unknowns x_j,

$$x = x_1 + x_2 + \ldots + x_n.$$

Then it is easy to solve for each x_j in terms of x and b_j,

$$\begin{aligned} x - b_1 &= (x_1 + x_2 + \ldots + x_n) - (x_2 + x_3 + \ldots + x_n) = x_1, \\ x - b_2 &= (x_1 + x_2 + \ldots + x_n) - (x_1 + x_3 + \ldots + x_n) = x_2, \\ &\vdots \\ x - b_n &= (x_1 + x_2 + \ldots + x_n) - (x_1 + x_2 + \ldots + x_{n-1}) = x_n. \end{aligned}$$

We can then solve for x,

$$\begin{aligned} x &= x_1 + x_2 + \ldots + x_n \\ &= (x - b_1) + (x - b_2) + \ldots + (x - b_n) \\ &= nx - (b_1 + b_2 + \ldots + b_n), \end{aligned}$$

and therefore,

$$x = \frac{1}{n-1}(b_1 + b_2 + \ldots + b_n).$$

Thus the problem will always have a unique solution ($n \geq 2$), and the unknowns will be integers if and only if $n-1$ divides $x = b_1 + b_2 + \ldots + b_n$ (assuming b_i are all integers). The corresponding augmented matrix is the following:

$$\begin{bmatrix} 0 & 1 & \ldots & 1 & b_1 \\ 1 & 0 & \ddots & \vdots & b_2 \\ \vdots & \ddots & \ddots & 1 & \vdots \\ 1 & \ldots & 1 & 0 & b_n \end{bmatrix}.$$

Problems 18 and 19

Problems 18 and 19 are quite similar to the preceding problems 16 and 17; problem 18 has 3 equations in 3 unknowns, and problem 19 has 4 equations in 4 unknowns. Problem 19 states:

> Find four numbers such that the sum of any three exceeds the fourth by a given number [$x_2+x_3+x_4 = x_1+b_1$, $x_1+x_3+x_4 = x_2+b_2$, $x_1+x_2+x_4 = x_3+b_3$, $x_1 + x_2 + x_3 = x_4 + b_4$].

Again, the given numbers are specified before a solution is calculated:

> Given are the differences 20, 30, 40, and 50 [$b_1 = 20$, $b_2 = 30$, $b_3 = 40$, $b_4 = 50$].

And again, the 4 equations in 4 unknowns are reduced to 1 equation in 1 unknown:

> Let $2x$ be the sum of the four unknown numbers [$2x = x_1 + x_2 + x_3 + x_4$]. Therefore, the four numbers are $x - 10$, $x - 15$, $x - 20$, $x - 25$ [$x_1 = x - \frac{1}{2}b_1$, $x_2 = x - \frac{1}{2}b_2$, $x_3 = x - \frac{1}{2}b_3$, $x_4 = x - \frac{1}{2}b_4$]. Therefore, $4x - 70 = 2x$ [$x_1 + x_2 + x_3 + x_4 = 2x$], and $x = 35$.[9]

That is, we first define a new unknown x such that $2x = x_1 + x_2 + x_3 + x_4$. We solve the initial equations for x_1, x_2, x_3, and x_4, respectively,

$$\begin{aligned} x_1 &= x_2 + x_3 + x_4 - b_1, \\ x_2 &= x_1 + x_3 + x_4 - b_2, \\ x_3 &= x_1 + x_2 + x_4 - b_3, \\ x_4 &= x_1 + x_2 + x_3 - b_4, \end{aligned}$$

and then adding, respectively, x_1, x_2, x_3, and x_4 to both sides, we have

$$\begin{aligned} 2x_1 &= x_1 + x_2 + x_3 + x_4 - b_1 = 2x - b_1, \\ 2x_2 &= x_1 + x_2 + x_3 + x_4 - b_2 = 2x - b_2, \\ 2x_3 &= x_1 + x_2 + x_3 + x_4 - b_3 = 2x - b_3, \\ 2x_4 &= x_1 + x_2 + x_3 + x_4 - b_4 = 2x - b_4, \end{aligned}$$

so that

$$x_1 = x - \frac{b_1}{2}, \quad x_2 = x - \frac{b_2}{2}, \quad x_3 = x - \frac{b_3}{2}, \quad x_4 = x - \frac{b_4}{2}.$$

Now we can write the 4 equations in 4 unknowns as 1 equation in 1 unknown,

$$2x = x_1 + x_2 + x_3 + x_4 = 4x - \frac{1}{2}(b_1 + b_2 + b_3 + b_4),$$

and solving for x, we have

$$x = \frac{1}{4}(b_1 + b_2 + b_3 + b_4).$$

Again, this can very easily be generalized for any $n \geq 3$. That is, assume we are asked to find n numbers $x_1, x_2, \ldots, x_n$ such that the sum of any combination of

[9] Statement, conditions, and solution adapted from Heath 1885, 167. In Heath's numbering, these are problems 18 and 20.

$n-1$ exceeds the remaining number x_j by a given number b_j,

$$\begin{aligned}
x_2 + x_3 + \ldots + x_n &= x_1 + b_1,\\
x_1 + x_3 + \ldots + x_n &= x_2 + b_2,\\
&\vdots\\
x_1 + x_2 + \ldots + x_{n-1} &= x_n + b_n.
\end{aligned}$$

We then define a new unknown x such that

$$2x = x_1 + x_2 + \ldots + x_n.$$

As above, we solve each of the initial equations for x_j, and to both sides of each equation we add, respectively, x_j, and then solve for each x_j in terms of x and b_j,

$$\begin{array}{rclcrclcrcl}
x_2 + x_3 + \ldots + x_n &=& x_1 + b_1 & \Longrightarrow & 2x &=& 2x_1 + b_1 & \Longrightarrow & x - \frac{1}{2}b_1 &=& x_1\\
x_1 + x_3 + \ldots + x_n &=& x_2 + b_2 & \Longrightarrow & 2x &=& 2x_2 + b_2 & \Longrightarrow & x - \frac{1}{2}b_2 &=& x_2\\
\multicolumn{11}{c}{\dotfill}\\
x_1 + x_2 + \ldots + x_{n-1} &=& x_n + b_n & \Longrightarrow & 2x &=& 2x_n + b_n & \Longrightarrow & x - \frac{1}{2}b_n &=& x_n
\end{array}$$

Now we can write the n equations in n unknowns as a single equation in a single unknown,

$$\begin{aligned}
2x = x_1 + x_2 + \ldots + x_n &= x - \frac{b_1}{2} + x - \frac{b_2}{2} + \ldots + x - \frac{b_n}{2}\\
&= nx - \frac{1}{2}(b_1 + b_2 + \ldots + b_n),
\end{aligned}$$

thus,

$$(n-2)x = \frac{1}{2}(b_1 + b_2 + \ldots + b_n),$$

and

$$x = \frac{1}{2n-4}(b_1 + b_2 + \ldots + b_n).$$

Again, if we write this as an augmented matrix, we have

$$\begin{bmatrix}
-1 & 1 & \ldots & 1 & b_1\\
1 & -1 & \ddots & \vdots & b_2\\
\vdots & \ddots & \ddots & 1 & \vdots\\
1 & \ldots & 1 & -1 & b_4
\end{bmatrix}.$$

Problems 22 and 23

Problem 22 has 3 equations in 4 unknowns, and problem 23 has 4 equations in 5 unknowns, and thus both are indeterminate. Problem 23 states:

> Find four numbers that become equal if each gives to the following number a given fraction of itself.

This problem, which is somewhat awkwardly expressed in words, means simply this: Let the four unknown numbers to be found be x_1, x_2, x_3, x_4, respectively; and let $\frac{1}{k_1}, \frac{1}{k_2}, \frac{1}{k_3}, \frac{1}{k_4}$ be the four given fractions. To take the first unknown x_1 as an example, we subtract $\frac{1}{k_1}x_1$ from x_1 to add to x_2, but we add to x_1 the quantity $\frac{1}{k_4}x_4$, which was subtracted from x_4. After doing similarly for each of the unknowns, the results are equal, giving the following equations:

$$x_1 - \frac{1}{k_1}x_1 + \frac{1}{k_4}x_4 = x_2 - \frac{1}{k_2}x_2 + \frac{1}{k_1}x_1 = x_3 - \frac{1}{k_3}x_3 + \frac{1}{k_2}x_2 = x_4 - \frac{1}{k_4}x_4 + \frac{1}{k_3}x_3. \quad \text{(A.1)}$$

Specific numerical values are assigned before the problem is solved:

> Let the first give $\frac{1}{3}$ of itself to the second, the second $\frac{1}{4}$ of itself to the third, the third $\frac{1}{5}$ of itself to the fourth, and the fourth $\frac{1}{6}$ of itself to the first.

To solve this indeterminate problem, first the value of one of the unknowns is given, transforming the problem into a system of 4 equations in 4 unknowns, and a new unknown x is defined:

> Assume the second unknown is 4. Set $3x$ equal to the first unknown.

Finally, this system of 4 equations in 4 unknowns is transformed into a system of 1 equation in 1 unknown:

> After giving and taking, the second becomes $x+3$. Since giving and taking from the first is $x+3$, the fourth unknown is $18-6x$. Since giving and taking from the fourth is $x+3$, the third unknown is $30x-60$. Since giving and taking from the third unknown is $24x-47$, which is also equal to $x+3$, $x = \frac{50}{23}$. Therefore the unknown numbers are $\frac{150}{23}$, 4, $\frac{120}{23}$, and $\frac{114}{23}$, or, multiplying by the denominator, 150, 92, 120, and 114.[10]

That is, the calculations proceed as follows. First, it is stipulated that $x_2 = 4$, and a new unknown x is defined such that $3x = x_1$, and we substitute these values of x_1 and x_2 into the initial equation for x_2 to find the common value of them all,

$$x_2 - \frac{1}{k_2}x_2 + \frac{1}{k_1}x_1 = 3 + x.$$

[10] Statement, conditions, and solution adapted from Heath 1885, 168–69 (his problem numbers 25 and 26).

We then rewrite the remaining equations as follows:

$$x_1 - \frac{1}{k_1}x_1 + \frac{1}{k_4}x_4 = 3 + x, \tag{A.2}$$

$$x_3 - \frac{1}{k_3}x_3 + \frac{1}{k_2}x_2 = 3 + x, \tag{A.3}$$

$$x_4 - \frac{1}{k_4}x_4 + \frac{1}{k_3}x_3 = 3 + x. \tag{A.4}$$

First, we substitute $x_1 = 3x$ into equation A.2,

$$\begin{aligned} 3 + x &= x_1 - \frac{1}{k_1}x_1 + \frac{1}{k_4}x_4 \\ &= 3x - x + \frac{1}{6}x_4, \end{aligned}$$

and solving, we have

$$x_4 = 18 - 6x.$$

Next, we substitute in $x_4 = 18 - 6x$ into equation A.4,

$$\begin{aligned} 3 + x &= x_4 - \frac{1}{k_4}x_4 + \frac{1}{k_3}x_3 \\ &= 18 - 6x - \frac{1}{6}(18 - 6x) + \frac{1}{5}x_3 \end{aligned}$$

and then, solving, we have

$$x_3 = 30x - 60.$$

Finally, we substitute $x_3 = 30x - 60$ and $x_2 = 4$ into equation A.3, which yields, from the original system of 4 equations in 5 unknowns, a single equation in a single unknown x,

$$\begin{aligned} 3 + x &= x_3 - \frac{1}{k_3}x_3 + \frac{1}{k_2}x_2 \\ &= 30x - 60 - \frac{1}{5}(30x - 60) + \frac{1}{4} \times 4 \\ &= 24x - 47, \end{aligned}$$

and solving for x, we have

$$x = \frac{50}{23}.$$

Therefore, the solution is

$$x_1 = \frac{150}{23}, \quad x_2 = 4, \quad x_3 = \frac{120}{23}, \quad x_4 = \frac{114}{23}.$$

To obtain a solution in which all the unknowns are integers, we multiply by 23, giving,

$$x_1 = 150, \quad x_2 = 92, \quad x_3 = 120, \quad x_4 = 114.$$

This problem is also easy to generalize, although calculating its solution in this general form is somewhat complicated. We wish to find n numbers such that

$$x_1 - \frac{1}{k_1}x_1 + \frac{1}{k_n}x_n = x_2 - \frac{1}{k_2}x_2 + \frac{1}{k_1}x_1 = \ldots = x_n - \frac{1}{k_n}x_n + \frac{1}{k_{n-1}}x_{n-1}.$$

We can solve this, following the solution to problem 23, as follows. First we transform this indeterminate system of n equations in $n+1$ unknowns into a determinate system of n equations in n unknowns by setting the value of $x_2 = k_2$. Next, we define a new unknown x as follows:

$$k_1 x = x_1.$$

We now calculate the value common to all the equations, substituting $x_1 = k_1 x$ and $x_2 = k_2$,

$$\begin{aligned} x_2 - \frac{1}{k_2}x_2 + \frac{1}{k_1}x_1 &= \frac{(k_2-1)x_2}{k_2} + x \\ &= (k_2 - 1) + x. \end{aligned}$$

Then we write the remaining $n-1$ equations as follows:

$$\begin{aligned} x_1 - \frac{1}{k_1}x_1 + \frac{1}{k_n}x_n &= (k_2-1)+x, \\ x_3 - \frac{1}{k_3}x_3 + \frac{1}{k_2}x_2 &= (k_2-1)+x, \\ x_4 - \frac{1}{k_4}x_4 + \frac{1}{k_3}x_3 &= (k_2-1)+x, \\ &\vdots \\ x_{n-1} - \frac{1}{k_{n-1}}x_{n-1} + \frac{1}{k_{n-2}}x_{n-2} &= (k_2-1)+x, \\ x_n - \frac{1}{k_n}x_n + \frac{1}{k_{n-1}}x_{n-1} &= (k_2-1)+x. \end{aligned}$$

We can rewrite this in the following manner:

$$x_n = k_n\Big((k_2 - 1 + x) - \frac{(k_1-1)}{k_1}x_1\Big), \tag{A.5}$$

$$x_2 = k_2\Big((k_2 - 1 + x) - \frac{(k_3-1)}{k_3}x_3\Big), \tag{A.6}$$

$$x_3 = k_3\Big((k_2 - 1 + x) - \frac{(k_4-1)}{k_4}x_4\Big), \tag{A.7}$$

$$\vdots$$

$$x_{n-2} = k_{n-2}\big((k_2 - 1 + x) - \frac{(k_{n-1} - 1)}{k_{n-1}} x_{n-1}\big), \tag{A.8}$$

$$x_{n-1} = k_{n-1}\big((k_2 - 1 + x) - \frac{(k_n - 1)}{k_n} x_n\big). \tag{A.9}$$

We now substitute the values calculated at each step into the previous equation. First, substituting $x_1 = k_1 x$ into equation A.5, and rearranging terms, we have

$$x_n = -k_n\big((k_1 - 1)x - ((k_2 - 1) + x)\big).$$

Substituting this value of x_n into equation A.9, and rearranging terms, gives us

$$x_{n-1} = k_{n-1}\big((k_n - 1)(k_1 - 1)x - ((k_n - 1) - 1)((k_2 - 1) + x)\big).$$

Substituting this value of x_{n-1} into equation A.8, we then have

$$\begin{aligned} x_{n-2} = -k_{n-2}\big(&(k_{n-1} - 1)(k_n - 1)(k_1 - 1)x \\ &- ((k_{n-1} - 1)(k_n - 1) - (k_{n-1} - 1) + 1)((k_2 - 1) + x)\big). \end{aligned}$$

We continue in this fashion, until we substitute x_4 into equation A.7, and we arrive at an expression for x_3 as follows:

$$\begin{aligned} x_3 = \pm k_3\Big(&(k_4 - 1)(k_5 - 1)\cdots(k_n - 1)(k_1 - 1)x \\ &- \big((k_4 - 1)(k_5 - 1)\cdots(k_n - 1) - (k_4 - 1)(k_5 - 1)\cdots(k_{n-1} - 1) + \cdots \\ &\mp (k_4 - 1)(k_5 - 1) \pm (k_4 - 1) \mp 1\big)((k_2 - 1) + x)\Big), \end{aligned}$$

where $\pm$ and $\mp$ are $+$ and $-$ respectively for n even, and $-$ and $+$ respectively for n odd. Finally, we substitute this value into the equation for x_3,

$$\begin{aligned} (k_2 - 1) + x &= x_3 - \frac{1}{k_3} x_3 + \frac{1}{k_2} x_2 \\ &= \frac{(k_3 - 1)}{k_3} x_3 + 1. \end{aligned}$$

Following simplification, omitted here, the solution is given by

$$x = \frac{\begin{array}{r}(k_n-1)(k_{n-1}-1)(k_{n-2}-1)\cdots(k_3-1)(k_2-1)\\ -(k_{n-1}-1)(k_{n-2}-1)\cdots(k_3-1)(k_2-1)\\ +(k_{n-2}-1)\cdots(k_3-1)(k_2-1)-\dots\\ \mp(k_3-1)(k_2-1)\pm(k_2-1)\mp 1\end{array}}{\begin{array}{r}(k_1-1)(k_n-1)(k_{n-1}-1)\cdots(k_4-1)(k_3-1)-\\ (k_n-1)(k_{n-1}-1)\cdots(k_4-1)(k_3-1)+\\ (k_{n-1}-1)\cdots(k_4-1)(k_3-1)-\dots\\ \mp(k_4-1)(k_3-1)\pm(k_3-1)\mp 1\end{array}}. \tag{A.10}$$

After rearranging terms and simplifying, omitted here, we have

$$x = \frac{\begin{array}{r}(\cdots((((k_n-1)-1)(k_{n-1}-1)+1)(k_{n-2}-1)-1)\cdots\\ (k_3-1)\pm 1)(k_2-1)\mp 1\end{array}}{\begin{array}{r}(\cdots((((k_1-1)-1)(k_n-1)+1)(k_{n-1}-1)-1)\cdots\\ (k_4-1)\pm 1)(k_3-1)\mp 1\end{array}}, \tag{A.11}$$

where, in order to make the pattern more clear, the initial $((k_i-1)-1)$ in the numerator and denominator has not been simplified to (k_i-2). Note that there will always be a solution, unless the denominator in equation A.11 is zero, which can occur, for example, for $n=4$, if $k_1=k_3=2$. Also, all the unknowns x_j will be integers if x is an integer, and therefore if we assign to x_2 the value

$$x_2 = k_2((\cdots((((k_1-1)-1)(k_n-1)+1)(k_{n-1}-1)-1)\cdots(k_4-1)\pm 1)(k_3-1)\mp 1),$$

then the solutions will all be integers. Although the solution to this problem is found in *Arithmetica* by solving for x and then substituting this into expressions for the x_j, we can also write the solutions as follows:

$$\begin{aligned}
x_1 &= k_1((\cdots((((k_n-1)-1)(k_{n-1}-1)+1)(k_{n-2}-1)-1)\cdots\\
&\quad\cdot(k_3-1)\pm 1)(k_2-1)\mp 1),\\
x_2 &= k_2((\cdots((((k_1-1)-1)(k_n-1)+1)(k_{n-1}-1)-1)\cdots\\
&\quad\cdot(k_4-1)\pm 1)(k_3-1)\mp 1),\\
x_3 &= k_3((\cdots((((k_2-1)-1)(k_1-1)+1)(k_n-1)-1)\cdots\\
&\quad\cdot(k_5-1)\pm 1)(k_4-1)\mp 1),\\
&\vdots\\
x_n &= k_n((\cdots((((k_{n-1}-1)-1)(k_{n-2}-1)+1)(k_{n-3}-1)-1)\cdots\\
&\quad\cdot(k_2-1)\pm 1)(k_1-1)\mp 1).
\end{aligned}$$

If we write this problem in its generalized form as an augmented matrix, we have

$$\begin{bmatrix} 1-\frac{1}{k_1} & 0 & \cdots & 0 & \frac{1}{k_n} & y \\ \frac{1}{k_1} & 1-\frac{1}{k_2} & 0 & \cdots & 0 & y \\ 0 & \frac{1}{k_2} & \ddots & \ddots & \vdots & \vdots \\ \vdots & \ddots & \ddots & 1-\frac{1}{k_{n-1}} & 0 & y \\ 0 & \cdots & 0 & \frac{1}{k_{n-1}} & 1-\frac{1}{k_n} & y \end{bmatrix}.$$

After a change of variables, writing $z_j = \frac{1}{k_j} x_j$ and $l_j = k_j - 1$, we have

$$\begin{bmatrix} l_1 & 0 & \cdots & 0 & 1 & y \\ 1 & l_2 & 0 & \cdots & 0 & y \\ 0 & 1 & \ddots & \ddots & \vdots & \vdots \\ \vdots & \ddots & \ddots & l_{n-1} & 0 & y \\ 0 & \cdots & 0 & 1 & l_n & y \end{bmatrix}.$$

From a modern, formal point of view, this problem is then in intriguing ways similar to the form of matrices analyzed in chapters 7 and 8 of this book.

Conclusions

This analysis also suggests interesting differences between these problems from *Arithmetica* and problems 3 and 12–15 from chapter 8 of the *Nine Chapters*, including the following:

1. The most important difference is the approach to the solution. In *Arithmetica*, the solution to problems of the form n equations in m unknowns was found by transforming them into a single equation in a single unknown.
2. The problems in *Arithmetica* were not visualized in an array, and this visualization is arguably a key step in the development of elimination and determinantal methods.
3. Most of the problems from *Arithmetica* are what we now term "dense" matrices, that is, matrices with few zeros, which are complicated to solve using determinants when $n \geq 4$ because of the number of terms involved. The forms given in equations 7.2 and 7.1 are "sparse" matrices, which can be more easily solved using determinants.

In sum, though solutions to what in modern terms we would call particular forms of systems of n equations in n unknowns were known quite early in many areas, records of a general method for solving these problems, such as elimination by row reductions, have been found only in China. In *Arithmetica*, solutions are found to problems that could be solved by transforming a system of n equations in m unknowns to a system with a single equation in a single unknown, and this method is extended to a variety of similar problems. In the same way, I have argued, problems 3 and 12–15 in chapter 8 of the *Nine Chapters* represent a set of problems for which a determinantal solution had been found for one example and extended to several similar problems. The key difference between these problems in *Arithmetica* and those we have analyzed from the *Nine Chapters* is the visualization of these problems in the form of arrays: these differences in notation are associated with the different methods of solution.

Examples from Modern Works on Linear Algebra

Examples of the form of matrices analyzed in chapters 7 and 8 of this book are quite difficult to find in modern mathematical works: I have found only one example similar to the general form given in equation 7.2. Among modern studies, the most encyclopedic collection of results on determinants is Muir 1906–1923, a four-volume work (reprinted as two), together approximately 2000 pages, which summarizes the abstracts of hundreds of papers on determinants from 1861 to 1900, roughly organized by categories. Muir also later published a comprehensive work on the theory of matrices, Muir 1933, approximately 750 pages. I have examined both works, and have found no mention or analysis of this type of matrix. The closest example Muir gives that even resembles this form of matrices—either the augmented matrix or the square matrix of coefficients of the unknowns—is the following problem, from Muir 1933, 22, to find the determinant of the following:

$$\begin{bmatrix} x & 0 & 0 & 0 & y \\ y & x & 0 & 0 & 0 \\ 0 & y & x & 0 & 0 \\ 0 & 0 & y & x & 0 \\ 0 & 0 & 0 & y & x \end{bmatrix}.$$

I have also examined numerous other works, including Cullen 1990; Hoffman and Kunze [1961] 1971; Horn and Johnson 1985; Horn and Johnson 1991; Serre 2002; Stewart 1973; Stewart 1998; Strang [1976] 1988; and Zhang 1999; to list only a few.

The only example I have found that is similar to the form in equation 7.2 is the following:

$$\begin{aligned} x_1 - a_{12}x_2 &= b_1, \\ x_2 - a_{23}x_3 &= b_2, \\ x_3 - a_{31}x_1 &= b_3, \end{aligned}$$

which can be written in matrix form as

$$\begin{pmatrix} 1 & -a_{12} & 0 \\ 0 & 1 & -a_{23} \\ -a_{31} & 0 & 1 \end{pmatrix} \begin{pmatrix} x_1 \\ x_2 \\ x_3 \end{pmatrix} = \begin{pmatrix} b_1 \\ b_2 \\ b_3 \end{pmatrix},$$

with the solution

$$x_1 = \frac{1}{\det A}(b_1 + b_2 a_{12} + b_3 a_{12} a_{23}),$$
$$x_2 = \frac{1}{\det A}(b_2 + b_3 a_{23} + b_1 a_{23} a_{31}),$$
$$x_3 = \frac{1}{\det A}(b_3 + b_1 a_{31} + b_2 a_{31} a_{12}).$$

The example is from a section titled, "Solution of Systems of Equations by Using Determinants," and the point in introducing this example is that, unless the number of unknowns is small, determinants are practicable only for solving problems with sparse matrices, where many of the entries are zero (Hohn 1973, 274–77).

Beyond this one example, I have found no other examples that are similar. To underscore how unusual the forms of matrices in equations 7.2 and 7.1 are, below are the two examples that most closely resemble these forms. The first example is the following:

$$A = \begin{bmatrix} -1 & 1 & 0 & 0 \\ 0 & -1 & 1 & 0 \\ 0 & 0 & -1 & 1 \\ 1 & 0 & 0 & -1 \end{bmatrix},$$

where, because this is an edge-node incidence matrix, all of the entries must be either 1, 0, or -1 (Strang [1976] 1988, 105). The second matrix is presented as an example of an M-matrix,[11]

$$\begin{bmatrix} 1 & -t & 0 & 0 & \cdots & 0 \\ 0 & 1 & -t & 0 & \cdots & 0 \\ 0 & 0 & 1 & -t & \ddots & \vdots \\ \vdots & \vdots & \ddots & \ddots & \ddots & 0 \\ 0 & 0 & \cdots & 0 & \ddots & -t \\ -t & 0 & \cdots & \cdots & 0 & 1 \end{bmatrix},$$

where $0 \leq t < 1$ (Horn and Johnson 1991, 374).

[11] An M-matrix is a matrix that is positive stable, that is, the real part of all eigenvalues is positive, and all of the nondiagonal entries are less than or equal to zero.

Appendix B
Chinese Mathematical Treatises

Bibliographies of Chinese Mathematical Treatises

This appendix represents a preliminary attempt to assemble a comprehensive bibliography of all of the Chinese mathematical treatises that have been recorded in Chinese bibliographies throughout history.

At present, there are no comprehensive bibliographies of Chinese mathematical treatises. In fact, the standard bibliography for Chinese mathematical treatises, *Complete Record of the Four Treasuries, Mathematics Volume* (*Sibu zong lu suanfa bian* 四部總錄算法編), edited by Ding Fubao 丁福保 and Zhou Yunqing 周雲青 (Shanghai: Shang wu yin shu guan 商務印書館, 1957), is very incomplete and out of date, listing only a fraction of the mathematical treatises that are recorded in Chinese bibliographies. Similarly, none of the recent Western historical studies of Chinese mathematics provide an adequate bibliography of Chinese mathematical treatises; in fact, bibliographies are often limited to several of the more important texts.

This appendix is divided into the following sections:

1. The first presents a comprehensive list of Chinese bibliographies that record titles of Chinese mathematical treatises. These bibliographies are ordered by the dynasty of the treatises recorded, earliest first.
2. The second section, given in the form of a table, presents a comprehensive list of all titles of Chinese mathematical treatises recorded in all available Chinese bibliographic treatises. Most of these treatises are no longer extant. In some cases, different treatises may have the same title; similarly, one treatise may be given different titles. In addition, the titles of Chinese treatises often included the number of chapters (*juan*). Differences in the recorded numbers of chapters for a title may indicate that portions of the text were lost, or that various titles were incorporated as one title. Therefore, each different title, as well as each variation in the number of chapters, is listed in this table as a separate entry. In order that information on titles can be compared, the table is arranged in columns, each column representing one of the most

important bibliographies for Chinese mathematical treatises of the late imperial period. From this table, then, it can easily be seen whether a title is listed in several bibliographies.

3. The final section, "Notes," then records each listing exactly as found in each bibliography, numbered as a reference to the table. This allows me to record any further information found in the bibliography, for example whether a treatise is listed with other treatises, whether an author or date is provided, or whether a notation states that the treatise is corrupt or incomplete. In cases where a treatise is known to be extant, I have included that fact in these notes.

Collections and Abbreviations

BJTCS Baojing tang congshu 抱經堂叢書.
BSJJZ Ba shi jingji zhi 八史經籍志.
CSJCCB Congshu jicheng chubian 叢書集成初編.
ESWSBB Ershi wu shi bu bian 二十五史補編.
GGTSKS Guangu tang suo kan shu 觀古堂所刊書.
GGTSMCK Guangu tang shumu congkan 觀古堂書目叢刊.
GXJBCS Guoxue jiben congshu 國學基本叢書.
HSYWZ Han shu yiwen zhi, she bu 漢書藝文志， 拾補.
JDJL Jing dian ji lin 經典集林.
JSGJBS Jishan guan ji bu shu 稷山館輯補書.
KGSSSFCS Kuai ge shi shi shan fang cong shu 快閣師石山房叢書.
LJYYWZ Liao Jin Yuan yiwen zhi 遼金元藝文志.
MDSMTBCK Ming dai shumu ti ba congkan 明代書目題跋叢刊.
MSYWZ Ming shi yiwen zhi, bubian, fubian 明史藝文志， 補編， 附編.
PXLCS Puxue lu congshu 樸學廬叢書.
QMACS Qianmo an congshu qi juan 千墨菴叢書七卷.
SBBY Sibu bei yao 四部備要.
SBCK Sibu cong kan 四部叢刊.
SKQS Wenyuan ge Siku quanshu 文淵閣四庫全書.
SLCSJB Song lin congshu jia bian 松鄰叢書甲編.
SSJJZ Sui shu jingji zhi, bu 隋書經籍志， 補.
SSYWZ Song shi yiwen zhi, bu fu bian 宋史藝文志， 補， 附編.
SXCS Shi xue cong shu 史學叢書.
TSJJYWHZ Tang shu jingji yi wen he zhi, bu lu 唐書經籍藝文合志， 附錄.
WJTCS Wenjing tang congshu 問經堂叢書　經典集林.
WJTSSMCK Wujin Tao shi shumu congkan 武進陶氏書目叢刊.
XYXSQS Xi yuan xiansheng quan shu 郎園先生全書.
YHSFJYS Yuhanshan fang ji yi shu, shi bian mulu lei 玉函山房輯佚書， 史編目錄類.
YHSFJYSXB Yuhanshan fang ji yi shu xu bian, shi bian mulu lei 玉函山房輯佚書續編， 史編目錄類.

YJZCSEJ Yujian zhai congshu er ji 玉簡齋叢書二集.
YYTCS Yueya tang congshu 粵雅堂叢書.
ZBZZCS Zhibuzu zhai congshu 知不足齋叢書.
ZGLDSMCK Zhongguo lidai shumu congkan 中國歷代書目叢刊.
ZGLDYWZ Zhongguo lidai yiwen zhi 中國歷代藝文志.
ZXCZZ Zhou xu chi za zhu 籀鄦誃雜著.

Bibliographies of the Earlier Han (206 B.C.E.*–9* C.E.*)*

Ban Gu 班固 and Yan Shigu 顏師古. *Qian Han shu yiwen zhi, yi juan* 前漢書藝文志一卷. In *BSJJZ, CSJCCB.*
Liu Guangfen 劉光蕡. *Qian Han shu yiwen zhi zhu, yi juan* 前漢書藝文志注一卷. In *ESWSBB.*
Liu Xiang 劉向 and Hong Yixuan 洪頤煊. *Bie lu yi juan* 別錄一卷. In *WJTCS, JDJL.*
———. *Qi lue yi juan* 七略一卷. In *WJTCS, JDJL.*
Liu Xiang and Ma Guohan 馬國翰. *Qi lue bie lu yi juan* 七略別錄一卷. In *YHSFJYS.*
Liu Xiang and Tao Junxuan 陶濬宣. *Bie lu yi juan* 別錄一卷. In *JSGJBS.*
———. *Qi lue bie lu er shi juan* 七略別錄二十卷. In *JSGJBS.*
Liu Xiang and Wang Renjun 王仁俊. *Bie lu bu yi yi juan* 別錄補遺一卷. In *YHSFJYSXB.*
———. *Qi lue bie lu yi juan* 七略別錄一卷. In *YHSFJYSXB.*
Liu Xiang and Yao Zhenzong 姚振宗. *Qi lue bie lu yi wen yi juan* 七略別錄佚文一卷. In *KGSSSFCS.*
Liu Xin 劉歆 and Tao Junxuan 陶濬宣. *Qi lue, yi juan* 七略一卷. In *JSGJBS.*
Liu Xin and Yao Zhenzong 姚振宗. *Qi lue yi wen, yi juan* 七略佚文一卷. In *KGSSSFCS.*
Wang Renjun 王仁俊. *Han shu yiwen zhi kaozheng jiao bu shi juan* 漢書藝文志攷證校補十卷. In *ZXCZZ.*
Wang Yinglin 王應麟. *Han yiwen zhi kaozheng shi juan* 漢藝文志考證十卷. In *ESWSBB, SKQS.*
Yao Zhenzong 姚振宗. *Han shu yiwen zhi tiao li ba juan shou yi juan* 漢書藝文志條理八卷首一卷. In *ESWSBB.*
———. *Han shu yiwen zhi she bu liu juan* 漢書藝文志拾補六卷. In *ESWSBB.*

Bibliographies of the Later Han (25–220 C.E.*)*

Gu Xiangsan 顧櫰三. *Bu Hou Han shu yiwen zhi sanshi yi juan* 補後漢書藝文志三十一卷. In *Xiaofanghu zhai congshu erji* 小方壺齋叢書二集; 10 *juan* edition in *ESWSBB.*

Hou Kang 侯康. *Bu Hou han shu yiwen zhi si juan* 補後漢書藝文志四卷. In *CSJCCB, ESWSBB.*

Qian Dazhao 錢大昭. *Bu xu Hou Han shu yiwen zhi yi juan* 補續後漢書藝文志一卷. In *ESWSBB, CSJCCB, SXCS,* and two *juan* edition in *Jixue zhai congshu* 積學齋叢書.

Yao Zhenzong 姚振宗. *Hou Han shu yiwen zhi si juan* 後漢書藝文志四卷. In *ESWSBB.*

Zeng Pu 曾樸. *Bu Hou Han shu yiwen zhi yi juan kao shi juan* 補後漢書藝文志一卷考十卷. In *ESWSBB.*

Bibliographies of the Three Kingdoms (220–280 C.E.*)*

Hou Kang 侯康. *Bu San guo yiwen zhi si juan* 補三國藝文志四卷. In *ESWSBB CSJCCB, SXCS.*

Yao Zhenzong 姚振宗. *San guo yiwen zhi si juan* 三國藝文志四卷. In *ESWSBB.*

Bibliographies of the Jin (280–420 C.E.*)*

Ding Guojun 丁國鈞 and Ding Chen 丁辰. *Bu Jin shu yiwen zhi si juan, bu yi yi juan, fulu yi juan, kan wu yi juan* 補晉書藝文志四卷補遺一卷附錄一卷刊誤一卷. In *ESWSBB, CSJCCB.*

Huang Fengyuan 黃逢元. *Bu Jin shu yiwen zhi si juan* 補晉書藝文志四卷. In *ESWSBB.*

Qin Rongguang 秦榮光. *Bu Jin shu yiwen zhi si juan* 補晉書藝文志四卷. In *ESWSBB.*

Wen Tingshi 文廷式. *Bu Jin shu yiwen zhi liu juan* 補晉書藝文志六卷. In *ESWSBB.*

Wu Shijian 吳士鑑. *Bu Jin shu jingjizhi si juan* 補晉書經籍志四卷. In *ESWSBB.*

Xun Xu 荀勗 and Wang Renjun 王仁俊. *Zhong jing bo yi juan* 中經簿一卷. In *Yuhanshan fang ji yi shu bu bian* 玉函山房輯佚書補編.

Bibliographies of the North-South Disunion (420–588 C.E.*)*

Chen Shu 陳述. *Bu Nan Qi shu yiwen zhi si juan* 補南齊書藝文志四卷. In *ESWSBB.*

Fu Liang 傅亮 and Fu Yili 傅以禮. *Xu wen zhang zhi yi juan* 續文章志一卷. In *Fu shi jia shu* 傅氏家書.

Liang Yuan Di 梁元帝 and Wang Renjun 王仁俊. *Jin lou zi cang shu kao yi juan* 金樓子藏書攷一卷, in *YHSFJYSXB.*

Nie Chongqi 聶崇岐. *Bu Song shu yiwen zhi yi juan* 補宋書藝文志一卷. In *ESWSBB.*

Ruan Xiaoxu 阮孝緒 and Wang Renjun 王仁俊. *Qi lu yi juan* 七錄一卷. In *YHS-FJYSXB.*

———. *Qi lu yi juan* 七錄一卷. Not consulted. Qing *chaoben,* in Fudan University Library and Beijing Library.

Wang Renjun 王仁俊. *Bu Liang shu yiwen zhi yi juan* 補梁書藝文志一卷. In *ZXCZZ.*

———. *Bu Song shu yiwen zhi yi juan* 補宋書藝文志一卷. In *ZXCZZ.*

Xu Chong 徐崇. *Bu Nan Bei shi yiwen zhi san juan* 補南北史藝文志三卷. In *ESWSBB.*

Bibliographies of the Sui (589–618 C.E.*)*

Wei Zheng 魏徵 and Zhangsun Wuji 長孫無忌. *Sui shu jingjizhi si juan* 隋書經籍志四卷. In *BSJJZ, CSJCCB.*

Yao Zhenzong 姚振宗. *Sui shu jingji zhi kaozheng wushi er juan shou yi juan* 隋書經籍志考證五十二卷首一卷. In *ESWSBB.*

Zhang Pengyi 張鵬一. *Sui shu jingjizhi bu er juan* 隋書經籍志補二卷. In *ESWSBB.*

Zhang Zongyuan 章宗源. *Sui jingjizhi kaozheng shisan juan* 隋經籍志考證十三卷. In *ESWSBB; Chongwen shuju huike shu* 崇文書局彙刻書; also *Sui shu jingji zhi kaozheng shisan juan* 隋書經籍志考證十三卷.

Bibliographies of the Tang (618–906 C.E.*)*

Liu Xu 劉昫. *Jiu Tang shu jingji zhi er juan* 舊唐書經籍志二卷. In *BSJJZ, CSJCCB.*

Ouyang Xiu 歐陽修 and Song Qi 宋祁. *Tang shu yiwen zhi si juan* 唐書藝文志四卷. In *BSJJZ, CSJCCB.*

Bibliographies of the Five Dynasties (907–960 C.E.*)*

Gu Huaisan 顧懷三. *Bu Wu dai shi yiwen zhi yi juan* 補五代史藝文志一卷. In *SXCS, CSJCCB, ESWSBB.*

Song Zujun 宋祖駿. *Bu Wu dai shi yiwen zhi yi juan* 補五代史藝文志一卷. In *PXLCS.*

Bibliographies of the Song (960–1279)

Chen Hanzhang 陳漢章. *Chong wen zong mu ji shi bu zheng si juan* 崇文總目輯釋補正四卷, *Zhuixue tang cong gao chu ji ben* 綴學堂叢稿初集本, in *ZGLDSMCK.*

Chen Kui 陳騤 and Zhao Shiwei 趙士煒. *Zhong xing guan ge shu mu* 中興館閣書目. In *SSYWZ.*

———. *Zhong xing guan ge shumu ji kao wu juan Gu yi shu lu cong ji ben* 中興館閣書目輯考五卷古逸書錄叢輯本. In *ZGLDSMCK.*

Chen Zhensun 陳振孫. *Zhi zhai shu lu jie ti ershi er juan* 直齋書錄解題二十二卷. In *ZGLDSMCK.* Also in *SKQS, CSJCCB.* Also *Zhi zhai shu lu jie ti wushi liu juan cun sishi qi juan* 直齋書錄解題五十六卷存四十七卷 [1–7, 17–56], at the Shanghai Library.

Gao Sisun 高似孫. *Zi lue si juan mu yi juan* 子略四卷目一卷. In *SKQS, CSJCCB, SBBY.* Also: Qing *chaoben* in the Shanghai Library and the Nanjing Library.

Huang Yuji 黃虞稷, Ni Can 倪燦, and Lu Wenchao 盧文弨. *Song shi yiwen zhi bu yi juan* 宋史藝文志補一卷. In *SSYWZ.* Also in *BJTCS, BSJJZ, SXCS, CSJCCB, ESWSBB.*

Shao Xingzhong 紹興中 and Xu Song 徐松. *Siku que shu mu* 四庫闕書目. In *SSYWZ.*

Shao Xingzhong and Ye Dehui 葉德輝. *Mishusheng xu bian dao siku que shu mu er juan* 祕書省續編到四庫闕書目二卷. In *SSYWZ.* Also in *ZGLDSMCK, GGTSZS, GGTSMCK, XYXSQS.*

Tuotuo 脫脫 et al. *Song shi yiwen zhi ba juan* 宋史藝文志八卷. In *SSYWZ.* Also in *BSJJZ, CSJCCB.*

Wang Renjun 王仁俊. *Xi Xia yiwen zhi yi juan* 西夏藝文志一卷. In *ESWSBB.*

Wang Yaochen 王堯臣. *Chongwen zong mu ershi juan* 崇文總目二十卷. In *SKQS.* Also *Chong wen zong mu liushi liu juan* 崇文總目六十六卷, Ming *chaoben* 明少本, at Tianyi ge wenwu baoguan suo 天一閣文物保管所; Qing *chaoben* at Nanjing Library 南京圖書館; and Qing *chaoben* 清抄本 at Hunan Provincial Library 湖南省圖書館.

Wang Yaochen et al., annotated by Qian Dongyuan 錢東垣 et al. *Chongwen zong mu wu juan bu yi yi juan fu lu yi juan* 崇文總目五卷補遺一卷附錄一卷. In *ZGLDSMCK,* reprint of *YYTCS.*

You Mou 尤袤. *Sui chu tang shu mu yi juan* 遂初堂書目一卷. In *ZGLDSMCK.* Also in: *SKQS, CSJCCB.* Also: *You shi Sui chu tang shu mu yi juan*尤氏遂初堂書目一卷, Qing *chaoben,* at the Nanjing Library.

Zhang Pan 張攀. *Zhongxing guan ge xu shu mu ji kao yi juan* 中興館閣續書目輯考一卷. In *ZGLDSMCK.*

Zhang Pan and Zhao Shiwei 趙士煒. *Zhongxing guan ge shu xu mu* 中興館閣書續目. In *SSYWZ.*

Zhao Gongwu 晁公武. *Zhaode xiansheng Jun zhai du shu zhi si juan hou zhi er juan fu zhi yi juan, kao yi yi juan* 昭德先生郡齋讀書志四卷後志二卷附志一卷考異一卷, *zhi* and *kao yi* by Zhao Xibian 趙希弁. In *SKQS.* Also: *Zhaode xiansheng Jun zhai du shu zhi si juan fu zhi yi juan, hou zhi er juan, er ben si juan, kao yi yi juan* 昭德先生郡齋讀書志四卷附志一卷後志二卷二本四卷考異一卷, in *Sibu cong kan san bian* 四部叢刊三編, *Xu gu yi congshu* 續古逸叢書.

Zhao Gongwu and Yao Yingji 姚應績. *Qu ben Jun zhai du shu zhi* 衢本郡齋讀書志二十卷. In *Wan wei bie cang* 宛委別藏. In *ZGLDSMCK*

Zhao Shiwei 趙士煒. *Zhongxing guan ge shumu ji kao* 中興館閣書目輯考. Beijing, 1933.

———. Song guo shi yiwen zhi 宋國史藝文志. In *SSYWZ*. Also *Song guo shi yiwen zhi ji ben* 宋國史藝文志輯本 (Beijing, 1933).

Zhongxing si chao guo shi 中興四朝國史. Fragments preserved in 文獻通考, collected by Zhao Shihui 趙士煒 in *Guo li Beiping tushuguan guan kan* 國立北平圖書館館刊 6, 1932, 413–436.

Bibliographies of the Liao and Jin (916–1234)

Huang Renheng 黃任恆. *Bu Liao shi yiwen zhi yi juan* 補遼史藝文志一卷. In *ESWSBB.*

Miao Quansun 繆荃孫. *Liao yiwen zhi* 遼藝文志一卷. In *ESWSBB.*

Ni Can 倪燦 and Lu Wenchao 盧文弨. *Bu Liao Jin Yuan yiwen zhi yi juan* 補遼金元藝文志一卷. In *BJTCS, BSJJZ, SXCS, CSJCCB, ESWSBB.*

Wang Renjun 王仁俊. *Liao shi yiwen zhi bu zheng yi juan* 遼史藝文志補證一卷. In *ESWSBB.*

Comprehensive pre-Yuan bibliographies (up to 1279)

Jin Menzhao 金門詔. *Bu San shi yiwen zhi yi juan* 補三史藝文志一卷. In *BSJJZ, SXCS, CSJCCB, ESWSBB.*

Lu Wenchao 盧文弨. *Wenxian tongkao jingji jiao bu yi juan* 文獻通考經籍校補一卷. In *BJTCS* [*Qun shu she bu chu bian* 群書拾補初編] and *CSJC* [*Qun shu she bu* 群書拾補].

Ma Duanlin 馬端臨. *Wen xian tongkao* 文獻通考. Edited by Hua nan shi da guji yanjiusuo 華東師大古籍研究所. Hua dong shifan daxue chubanshe 華東師范大學出版社.

Zhuang Shuzu 莊述祖. *Lidai zai ji zu zheng lu yi juan* 歷代載籍足徵錄一卷. In *Zhen yi huan yi shu* 珍埶宧遺書, *Xunzuan tang congshu* 訓纂堂叢書.

Bibliographies of books in the Yuan Dynasty (1260–1368)

Hu Shi'an 胡師安. *Yuan Xihu shu yuan chong zheng shu mu* 元西湖書院重整書目一卷. In *SLCSJB*.
Liao Jin Yuan yiwen zhi 遼金元藝文志.
Qian Daxin 錢大昕. *Yuan shi yiwen zhi si juan* 元史藝文志四卷. In *BSJJZ*. Also: *Bu Yuan shi yiwen zhi si juan* 補元史藝文志四卷 in *Er shi wu shi* 二十四史.

Bibliographies of books in the Ming Dynasty (1368–1644)

Bibliographies compiled during the Ming Dynasty

Chen Bingyao 陳秉鑰. *Jigu ge sho ke shumu yi juan* 汲古閣所刻書目一卷; n.a., *Bu yi yi juan, shu ban cun wang kao yi juan* 補遺一卷書板存亡考一卷. At the Beijing Library; not consulted.
Chen Di 陳第. *Shishan tang cang shu mulu er juan* 世善堂藏書目錄二卷. In *MDSMTBCK*. Also in *ZBZZCS*; in *CSJCCB*.
Cheng Dawei 程大位. *Suanfa tongzong* 算法統宗. See SFTZ, in the Bibliography, Primary Sources—Chinese.
Chong zheng Nei ge da ku canben shu ying 重整內閣大庫殘本書影. Beijing: Gugong bowuyuan wenxian guan 故宮博物院文獻館, 1933.
Du Mu 都穆. *Nan hao ju shi wen ba si juan* 南濠居士文跋四卷. Ming *keben*, at Beijing Library; *Nan hao wen ba si juan* 南濠文跋四卷, Qing *chaoben*, at Nanjing Library; not consulted.
Gao Ru 高儒. *Bai chuan shu zhi* 百川書志. In *Bai chuan shu zhi, Gu jin shu ke* 百川書志 古今書刻. In *MDSMTBCK*. Also in *GGTSMCK*, *XYXSQS*. Modern edition: Gudian wenxue chubanshe 古典文學出版社 (1957). Also: *Bai chuan shu zhi ershi juan* 百川書志二十卷 in *QMACS*, at Beijing Library and Shanghai Library.
Guo Pan 郭磐. *Ming tai xue jingjizhi* 明太學經籍志. In *MDSMTBCK*. Also in *Tan yin lu congshu* 蟫隱廬叢書.
Hao Qingbo 郝慶柏 *Yongle da dian shumu kao si juan* 永樂大典書目考四卷. In *Liao hai congshu di ba ji* 遼海叢書第八集.
Huang chao shu ke bu fen juan 皇朝書刻不分卷. Qing *chaoben*, at Shanghai Library; not consulted.
Jiao Hong 焦竑. *Guo shi jingjizhi wu juan fulu yi juan* 國史經籍志五卷附錄一卷. In *MSYWZ*, *MDSMTBCK*. Also in *CSJCCB*, *YYTCS* .
Jigu ge Mao shi cang shu mulu 汲古閣毛氏藏書目一卷. Qing *chaoben*. At the Chinese Academica Sinica Library 中國科學院圖書館; not consulted.
Jingu tang shumu er juan 近古堂書目二卷. In *MDSMTBCK*. Also in *YJZCSEJ*.
Kuaiqi Niu shi shi xue lou zhen cang tu shu mu 會稽鈕氏世學樓珍藏圖書目. In *MDSMTBCK*.

Li Echong 李鶚翀. *Jiangyin Li shi deyue lou shumu zhailu* 江陰李氏得月樓書目摘錄一卷. In *MDSMTBCK*. Also in *Jiangyin congshu* 江陰叢書, *Suxiang shi congshu* 粟香室叢書, *Changzhou xian zhe yishu di yi ji, shi bian* 常州先哲遺書第一集，史編.

Li Tingxiang 李廷相. *Puyang Puting Li xiansheng jia cang mulu* 濮陽蒲汀李先生家藏目錄. In *MDSMTBCK*. Also in *Yujian zhai congshu er ji* 玉簡齋叢書二集.

Liu Ruoyu 劉若愚. *Nei ban jing shu ji lue* 內板經書紀略. In *MDSMTBCK*. Also in *SLCSJB*.

Liu Yuanliang 劉元亮. *Yu xuan xin zuan gu jin shu mu er juan* 玉軒新纂古今書目二卷. Ming *chaoben*, at Shandong Provincial Museum 山東省博物館; not consulted.

Mao Jin 毛晉. *Jigu ge jiao ke shu mu* 汲古閣校刻書目. In *MDSMTBCK*.

———. *Jigu ge Mao shi cang shu mulu* 汲古閣毛氏藏書目錄. Harvard University-Yenching University Library, *chaoben*, 1950.

———. *Jigu ge shuba* 汲古閣書跋. Gudian wenxue chubanshe 古典文學出版社, 1958.

———. *Ti ba er juan* 題跋二卷. Ming *keben*, at the Beijing Library, Shanghai Library, and Nanjing Library; Qing *chaoben*, at the Zhejiang Library; not consulted.

———. *Ti ba xu er juan* 題跋續二卷. Ming *keben*. At the Shanghai Library; not consulted.

———. *Yin hu ti ba er juan* 隱湖題跋二卷. In *MDSMTBCK*. Also in *Yu shan cong ke* 虞山叢刻.

Mao Yi 毛扆. *Jigu ge kan shu xi mu yi juan, Jigu ge zhen cang mi ben shumu yi juan* 汲古閣刊書細目一卷汲古閣珍藏祕本書目一卷. At the Sichuan Provincial Library 四川省圖書館; not consulted.

———. *Jigu ge zhen cang mi ben shu mu* 汲古閣珍藏祕本書目一卷. In *CSJCCB*, *Shi li ju Huang shi congshu* 士禮居黃氏叢書. Also at the Beijing Library, Nanjing Library.

Mei Zhuo 梅鷟. *Nan yong jing ji kao er juan* 南廱志經籍考二卷. In *MDSMTBCK*. Also in *GGTSKS, GGTSMCK, SLCSJB, XYXSQS*.

Nanhao ju shi wen ba si juan 南濠居士文跋四卷. In *MDSMTBCK*.

Qi Chengye[?] 祁承㸁.[1] *Dansheng tang cang shu mu* 澹生堂藏書目十四卷. In *MDSMTBCK*. Also in *Shaoxing xian zheng yi shu di san ji* 紹興先正遺書第三集.

———. *Dan sheng tang cang shu mu ba juan, cang shu yue yi juan, gengshen zheng shu xiao ji yi juan, gengshen zheng shu li lue yi juan* 澹生堂藏書目八卷藏書約一卷庚申整書小記一卷庚申整書例略一卷. Qing *chaoben*, at the Beijing Library; not consulted.

———. *Dansheng tang cang shu mu bu fen juan, cang shu xun yi juan, cang shu yue yi juan* 澹生堂藏書目不分卷藏書訓一卷藏書約一卷, Ming *chaoben*, at

[1] The third character in Qi's name is idiosyncratic, and does not appear in standard dictionaries; in fact, the component on the right does not appear in standard dictionaries. In the character I have composed here, 㸁, I have used the character *ye* 業 as the right-hand component, which I have also used as the most likely pronunciation.

the Shanghai Library (only the sections *jing* 經, *shi* 史, *zi* 子, *cang shu xun* 藏書訓, and *cang shu yue* 藏書約 are extant); not consulted.

———. *Dansheng tang cang shu mu bu fen juan* 澹生堂藏書目不分卷, at the Qingdao Municipal Museum 清島市博物館; Qing *chaoben*, at the Beijing Library; *Dansheng tang cang shu mu bu fen juan* 淡生堂藏書目不分卷, *gaoben*, at the Nanjing Library; not consulted.

———. *Dansheng tang cang shu xun yue si juan kuang ting ji er juan* 澹生堂藏書訓約四卷曠亭集二卷. Ming *keben*, at the Beijing Library; not consulted.

———. *Dansheng tang cang shu yue, si juan* 澹生堂藏書約四卷. In *Ou xiang ling she* 藕香零拾; one *juan* edition in *ZBZZCS di wu ji* 第五集, and *Biji xiaoshuo daguan di ba ji* 筆記小說大觀第八輯.

Qian Pu 錢溥. *Mige shumu* 祕閣書目. In *MDSMTBCK*. Also: *Mige shumu bu fen juan* 祕閣書目不分卷, Qing *chaoben*, at the Chinese Academica Sinica Library 中國科學院圖書館.

Qian Qianyi 錢謙益. *Yiyun lou shumu bu yi yi juan* 繹雲樓書目補遺一卷. In Ye Dehui 葉德輝 *Guan gu tang hui ke shu* 觀古堂彙刻書, published in Xiangtan in Guangxu renyin 湘潭光緒壬寅 [1903], republished in Taibei 台北 in 1971, pp. 483–494.

Sun Fengyi 孫馮翼 *Siku quan shu Yongle da dian ben shumu yi juan* 四庫全書輯永樂大典本書目一卷. In *Liao hai congshu di ba ji* 遼海叢書第八集.

Wang Dunhua 王敦化. *Ming ban shu jing yan lu* 明板書經眼錄〔子部〕. Qilu daxue guoxue yanjiusuo qian yin ben 齊魯大學國學研究所鉛印本.

Wang Lian 王璉. *Li dai jingji kao ershi si juan* 歷代經籍考二十四卷. Ming Zhengde 12, *Shendu zhai keben* 明正德十二年慎獨齋刻本, at the Beijing Normal University Library 北京師範大學圖書館; not consulted.

Wu Kuan 吳寬. *Wu Wending gong cang shu mu yi juan* 吳文定公藏書目一卷. In *QMACS*; not consulted.

Xu Bo[?] 徐𤈦.[2] *Hongyu lou ti ba er juan* 紅雨樓題跋二卷. Qing *chaoben* in Beijing Library and Nanjing Library; not consulted.

———. *Xu shi jia cang shumu* 徐氏家藏書目. In *MDSMTBCK*. Also: *Xu shi jia cang shumu si juan* 徐氏家藏書目七卷, in Beijing Library; four *juan* in the Shanghai Library.

Xu Bo[?] and Miao Quansun 繆荃孫. *Chong bian Hongyu lou ti ba er juan* 重編紅雨樓題跋二卷. In *MDSMTBCK*. Also in *Qiaofan lou congshu* 峭帆樓叢書.

Xu Tu 徐圖. *Xing ren si chong ke shu mu* 行人司重刻書目. In *MDSMTBCK*. Also edition not divided into *juan* in *Jimao cong bian san ce* 己卯叢編三冊, 1939 edition.

Xuanshang zhai shumu ba juan 玄賞齋書目八卷. In *MDSMTBCK*.

Xun Nengchuan 孫能傳 and Zhang Xuan 張萱. *Neige cangshu mulu ba juan* 內閣藏書目錄八卷. In *MDSMTBCK*. Also in *Shi yuan cong shu di yi ji* 適園叢書第一集; modern edition published by Wenwu chubanshe 文物出版社 (1991) 9.

[2] The character for Xu's given name is idiosyncratic. In the character I have composed here, 火勃, the character *bo* 勃 is the right-hand component, which I have also used as the most likely pronunciation.

Yang Shiqi 楊士奇. *Wenyuan ge shumu ershi juan* 文淵閣書目二十卷, reprints of the *Duhua zhai congshu wu ji* 讀畫齋叢書戊集 edition in *MDSMTBCK*, and typeset edition in *MSYWZ*. Also in *CSJCCB* and *GXJBCS*. Four *juan* edition in *SKQS*; edition not divided into *juan*, Qing *chaoben* 清抄本 in Beijing Library and Nanjing Library.

Yao Guangxiao 姚廣孝. *Yongle da dian mulu liushi juan* 永樂大典目錄六十卷. In Yang Shangwen 楊尚文, ed., *Lian yun yi congshu* 連筠簃叢書.

Ye Sheng 葉盛. *Lüzhu tang shumu san juan* 菉竹堂書目三卷. In *MDSMTBCK*. Also six *juan* edition: in *CSJCCB*, *Yueya tang congshu er bian di wu ji* 粵雅堂叢書二編第十五集. Also: six *juan* edition at the Shanghai Library and Beijing Library [only 1–2 and 4–6 extant]; edition not divided into *juan* [*bu fen juan* 不分卷], Qing *chaoben* at the Beijing Library and Shanghai Library.

Zhao Li 晁瑮. *Zhao shi Baowen tang shumu san juan* 晁氏寶文堂書目三卷. In *MDSMTBCK*.

Zhao Qimei 趙琦美. *Guawang guan shumu* 脈望館書目. In *MDSMTBCK*. Also in *YJZCSEJ*. Edition not divided into *juan* in *Hanfen lou mi ji di liu ji* 涵芬樓祕笈第六集. Also: edition not divided into *juan*, Qing *chaoben*, and four *juan* edition, in Beijing Library.

Zhao Yongxian 趙用賢. *Zhao Dingyu shumu* 趙定宇書目. In *MDSMTBCK*. Also edition not divided into *juan*, Qing *chaoben*, at the Shanghai Library (*ZGGJS-BSM*, p. 1386). Modern edition published by Gudian wenxue chubanshe 古典文學出版社, 1957.

Zhou Hongzu 周弘祖. *Gu jin shu ke er juan* 古今書刻二卷. In *GGTSKS, GGTSM, XYSS*. Also in *Bai chuan shu zhi, Gu jin shu ke* 百川書志　古今書刻, Gudian wenxue chubanshe 古典文學出版社, 1957. In *MDSMTBCK*.

Zhu Mujia[?] 朱睦㮮.[3] *Ju le tang yi wen mulu shi qi juan* 聚樂堂藝文目錄十七卷, Qing *chao ben*, at the Beijing Library. Also six *juan* edition, Qing *chaoben*, also at the Beijing Library; not consulted.

———. *Wanjuan tang shumu* 萬卷堂書目四卷. In *MDSMTBCK*. Also in: *GGT-SKS, GGTSMCK, YJZCSEJ, XYXSQS*. Also: edition not divided into *juan*, Qing *chaoben*, at Nanjing Library, Qinghua University Library.

———. *Wanjuan tang jia cang yi wen mu bu fen juan* 萬卷堂家藏藝文目不分卷, Qing *chaoben* at Nanjing Library; not consulted.

———. *Wanjuan tang yi wen mu lu shi juan* 萬卷堂藝文目錄十卷. In *QMACS*; two *juan* edition, Qing *chaoben*, at the Beijing Library; not consulted.

Bibliographies compiled after the Ming Dynasty

Fan Qin 范欽. *Tianyi ge shumu si juan* 天一閣書目四卷. Also, edition not divided into *juan* at the Beijing Library. See also Ruan Yuan 阮元, *Tianyi ge shumu shi*

[3] The third character in Zhu's name is idiosyncratic. The character I have composed, 㮮, has the character *jia* 挈 as the right-hand component, which I have used as the most likely pronunciation of the character.

juan 天一閣書目十卷, Qing *keben* at Fudan University; Fan Maozhu 范懋柱 *Tianyi ge jin cheng shu bu fen juan* 天一閣進呈書不分卷. Qing *chaoben* at Shandong sheng Bowuguan 山東省博物館.

Fu Weilin 傅維麟. *Ming shu yiwen zhi* 明書經籍志. Jiangsu Guangling guji keyinshe 江蘇廣陵古籍刻印社, 1988.

Gu Tinglong 顧廷龍 and Pan Chengbi 潘承弼, ed. *Ming dai banben tu lu chu bian* 明代版本圖錄初編. Kaiming shudian 開明書店, n.d.

Hu Yinglin 胡應麟. *Sibu zheng wei san juan* 四部正偽三卷. In *Shao shi shan fang si ji, bi cong* 少室山房四集, 筆叢; *Guangya shuju congshu, za zhu, Shao shi shan fang ji, Shao shi shan fang bi cong* 廣雅書局叢書, 雜著, 少室山房集, 少室山房筆叢; and *Ming Qing biji congkan, Shao shi shan fang bi cong* 明清筆記叢刊, 少室山房筆叢.

Huang Yuji 黃虞稷. *Qianqing tang shumu sanshi er jaun* 千頃堂書目三十二卷. In *SKQS, Shi yuan congshu di er ji* 適園叢書第二集. Also edition not divided into *juan*.

———, collated and edited by Wang Hongxu 王鴻緒, Zhang Yu 張廷玉, et al. *Ming shi yiwen zhi* 明史藝文志. Zhonghua shuju edition; also in *BSJJZ, CSJCCB;* in *MYBF.*

Jin Menzhao 金門詔. *Ming shi jingjizhi yi juan* 明史經籍志一卷. In *Jin tai shi quan ji* 金太史全集.

Li Jinhua 李晉華. *Ming dai chi zhuan shu kao* 明代敕撰書考 (Official publications ordered by the Ming Emperors). Harvard-Yenching Institute Sinological Index Series, Supplement No. 3, Beijing, 1932.

Mei Wending 梅文鼎. *Wuyan li suan shu ji yi juan* 勿庵歷算書記一卷. In *SKQS.* Also: *Wuyan li suan shu mu yi juan* 勿庵歷算書目一卷 in *ZBZZCS, CSJCCB.*

Ming ji zhushu shuji bu fen juan 明季著述書籍不分卷. Qing *chaoben,* at Shanghai Library; not consulted.

Ming shi yiwen zhi bubian 明史藝文志補編. In *MSYWZ.*

Ming shu jingjizhi 明書經籍志. In *MSYWZFB.*

Qin ding Siku quan shu zong mu erbai juan shou si juan 欽定四庫全書總目二百卷首四卷. In *Wuying dian ju zhen ban shu* 武英殿聚珍版書.

Qin ding Siku quan shu kaozheng yibai juan 欽定四庫全書考證一百卷. In *Wuying dian ju zhen ban shu* 武英殿聚珍版書; also *Siku quan shu kaozheng yibai juan* 四庫全書考證一百卷 in *CSJCCB.*

Si Ming Tianyi ge cang shu mulu 四明天一閣藏書目錄. In *Yujian zhai congshu er ji* 玉簡齋叢書二集.

Tao Xiang 陶湘. *Ming Mao shi Jigu ge ke shu mulu yi juan,* 明毛氏汲古閣刻書目錄一卷. In *Wujin Tao shi shumu congkan* 武進陶氏書目叢刊.

———. *Ming Wu Xingmin ban shumu* 明吳興閔板書目一卷. In *Wujin Tao shi shumu congkan* 武進陶氏書目叢刊.

———. *Ming dai Neifu jing chang ben shumu yi juan* 明代內府經廠本書目一卷. In *Wujin Tao shi shumu congkan* 武進陶氏書目叢刊.

Wang Hongxu 王鴻緒. *Ming shi gao* 明史稿. Wenhai chubanshe 文海出版社.

Xue Fucheng 薛福成. *Tianyi ge jian cun shumu si juan* 天一閣見存書目四卷, *gaoben,* at the Beijing Library; *Tianyi ge jian cun shumu si juan, shou yi juan,*

mo yi juan 天一閣見存書目四卷首一卷末一卷, *gaoben*, at the Zhejiang Library; not consulted.
Yuan Tongli 袁同禮. *Yongle da dian xiancun juan mu biao* 永樂大典現存卷目表, 1929.
Zheng Demao 鄭德懋. *Jigu ge jiao ke shumu yi juan, bu yi yi juan, ke ban cun wang kao yi juan* 汲古閣校刻書目一卷補遺一卷刻版存亡考一卷. In *Xiaoshishan fang congshu, di jiu ce* 小石山房叢書第九冊.
Zheng fang Ming ji yi shu mu 徵訪明季遺書目.
Zhu Yizun 朱彝尊. *Ming shi zong cai ji shumu yi juan* 明詩綜采輯書目一卷. In Shijing ge congshu 石經閣叢書.

Selected bibliographies of books in the Qing Dynasty (1644–1911)

Chijing zhai shumu 持靜齋書目 (Catalogue of books in the Chijing Studio).
Ding Shen 丁申. *Wulin cangshu lu* 武林藏書錄 (Catalogue of books preserved at Wulin). Gudian wenxue chubanshe 古典文學出版社, 1957.
Lu Xinyuan 陸心源. *Bisong lou cangshu zhi* 皕宋樓藏書志 (Record of books preserved at the Bisong Hall).
———. *Bisong lou cangshu xuzhi* 皕宋樓藏書續志 (Continuation of the record of books preserved at the Bisong Hall).
Ma Ying 馬瀛 and Pan Jingzheng 潘景鄭. *Yinxiangxian guan shumu* 吟香仙館書目 (Catalogue of books of the Yinxiangxian Hall). Gudian wenxue chubanshe 古典文學出版社, 1958.
Mo Youzhi 莫友芝. *Lü ting zhi jian zhuan ben shumu* 郘亭知見傳本書目.
Qi Lixun 祁理孫. *Yiqing cangshulou shumu* 奕慶藏書樓書目 (Catalogue of books preserved at the Yiqing Library). Gudian wenxue chubanshe 古典文學出版社, 1958.
Qian Zeng 錢曾. *Yeshi yuan cang shumu* 也是園藏書目 (Catalogue of books preserved at the Yeshi Garden).
Quan hui shumu 全燬書目 (Catalogue of books completely destroyed), in *Qing dai jin hui shumu si zhong* 清代禁燬書目四種.
Rong Zhu 榮柱 and Yao Jinyuan 姚覲元. *Wei ai shumu* 違礙書目, (Catalogue of prohibited books), in *Qing dai jinhui shumu sizhong* 清代禁燬書目四種.
Shao Yichen 邵懿辰 *Zeng ding Siku jianming mulu biao zhu ershi juan* 增訂四庫簡明目錄標注二十卷.
Shen Deshou 沈德壽. *Baojing lou cangshu zhi* 抱經樓藏書志 (Record of books preserved in the Baojing Hall). Zonghua shuju 中華書局, 1990.
Wu Guang 吳光. *Huang Zongxi zhuzuo huikao* 黃宗羲著作彙考 (Combined investigations on the works of Huang Zongxi). Taiwan xuesheng shuju 臺灣學生書局, 1990.
Wu Weizu 吳慰祖. *Siku caijin shumu* 四庫採進書目 (Catalogue of books presented for the four treasuries). Shangwu yinshu guan 商務印書館, 1960.
Yao Ruo 姚若. *Lü ting shumu* 郘亭書目 (Catalogue of books of the Lü Pavilion).

Yong Rong 永瑢. *Siku quanshu zongmu* 四庫全書總目 (General catalogue of the complete works of the four treasuries). Zonghua shuju 中華書局, 1992.

Zhao Li 晁瑮. *Xu shi hongyu lou shumu* 徐氏紅雨樓書目 (Catalogue of books in the Hongyu Hall of Mr. Xu). Gudian wenxue chubanshe 古典文學出版社, 1957.

Zhao Wei 趙魏. *Zhuyan an zhuan chao shumu* 竹崦盦傳鈔書目 (Zhuyan Hut bibliographies).

Zheng tang dushu ji 鄭堂讀書記 (Records of reading books of the Zheng Hall).

Zheng Yuanqing 鄭元慶. *Bisong lou cangshu yuanliu kao* 皕宋樓藏書源流考 (Investigations of the origin of books preserved at the Bisong Hall).

Modern Bibliographies

General Bibliographies

Du Xinfu 杜信孚, Zhou Guangpei 周光培, and Jiang Xiaoda 蔣孝達. *Ming dai banke zonglu* 明代版刻綜錄 (Comprehensive catalogue of Ming Dynasty publications). Jiangsu Guangling guji keyinshe 江蘇廣陵古籍刻印社, 1983.

Feng Cheng 馮澂. *Qiangzili zhai xuji si zhong* 強自力齋續集四種 (Four continuations of the literary collection of Qiangzili Studio).

Guoli zhongyang tushuguan tecang zu 國立中央圖書館特藏組, ed. *Zhongguo lidai yiwen zongzhi* 中國歷代藝文總志 (Combined record of dynastic bibliographies). Taipei: Guoli zhongyang tushuguan 國立中央圖書館, 1989.

Hong Ye 洪業, Nie Chongqi 聶崇岐, Li Shuchun 李書春, and Ma Xiyong 馬錫用. *Shihuo zhi shiwu zhong zonghe yinde, yiwen zhi ershi zhong zonghe yinde* 食貨志十五種綜合引得 藝文志二十種綜合引得 (Combined indices to fifteen Records of government finance; Combined indices to twenty bibliographies), Shanghai guji chubanshe 上海古籍出版社, 1986.

Gu Xiu 顧修. *Hui ke shumu* 彙刻書目 (Catalogue of books combined in printing). Shanghai: Fuying shuju 福瀛書局, 1886.

Jiangsu shengli guoxue tushuguan zongmu bubian, liu juan 江蘇省立國學圖書館總目補編六卷 (Supplement to the General catalogue of the Jiangsu Province Library of Chinese Studies, 6 *juan*).

Jinshu zongmu 禁書總目 (General catalogue of prohibited books), in *Qing dai jin hui shumu si zhong* 清代禁燬書目四種.

Liu Yizheng 柳詒徵. *Jiangsu shengli guoxue tushuguan zongmu, sishi si juan* 江蘇省立國學圖書館總目四十四卷 (General catalogue of the Jiangsu Province Library of Chinese Studies, 44 *juan*).

Sun Dianqi 孫殿起. *Fan shu ou ji* 販書偶記 (Fortuitous records of books sold), Shanghai guji chubanshe 上海古籍出版社, 1982.

———. *Fan shu ou ji xu bian* 販書偶記續編 (Continuation of Fortuitous records of books sold), Shanghai guji chubanshe 上海古籍出版社, 1980.

Yao Jinguang 姚覲光. *Qing dai jinhui shumu si zhong* 清代禁燬書目四種 (Four catalogues of books prohibited and destroyed during the Qing Dynasty). In *Wanyou wenku* 萬有文庫.

Yao Jinyuan 姚覲元 and Sun Dianqi 孫殿起. *Qing dai jin hui shumu, Qing dai jinshu zhi jian lu* 清代禁毀書目 清代禁書知見錄 (Catalogue of books prohibited and destroyed during the Qing Dynasty; Record of books prohibited in the Qing Dynasty, known and seen). Shangwu yinshuguan 商務印書館, 1957.

Zhang Yu 章鈺 and Wu Zuocheng 武作成. *Qing shi gao yiwen zhi ji bubian* 清史稿藝文志及補編 (Bibliography section of the History of the Qing, with supplement). Zonghua shuju 中華書局, 1982.

Zhang Zhidong 張之洞 and Miao Quanxun 繆荃孫. *Shumu dawen buzheng wu juan* 書目答問補正五卷 (Supplement and corrections to Answers to questions on bibliography), Shanghai guji chubanshe 上海古籍出版社, 1986.

Bibliographies of Rare Editions

Beijing daxue tushuguan cang shanben shumu 北京大學圖書館藏善本書目 (Catalogue of rare editions preserved at the Beijing University Library), Beijing daxue tushuguan 北京大學圖書館, 1958.

Wang Zhongmin 王重民. *Shanben shuji jing yan lu* 善本書籍經眼錄 (Record of rare editions seen).

Bibliographies of Mathematical Treatises

Chen Li 陳澧. *Suan shu mulu yi ce* 算書目錄一冊 (Catalogue of mathematical treatises, one volume), in *Cao mu* 草目.

Ding Fubao 丁福保. *Suanxue shumu tiyao san juan* 算學書目提要三卷 (Annotated bibliography of mathematical treatises, three *juan*).

Ding Fubao 丁福保 and Zhou Yunqing 周雲青. *Sibu zong lu suanfa bian* 四部總錄算法編 (General record of the Four Sections, mathematics edition), Shangwu yinshuguan 商務印書館, 1957.

———. *Sibu zong lu suanfa bian buyi* 四部總錄算法編補遺 (Supplement to the General record of the Four Sections, mathematics edition).

———. *Sibu zong lu tianwen bian* 四部總錄天文編 (General record of the Four Sections, astronomy edition), Shangwu yinshuguan 商務印書館, 1956.

Feng Cheng 馮澂. *Suanxue kao chubian ershi juan* 算學考初編二十卷 (Investigation into mathematics, first edition, 20 *juan*), in *Qiangzili zhai xuji si zhong* 強自力齋續集四種.

Feng Cheng and Liu Duo 劉鐸. *Gujin suanxue shulu, suanxue kao chubian he zhu* 古今算學書錄、算學考初編合注 (Catalogue of treatises on mathematics, recent and ancient; and Investigation into mathematics, first edition; combined, with commentary), in *Si bu zong lu suanfa bian* 四部總錄算法編.

Fu Zengxiang 傅增湘. *Cang yuan qun shu jing yan lu* 藏園群書經眼錄, Zonghua shuju 中華書局, 1983.

Gu jin suanshu congshu bian mu 古今算術叢書編目.

Li Yan 李儼. *Beiping ge tushuguan suo cang Zhongguo suanxue shu lianhe mulu* 北平各圖書館所藏中國算學書聯合目錄 (Combined catalogue of Chinese mathematical treatises preserved in various libraries in Beijing).

———. *Dongfang tushuguan shanben suan shu jieti* 東方圖書館善本算書解題 (Explanation of rare edition mathematical treatises in the Oriental library), in 中算史論叢.

———. *Li Yan suo cang suanxue shumu* 李儼所藏算學書目 (Catalogue of mathematical treatises preserved by Li Yan). In *Kexue* 科學.

Liang Zhaokeng 梁兆鏗. *Tianwen suanfa kao shiliu ce* 天文算法考十六冊 (Investigation into astronomy and mathematics, 16 volumes).

Liu Chaoyang 劉朝陽. *Zai bu Qiu bian Zhongguo suanxue shumu* 再補裘編中國算學書目 (Another supplement to Qiu's catalogue of Chinese mathematical treatises), in *Yuyan lishi yanjiusuo zhoukan* 語言曆史研究所週刊.

Liu Yingxian 劉應先. *Zhongguo shuxue shuji kao* 中國數學書籍考 (Investigation into treatises on Chinese mathematics). In *Wu Gao shuli* 武高數理.

Qiu Chongman 裘沖曼. *Tianwen suanxue shumu huibian* 天文算學書目彙編 (Compilation of catalogues of treatises on astronomy and mathematics). In *Qinghua xuebao* 清華學報.

———. *Zhongguo suanxue shumu huibian* 中國算學書目彙編 (Compilation of catalogues of Chinese mathematical treatises). In *Qinghua xuebao* 清華學報.

Shao Zhang 邵章 and Shao Yichen 邵懿辰. *Zengding siku jianming mulu biaozhu* 增訂四庫簡明目錄標注 (Concise catalogue of the books of the Four treasuries, with annotations, revised and enlarged), Shanghai guji chubanshe 上海古籍出版社, 1979.

Tang Tiandong 湯天棟. *Zhongguo suanxue shumu huibian zhiyi* 中國算學書目彙編質疑 (Questions concerning Compilation of catalogues of Chinese mathematical treatises). In *Xue yi* 學藝.

Yan Dunjie 嚴敦傑. *Gaoben Zhong suan jiao xu mu* 稿本中算斠序目 (Draft of a catalogue of prefaces to Chinese mathematical treatises, corrected). In *Wen shi fukan* 文史副刊.

———. *Huihui lifa shumu* 回回曆法書目 (Catalogue of calendrical methods of the Muslims). In *Yi shi bao* 益世報.

———. *Qing Guangxu nian Shu ke suanshu* 清光緒年蜀刻算書 (Mathematical treatises published during the Guangxu period of the Qing Dynasty). In *Tushu yuekan* 圖書月刊.

———. *Shu zhong chou ke suan shu* 蜀中疇刻算書 (Mathematical treatises printed by astronomers in Sichuan). In *Zhenli zazhi* 真理雜誌.

———. *Sichuan tian suan yiwen zhi lüe* 四川天算藝文志略 (A summary of the bibliography of astronomy and mathematics of Sichuan). In *Shi shi xin bao xue deng* 時事新報學燈.

———. *Sichuan tongsu suan shu kao* 四川通俗算書考 (An investigation into popular mathematical treatises of Sichuan). In *Xue deng* 學燈.

Zeng Yuanrong 曾遠榮. *Zhongguo suanxue shumu huibian zengbu* 中國算學書目彙編增補 (Supplement to the Compilation of catalogues of Chinese mathematical treatises). In *Qinghua xuebao* 清華學報.

Zeng Zhaoan 曾昭安. *Shuxue jia xingming lu* 數學家姓名錄 (Record of the names of mathematicians). In *Like jikan* 理科季刊.

Zhao Zongjian 趙宗建. *Jiushan lou shumu* 舊山樓書目 (Catalogue of books from the Jiushan Hall).

Zhongguo jian cun suanxue congshu zongmu 中國見存算學叢書總目 (General catalogue of collections of mathematical treatises extant in China).

Critical studies of bibliographies

Chang Bide 昌彼得. "Jiao Hong *Guo shi jingji zhi* de pingjia" 焦竑《國史經籍志》的評價. In *Qu Wanli xiansheng qi zhi rong qing wenji* 屈萬里先生七秩榮慶文集, 307–317.

Chen Yuesu 陳樂素. "Lue lun *Zhi zhai shulu jieti*" 略論《直齋書錄解題》, in *Qiu shi ji* 求是集 2: 310–317.

———. "Yuan ben yu Qu ben *Jun zhai du shu zhi*" 袁本與衢本《郡齋讀書志》, in *Qiu shi ji* 求是集 2: 275–292.

Ding Yu 丁瑜. "*Zhi zhai shulu jieti* ji qi zuozhe jianlun" 《直齋書錄解題》及其作者簡論, in *Guji luncong* 古籍論叢 1: 235–245.

Liu Changrun 劉昌潤. "*Wanjuan tang shumu* xiaokao" 《萬卷堂書目》小考, in *Wen shi* 文史 18: 272.

Sun Meng 孫猛. "*Jun zhai du shu zhi* Qu Yuan er ben de bijiao yanjiu—jian lun 'Jun zhi du shu zhi de cheng shu guocheng' " 《郡齋讀書志》衢袁二本的比較研究－兼論《郡齋讀書志的成書過程》, in *Wen shi* 文史 20: 97–120.

Wang Zhongmin 王重民. "Dui yu *Sui shu jingji zhi* de chubu tanlun" 對於《隋書經籍志》的初步探論, in *Muluxue lunwen xuan* 目錄學論文選.

———. "Qianqing tang shumu kao" 千頃堂書目考, in *Guoxue jikan* 國學季刊 7.1 (1950): 69–90.

Wang Guoyuan 汪辟疆. *Muluxue yanjiu* 目錄學研究. Shanghai: Shangwu yinshuguan 上海商務印書館, 1934.

Wang Wenjin 王文進. "Ming Mao shi xie ben shumu" 明毛氏寫本書目. In *Zhou Shutao xiansheng liushi shengri jidai lunwenji* 周叔弢先生六十生日紀代論文集, 5–22.

Xie Dexiong 謝德雄. "Song dai muluxue de fazhan ji qi chengjiu" 宋代目錄學的發展及其成就, in *Muluxue lunwen xuan* 目錄學論文選.

Xie Guozheng 謝國楨. "Ming Qing shidai de muluxue" 明清時代的目錄學 in *Muluxue lunwen xuan* 目錄學論文選.

Yu Jiaxi 余嘉錫. "*Jule tang yi wen mulu* kao" 《聚樂堂藝文目錄》考. In *Zhongguo shuji kao lunji* 中國書籍考論集, 228–231.

Zhang Minghua 張明華. "Huang Yuji he ta de *Qianqing tang shumu*" 黃虞稷和他的《千頃堂書目》, in *Xue lin man lu* 學林漫錄 9: 118–124.

———. "*Qianqing tang shumu* de yuanliu" 《千頃堂書目》的源流, in *Wen shi* 文史 20: 121–145.

Mathematical Treatises Listed in Chinese Bibliographies

The following table and notes record each listing of a Chinese mathematical treatise exactly as it is found in each bibliography, and a number is assigned to each treatise for reference in the table.

The first column of the table gives the title, including the number of chapters (*juan* 卷). The second column gives the authors (if recorded). The third column, labeled "Pre-Ming," is for entries from bibliographies before the Ming Dynasty. The next fifteen columns are for entries from the most important Ming and early Qing dynasty bibliographies. The last column, labeled "Extant texts," records some of the Yuan and Ming dynasty texts that are extant, many of which, it should be noted, were not recorded in bibliographies of the time.

The notes (beginning on page 241, following the table) then record any further information found in the bibliography, including the following, for each treatise:

1. The number I have assigned to each treatise.
2. Any information on the author.
3. Any information on the date or dynasty.
4. Other treatises listed together with this treatise.
5. Notes on whether the treatise is extant, corrupt, or incomplete.

Treatise Title:	Bibliography: Author:	Pre-Ming	文淵閣書目	秘閣書目	晁氏寶文堂書目	算法統宗	國史經籍志	內閣藏書目錄	世善堂藏書目錄	脈望館書目	近古堂書目	玄賞齋書目	其它明代書目	千頃堂書目	明書	明史稿	明史藝文志	Extant texts
一位算法二卷	江本	1					2											
一鴻筭法	余楷					3												
丁巨算法	丁巨																	4
七經算術通義七卷	陰景愉	5					6											
九九古經歌一卷														7				
九九算術二卷	楊淑	8					9											
九章大全					10													
九章六曹算經一卷	劉徽	11					12											
九章比類筭法	吳敬					13												
九章別術二卷	徐岳、甄鸞	14					15											
九章重差一卷	劉向	16					17											
九章重差圖一卷	劉徽	18					19											
九章推圖經法一卷	張峻	20					21											
九章通明筭法	劉仕隆					22												
九章詳注比類算法					23									24				
九章詳通筭法	余進					25												
九章詳註比類均輸算法大全										26								
九章詳註筭法	許榮					27												
九章算法				28									29					

Treatise Title:	Author:	P-M	文淵	秘閣	晁氏	算法	國史	內閣	世善	胍望	近古	玄賞	其它	千頃	明書	明稿	明志	Ext.
九章算法大全九卷	吴敬											30		31				
九章算法比類大全	吴敬									32	33							34
九章算法詳註九卷	許榮、孟仁										35							
九章算術九卷	徐岳	36					37											
九章算術二卷	徐岳、甄鸞	38					39											
九章算術十卷	劉徽	40					41											
九章算術注九卷	李淳風	42					43											
九章算經			44												45			
九章算經九卷	甄鸞	46					47		48									
九章算經二十九卷	徐岳、甄鸞	49					15											
九章算經音義一卷	李籍	50					51											
九章雜算文一卷	劉祐	52					53											
九經術疏九卷	宋泉之	54					55											
三元化零歌						56												
三等數一卷	董泉、甄鸞	57					58											
中西數學圖說	李篤培																	59
五曹孫子等算經二十卷	李淳風	60					61											
五曹筭法		62		63		64												
五曹算經		65	66									67			68			
五曹算經五卷	甄鸞	69					70											71
五曹算經五卷	韓延、夏侯陽	72					70											
五經筭法	甄鸞	73				74												75
五經算術			76	77											78			
五經算術一卷	甄鸞	79																

Treatise Title:	Author:	P-M	文淵	秘閣	晁氏	算法	國史	內閣	世善	胍望	近古	玄賞	其它	千頃	明書	明稿	明志	Ext.
五經算術二卷	李淳風	79					80											
五經算術二卷	甄鸞						81											
元寶鈔通考節要			82	83											84			
六分論	唐順之																	85
六門筭法					86													
六問算法五卷	龍受益								87									
分法論	唐順之																	88
勾股術										89								
勾股筭術						90						91						
勾股算法													92					
勾股算術	顧應祥																	93
心機算術括一卷	一行、黃棲巖	94					95											96
方員句股圖解一卷														97				
方程論	梅文鼎																	98
方圓術	顧應祥									89	99							100
方圓算術												101						
方圓論	程大位																	102
句股容方圓論	唐順之																	103
句股測望論	唐順之																	104
句股等六論一卷	唐順之						105							106			107	
句股算術	李瓚									108	109			110				
句股機要一卷	楊雲翼								111					112				
四元玉鑑二卷	朱世傑													113				114
正明筭法	張爵					115												

Treatise Title:	Author:	P-M	文淵	秘閣	晁氏	算法	國史	內閣	世善	脈望	近古	玄賞	其它	千頃	明書	明稿	明志	Ext.
田畝比類乘除捷法	楊輝																	116
百家纂譜										117								
妙錄算經															118			
抄錄算法			119	120														
求一算法九例一卷							121											
求一算經一卷									122									
求指家元要一卷	李紹穀						123											
走盤集						124												
周易軌限算一卷							125											
周髀算經圖解	朱載堉																	126
弧矢弦術						127												
弧矢算術										89								
弧矢算術	唐順之																	128
弧矢算術二卷	顧應祥										129			130			131	132
弧矢論	唐順之																	133
明古筭法						134												
明源筭法						135												
法算細歷一卷							136											
金科筭法						137												
金蟬脫殼					138													
金蟬脫殼縱橫算法											139							
度測三卷	陳藎謨																	140
律呂算例					141									142				
指明筭法	夏源澤					143					144							145

Treatise Title:	Author:	P-M	文淵	秘閣	晁氏	算法	國史	內閣	世善	胍望	近古	玄賞	其它	千頃	明書	明稿	明志	Ext.
指南筭法						146												
泉志			147	148											149			
訂正筭法	林高					150												
重明筭法						151												
乘除通變本末						152												
乘除通變本末	楊輝																	153
乘除算例一卷		154					136											
夏侯陽筭法						155												
夏侯陽算經			156	157								158			159			
夏侯陽算經一卷		160					161											
夏侯陽算經一卷	韓延	162					163											164
孫子筭法						165												
孫子算經		166	167	168								169			170			
孫子算經二卷		171					172											173
海島筭法						174												
海島算經			175	176											177			
海島算經一卷	李淳風						178											
海島算經一卷	劉徽						178											179
益古衍段	李冶		180	181								182			183			184
益古衍段三卷	李冶							185						186				
益古衍義三卷	李冶						187											
益古筭法						188												
神道大編曆宗算會	周述學																	189
馬傑改正筭法	吳橋人					190												

Treatise Title:	Author:	P-M	文淵	秘閣	晁氏	算法	國史	內閣	世善	胍望	近古	玄賞	其它	千頃	明書	明稿	明志	Ext.
婆羅門陰陽算歷一卷		191					192											
婆羅門算法三卷		193					192											
婆羅門算經三卷		194					192											
庸章筭法	朱元溶					195												
張丘建筭法	張邱建					196												
張邱建算經二卷	張邱建	197					198					199						200
得一算經七卷	陳從運	201					202											
捷用算法			203	204											205			
啟蒙發明筭法	鄭高昇					206												
啟蒙算法四冊													207					
曹唐筭法						208												
眾家算陰陽法一卷		209					210											
通原算法				211											212			213
通原算法二卷							214							215				
通微集						216												
通源算法			217															
通機集						218												
通變算寶			219	220											221			
測圓海鏡	李冶					222				223	224							225
測圓海鏡	李冶		226	227				228				229			230			
測圓海鏡十二卷	李冶						187							186				
測圓海鏡分類釋術	顧應祥																	231
測圓海鏡分類釋術序	顧應祥										232	233						
測圓算術										234	235	236						

Treatise Title:	Author:	P-M	文淵	秘閣	晁氏	算法	國史	內閣	世善	脈望	近古	玄賞	其它	千頃	明書	明稿	明志	Ext.
測圓算術四卷	顧應祥													130			131	237
証古筭法						238												
量田要例算法一卷							136											
鈐經						239												
鈐釋						240												
開方說	陳蓋謨																	241
黃帝九章						242												
黃鍾算術												243						
黃鐘術										89								
黃鐘算法三十八卷		244					210											
圓方句股圖解	朱載堉																	245
新易一法算軌九例要訣一卷	龍受益	246					125											
楊輝九章			247	248											249			
楊輝九章一卷	楊輝						250							251			252	
萬物筭數					253													
詳明筭法	安止齋、何平子				254	255												256
詳明算法			257	258									259		260			
詳解九章算法	楊輝																	261
詳解日用筭法						262												
詳解黃帝九章						263												
賈憲九章	賈憲					264												
筭林拔法						265												
筭法大全										266			267					
筭法百課珠				268											269			

Treatise Title:	Author:	P-M	文淵	秘閣	晁氏	算法	國史	內閣	世善	脈望	近古	玄賞	其它	千頃	明書	明稿	明志	Ext.
筭理明解	陳必智					270												
筭術拾遺						271												
筭學通衍	劉洪					272												
嘉量算經三卷	朱載堉													273			274	275
摘奇算法			276	277											278			
算法斅古集二卷	賈憲						279											
算法	杜文亨																	280
算法二卷													281					
算法二卷	龍受益						125											
算法全能集	賈亨輯		282	283								284			285			286
算法指南	黃龍吟																	287
算法啟蒙一卷														288				
算法統宗	程大位																	289
算法袖訣				290														
算法通纂										291								
算法透簾			292	293											294			
算法透簾草一卷							214							295				296
算法補缺			297												298			
算法纂要四卷	程大位																	299
算律呂法一卷							210											
算疏一卷	張去斤	300					301											
算術二十六卷	許商						302											
算術十六卷	杜忠						303											
算術百顆珠			304															

Treatise Title:	Author:	P-M	文淵	秘閣	晁氏	算法	國史	內閣	世善	脈望	近古	玄賞	其它	千頃	明書	明稿	明志	Ext.
算術百顆珠一卷							214							305				
算經三卷	謝察微	306					307											308
算經易義一卷	張續	309					310											
算經表序一卷							311											
算經品一卷														312				
算經要用百法一卷	徐岳	313					314											
算經圖釋九卷	彭絲								315					316				
算學啓蒙	朱世傑																	317
算學新說	朱載堉																	318
算學源流			319	320											321			322
算學寶鑑	王文素																	323
綴術六卷	祖沖之	324					325											326
綴算舉例一卷	楊廉													327			328	
趙 歐算經一卷		329					330											
數書											331	332						
數書九章	秦九韶										333	334						335
數術紀遺	甄鸞〔註〕											336						337
數術記遺一卷	徐岳	338					339				340	341						
數學九章			342	343											344			
數學九章九卷	秦九韶						345	346						347				
數學通軌	柯尚遷																	348
數學圖訣發明一卷	楊廉													327			328	
盤珠集						349												
盤珠算法	徐心魯																	350

Treatise Title:	Author:	P-M	文淵	秘閣	晁氏	算法	國史	內閣	世善	胍望	近古	玄賞	其它	千頃	明書	明稿	明志	Ext.
範圍分類					351													
範圍歌訣					352													
緝古筭法						353												
緝古算術四卷	王孝通						354					355						
緝古算經	王孝通																	356
頴陽書三卷	邢和璞						357											
膩皬應用筭法					358													
辨古筭法						359												
錢式			360	361											362			
錢誌				363											364			
錢譜			365	366											367			
應用筭法						368												
謝經算術三卷							369											
雙珠算法二卷														370				
議古根源						371												
釋測圓海鏡十卷	顧應祥													130			131	
續古摘奇筭法						372												373

Notes to the Table of Chinese Mathematical Treatises

[1] *Yi wei suanfa,* two chapters (written by Jiang Ben) 一位算法二卷〔江本撰〕. *Xin Tang zhi, Song zhi.* No longer extant.

[2] *Yi wei suanfa,* two chapters, Jiang Ben 《一位算法》二卷〔江本〕.

[3] *Yi hong suanfa* (written by Yu Kai of Yin in 1584) 一鴻筭法〔萬曆甲申銀邑余楷作〕.

[4] In ZKJD.

[5] *Qi jing suanshu tong yi,* two chapters (written by Yin Jingyu) 七經算術通義七卷〔陰景愉撰〕. *Jiu Tang zhi, Xin Tang zhi.* No longer extant.

[6] *Qi jing suanshu tong yi,* seven chapters 《七經算術通義》七卷.

[7] *Jiu jiu gu jing ge,* one chapter 九九古經歌一卷. Listed under the heading Yuan Dynasty.

[8] *Jiu jiu suanshu,* two chapters (written by Yang Shu) 九九算術二卷〔楊淑撰〕. Also in *Sui zhi.* No longer extant.

[9] *Jiu jiu suanshu,* two chapters (Yang Shu) 《九九算術》二卷〔楊淑〕.

[10] *Jiuzhang da quan* 九章大全.

[11] *Jiuzhang liu cao suan jing,* one chapter 九章六曹算經一卷. *Sui zhi.* No longer extant.

[12] *Jiuzhang liu cao suan jing,* one chapter (Liu Hui); *Jiuzhang chong cha tu,* one chapter (Liu Hui) 《九章六曹算經》一卷〔劉徽〕 《九章重差圖》一卷〔劉徽〕.

[13] *Jiuzhang bi lei suanfa* (written by Mr. Wu of Qiantang in Ming Jingtai gengwu [1450]. Altogether eight volumes; includes multiplication and division; divided into the categories of the *Nine Chapters*; at the end of each chapter there are difficult problems; there are many places in this book where the categorization by chapters is chaotic and mistaken) 九章比類筭法〔景泰庚午錢塘吳氏作共八本有乘除分九章每章後有難題其書章類繁亂差訛者亦多〕. Harvard edition: 九章比類法〔景泰庚午錢塘吳信民作共八本分九章每章後有難題其書章類繁亂差訛〕.

[14] *Jiuzhang bie shu,* two chapters 九章別術二卷. *Sui zhi.* No longer extant.

[15] *Jiuzhang bie shu,* two chapters (Xu Yue and Zhen Luan); *Suan jing,* twenty-nine chapters (Xu Yue and Zhen Luan) 《九章別術》二卷〔徐岳甄鸞〕 《算經》二十九卷〔徐岳甄鸞〕.

[16] *Jiuzhang chong cha,* one chapter (written by Liu Xiang) 九章重差一卷〔劉向撰〕. *Sui zhi, Jiu Tang zhi, Xin Tang zhi,* etc. Original no longer extant. Later titled *Hai dao suan jing* 海島算經. Liu Xiang should probably be Liu Hui. The *Jiu Tang zhi* and the *Xin Tang zhi* both record this title in this form, along with *Hai dao suan jing.*

[17] *Jiuzhang chong cha,* one chapter (Liu Xiang) 《九章重差》一卷〔劉向〕.

[18] *Jiuzhang chong cha tu,* one chapter (written by Liu Hui) 九章重差圖一卷〔劉徽撰〕.

[19] *Jiuzhang chong cha tu,* one chapter (Liu Hui) 九章重差圖一卷〔劉徽〕.

[20] *Jiuzhang tui tu jing fa,* one chapter (written by Zhang Jun) 九章推圖經法一卷〔張峻撰〕. *Sui zhi.* No longer extant.

[21] *Jiuzhang tui tu jing fa*, one chapter (Zhang Jun) 《九章推圖經法》一卷〔張峻〕.

[22] *Jiuzhang tongming suanfa* (written by Liu Shilong of Linjiang in [Ming] Yongle 22 [1424]; includes nine chapters but not methods of multiplication, division, etc.; at the end 33 difficult problems are solved) 九章通明筭法〔永樂二十二年臨江劉仕隆作九章而無乘除等法後作難題三十三款〕.

[23] *Jiuzhang xiang zhu bi lei suanfa* 九章洋注比類筭法. *Yang* 洋 should be *xiang* 詳.

[24] *Jiuzhang xiang zhu bi lei suanfa* 九章詳注比類算法. Listed under the heading Yuan Dynasty.

[25] *Jiuzhang xiang tong suanfa* (written by Yu Jin of Poyang in [Ming] Chenghua guimao [1483], using the *xiang ming* and *tong ming* methods) 九章詳通筭法〔成化癸卯鄱陽余進作釆取詳明通明法〕.

[26] *Jiuzhang xiang zhu bi lei jun shu suanfa da quan*, six volumes 九章詳註比類均輸算法大全六本.

[27] *Jiuzhang xiang zhu suanfa* (written by Xu Rong of Jinling in [Ming] Chenghua wuxu [1478], using the methods of Mr. Wu) 九章詳註筭法〔成化戊戌金陵許榮作釆取吳氏之法〕.

[28] *Jiuzhang suanfa* (four) 九章算法〔四〕.

[29] 九章算法. Listed under the heading *shu fang* 書坊, in the category *za shu lei* 雜書類.

[30] Wu Xinmin, *Jiuzhang suanfa da quan* 吳信民九章算法大全.

[31] *Jiuzhang suanfa da quan*, nine chapters 九章算法大全九卷. Listed under the heading Yuan Dynasty. However, the listing from *Xuanshang zhai shumu* states that this work is by Wu Jing (*zi* Xinmin) of the Ming Dynasty.

[32] *Jiuzhang suanfa bi lei da quan*, eight volumes 九章算法比類大全八本.

[33] *Jiuzhang suanfa bi lei da quan* 九章算法比類大全.

[34] In ZKJD.

[35] *Jiuzhang suanfa xiang zhu*, nine chapters (re-edited by Xu Rong and Meng Ren of Jinling) 九章算法詳註九卷〔金陵許榮孟仁重編〕.

[36] Listed in ZGLDYWZ.

[37] *Jiuzhang suanshu*, nine chapters (Xu Yue) 《九章算術》九卷〔徐岳〕.

[38] *Jiuzhang suanshu*, two chapters 九章算術二卷〔徐岳甄鸞重述〕. *Sui zhi*. No longer extant.

[39] *Jiuzhang suanshu*, two chapters (rewritten by Xu Yue and Zhen Luan) 《九章算術》二卷〔徐岳甄鸞重述〕.

[40] *Jiuzhang suanshu*, ten chapters (written by Liu Hui) 九章算術十卷〔劉徽撰〕.

[41] *Jiuzhang suanshu*, ten chapters (Liu Hui) 《九章算術》十卷〔劉徽〕.

[42] Listed in ZGLDYWZ.

[43] *Jiuzhang suanshu zhu*, nine chapters (Li Chunfeng); *Yao lue*, one chapter (Li Chunfeng) 《九章算術注》九卷〔李淳風〕 《要略》一卷〔李淳風〕. The latter treatise, probably an appendix or postscript, has not been included in the table.

[44] *Jiuzhang suan jing*, one copy, four books, incomplete 九章算經〔一部四冊闕〕.
[45] *Jiuzhang suan jing* 九章算經.
[46] Listed in *Jiu Tang zhi, Xin Tang zhi.*
[47] *Jiuzhang suan jing*, nine chapters (Zhen Luan) 《九章算經》九卷〔甄鸞〕.
[48] *Jiuzhang suan jing* (Li Chunfeng's annotations state written by the Duke of Zhou) 九章算經九卷〔李淳風註云周公作〕. Listed under the heading "miscellaneous arts" (*za yi* 雜藝).
[49] *Jiuzhang suan jing*, twenty-nine chapters (written by Xu Yue and Zhen Luan) 九章算經二十九卷〔徐岳甄鸞撰〕.
[50] Listed in ZGLDYWZ.
[51] *Jiuzhang suan jing yin yi*, one chapter (Li Ji) 《九章算經音義》一卷〔李籍〕.
[52] *Jiuzhang za suan wen*, two chapters (written by Liu You) 九章雜算文二卷〔劉祐撰〕. *Jiu Tang zhi, Xin Tang zhi.* No longer extant.
[53] *Jiuzhang za suan wen*, one chapter (Liu You) 《九章雜算文》一卷〔劉祐〕.
[54] *Jiu jing shu shu*, nine chapters (written by Song Quanzhi) 九經術疏九卷〔宋泉之撰〕. *Jiu Tang zhi, Xin Tang zhi.* No longer extant.
[55] *Jiu jing shu shu*, nine chapters (Song Quanzhi) 《九經術疏》九卷〔宋泉之〕.
[56] *San yuan hua ling ge* 三元化零歌. This title does not appear in extant Ming bibliographies. The heading for these entries states: "During the Yuanfeng (1078–1085), Shaoxing (1131–1162), Chunxi (1174–1189) periods of the Song Dynasty and following, publications were numerous; recorded below are publications either heard of or seen" 元豐紹興淳熙以來刊刻者多且以見聞者著之.
[57] Dong Quan, *San deng shu*, one chapter 董泉三等數一卷. *Jiu Tang zhi, Xin Tang zhi.* No longer extant.
[58] *San deng shu*, one chapter (Dong Quan) 《三等數》一卷〔董泉〕.
[59] At the Institute for the History of Natural Science Library.
[60] *Tang zhi.*
[61] *Zhou bi suan jing*, two chapters (by Li Chunfeng); *Wu cao, Sunzi*, and other mathematical classics, twenty chapters (Li Chunfeng) 《周髀算經》二卷〔李淳風〕 五曹孫子等算經二十卷〔李淳風〕.
[62] This form of the title does not appear in extant bibliographies.
[63] *Wu cao suanfa* (one) 五曹算法〔一〕.
[64] *Wu cao suanfa* 五曹筭法. In Cheng's bibliography, the heading for these ten entries states: "The Ten Books, published in the seventh year of the Yuanfeng period in the Song Dynasty [1084], entered into the Palace Library, and reprinted by the Tingzhou Xuexiao" 宋元豐七年刊十書入秘書省又刻於汀州學校. The original edition in *Suanfa tongzong jiaoshi* 算法統宗校釋 has 宋元豐七年刊十書人秘書省人刻於汀州學校; p. 1002 n. 52 corrects both instances of *ren* 人.
[65] *Jiu Tang zhi, Xin Tang zhi.* Original edition no longer extant.
[66] *Wu cao suan jing*, one copy, one book, incomplete 五曹算經〔一部一冊闕〕.
[67] *Wu cao suan jing* 五曹算經.
[68] *Wu cao suan jing* 五曹算經.

[69] *Jiu Tang zhi, Xin Tang zhi.*

[70] *Wu cao suan jing,* five chapters (by Zhen Luan); also five chapters (by Han Yan and Xiahou Yang) 《五曹算經》五卷〔甄鸞〕 又五卷〔韓延、夏侯陽〕.

[71] In ZKJD.

[72] *Xin Tang zhi.*

[73] This exact title does not appear in extant early bibliographies.

[74] *Wu jing suanfa* 五經筭法. In the Ten Books, published in 1084. See note 64.

[75] In ZKJD.

[76] *Wu jing suanshu,* one copy, one book, incomplete 五經算術〔一部一冊闕〕.

[77] *Wu jing suanshu* (one) 五經算術〔一〕.

[78] *Wu jing suanshu* 五經算術.

[79] *Wu jing suanshu,* one chapter (written by Zhen Luan), also two chapters (written by Li Chunfeng) 五經算術一卷〔甄鸞撰〕又二卷〔李淳風撰〕. *Xin Tang zhi.*

[80] *Wu jing suanshu,* two chapters (Li Chunfeng) 《五經算術》二卷〔李淳風〕.

[81] *Wu jing suanshu,* two chapters (commentary by Zhen Luan) 《五經算術》二卷〔甄鸞注〕.

[82] *Yuanbao chao tongkao jieyao* (one copy, four books, incomplete) 元寶鈔通考節要〔一部四冊闕〕.

[83] *Yuanbao chao tongkao jieyao* (four) 元寶鈔通考節要〔四〕.

[84] *Yuanbao chao tongkao jieyao* 元寶鈔通考節要.

[85] In *Tang Shunchuan xiansheng wenji shiba juan* 唐荊川先生文集十八卷. At the Institute for the History of Natural Science Library.

[86] *Liu men suanfa* 六門算法.

[87] *Liu wen suanfa,* five chapters (by Tang dynasty Long Shouyi) 六問算法五卷〔唐龍受益〕. Listed under the heading *za yi* 雜藝.

[88] In *Tang Shunchuan xiansheng wenji shiba juan* 唐荊川先生文集十八卷. At the Institute for the History of Natural Science Library.

[89] *Hu shi suanshu, fang yuan shu, huang zhong shu, gou gu shu,* together one volume 弧矢算術方圓術黃鐘術勾股術共一本.

[90] *Gou gu suanshu* (written by Minister Ruoxi Gu Yingxiang of Wuxing in [Ming] Jiajing *guisi* [1533]. Without multiplication or division) 勾股筭術〔嘉靖癸巳吳興尚書箬溪顧應祥作無乘除〕. Reading *xing* 興 for what is apparently an idiosyncratic simplification.

[91] *Gou gu suanshu* 勾股算術.

[92] *Wanjuan tang shumu* 萬卷堂書目: *Gou gu suanfa,* one book 勾股算法一冊. Listed under the heading *nong jia* 農家.

[93] In ZKJD.

[94] *Xin Tang zhi.*

[95] *Xin ji suanshu kuo,* one chapter 《心機算術括》一卷.

[96] In ZKJD.

[97] *Fang yuan gou gu tu jie,* one chapter 方員句股圖解一卷. Listed under the heading Yuan Dynasty.

[98] In ZKJD.

[99] *Fang yuan shu* 方圓術.
[100] At the Institute for the History of Natural Science Library.
[101] *Fang huan suanshu* 方圜算術.
[102] Probably not extant.
[103] In SKQS.
[104] In *[Tang] Bingchuan wenji* 荊川文集, in SKQS.
[105] *Gou gu deng liu lun*, one chapter (by Tang Shunzhi) 《句股等六論》一卷〔唐順之〕.
[106] Tang Shunzhi, *Gou gu deng liu lun*, one chapter 唐順之句股等六論一卷. Listed under the heading Ming Dynasty.
[107] Tang Shunzhi, *Gou gu deng liu lun*, one chapter 唐順之《句股等六論》一卷. Listed under the category *xiaoxue* 小學.
[108] *Gou gu suanshu*, one volume 勾股算術一本.
[109] *Gou gu suanshu* 勾股算術.
[110] Li Zan (from Wucheng), *Gou gu suanshu* 李瓚句股算術〔烏程人〕. Listed under the heading Ming Dynasty.
[111] *Gou gu ji yao*, one chapter (by Yang Yunyi of the Jin dynasty) 句股機要一卷〔金楊雲翼〕. Listed under the heading *za yi* 雜藝.
[112] Yang Yunyi, *Gou gu ji yao* 楊雲翼句股機要. The only entry under the heading for the Jin Dynasty.
[113] Zhu Shijie, *Si yuan yu jian*, two chapters 朱世傑四元玉鑑二卷. Listed under the heading Yuan Dynasty.
[114] In ZKJD.
[115] *Zheng ming suanfa* (written by Zhang Jue of Jintai in [Ming] Jiajing *jihai* [1539]) 〔正明筭法〕嘉靖己亥金臺張爵作. Reading *ji* 己 for *si* 巳.
[116] In ZKJD.
[117] *Bai jia zuan pu*, one volume 百家纂譜一本.
[118] *Chao lu suan jing* 妙錄算經.
[119] *Chao lu suanfa*, one copy, one book, incomplete 抄錄算法〔一部一冊闕〕.
[120] *Chao lu suanfa* 抄錄算法.
[121] *Qiu yi suanfa jiu li*, one chapter 《求一算法九例》一卷.
[122] *Qiu yi suan jing*, one chapter 求一算經一卷.
[123] *Qiu zhi jia yuan yao*, one chapter (Li Shaogu) 《求指家元要》一卷〔李紹穀〕.
[124] *Zou pan ji* 走盤集. From after 1078. See note 56.
[125] Long Shouyi, *Suanfa*, two chapters; *Zhou yi gui xian suan*, one chapter; *Xin yi yi fa suan gui jiu li yao jue*, one chapter (Long Shouyi) 龍受益《算法》二卷《周易軌限算》一卷 《新易一法算軌九例要訣》一卷〔龍受益〕.
[126] In *Gu jin suanxue congshu* 古今算學叢書.
[127] *Hu shi xuan* (written by Gu Ruoxi Yingxiang in [Ming] Jiajing *renzi* [1552]) 弧矢弦〔嘉靖壬子顧箬溪作無乘除〕.
[128] Listed in ZGLDYWZ.
[129] *Hu shi suanshu* 弧矢算術.

[130] Gu Yingxiang, *Ce yuan suanshu,* four chapters; also *Hu shi suanshu,* two chapters; also *Shi Ce yuan hai jing,* ten chapters. 顧應祥測圓算術四卷又弧矢算術二卷又釋測圓海鏡十卷. Listed under the Ming Dynasty.

[131] Gu Yingxiang, *Ce yuan suanshu,* four chapters; *Hu shi suanshu,* two chapters; *Shi Ce yuan hai jing,* ten chapters. 顧應祥《測圓算術》四卷、《弧矢算術》二卷、《釋測圓海鏡》十卷. Listed under the category *xiaoxue* 小學.

[132] In SKQS, ZKJD.

[133] In SKQS.

[134] *Ming gu suanfa* 明古筭法. From after 1078. See note 56.

[135] *Ming yuan suanfa* 明源筭法. From after 1078. See note 56.

[136] *Cheng chu suan li,* one chapter; *Fa suan xi li,* one chapter; *Liang tian yao li suanfa,* one chapter 《乘除算例》一卷 《法算細歷》一卷 《量田要例算法》一卷.

[137] *Jin ke suanfa* 金科筭法. Published between 1078 and 1189. See note 56.

[138] *Jin chan tuo ke* 金蟬脫殼.

[139] *Jin chan tuo ke zong heng suanfa* (author unknown) 金蟬脫殼縱橫算法一卷〔不知作者〕.

[140] At the Institute for the History of Natural Science Library.

[141] *Lülü suan li* 律呂算例.

[142] *Lülü suan li* 律呂算例. Listed under the heading Yuan Dynasty.

[143] *Zhi ming suanfa* (written by Xia Yuanze of Jiangning in [Ming] Zhengtong *jiwei* [1439]; the nine chapters are incomplete) 指明筭法〔正統巳未江寧夏源澤作而九章不全〕. *Si* 巳 should be *ji* 己; reading *ning* 寧 for idiosyncratic variant.

[144] *Zhi ming suanfa,* two chapters (author unknown, twenty four examples) 指明算法二卷〔不知作者二十四則〕.

[145] Apparently a hand-copied version is at Cambridge; also an edition is preserved in Japan. Not seen.

[146] *Zhi nan suanfa* 指南筭法. From after 1078. See note 56.

[147] *Quan zhi,* one copy, two book, incomplete 泉志〔一部二冊闕〕.

[148] *Quan zhi* 泉志.

[149] *Quan zhi* 泉志.

[150] *Ding zheng suanfa* (written by Lin Gao of Zhedong Kuaiqi in Ming Jiajing *gengzi* [1600], with detailed annotations, fixing positions) 繘正筭法〔嘉靖庚子浙東會稽林高作詳解定位〕.

[151] *Chong ming suanfa* 重明筭法. No date is given, but it is listed among treatises from the Jiajing period.

[152] *Cheng chu tong bian ben wei* 乘除通變本未 [the final character *wei* 未 should be *mo* 末]. Listed under the heading, "During the periods Jiading [1208–1224], Xianchun [1265–1274], and Deyou [1275–1276], many books were also published" 嘉定咸淳德祐等年又刊各書. Following these entries is the statement, "The above all come from Yang Hui's *Zhai qi*" 巳上俱出楊輝摘奇內 (the character *si* 巳 should be *yi* 已).

[153] In ZKJD.

[154] Listed in *Mi shu xu mu* 秘書續目. No longer extant.

[155] *Xiahou Yang suanfa* 夏侯陽筭法. In the Ten Books, published in 1084. See note 64.

[156] *Xiahou Yang suan jing,* one copy, one book, incomplete 夏侯陽算經〔一部一冊闕〕.

[157] *Xiahou Yang Suan jing* (one) 夏侯陽算經〔一〕.

[158] *Xiahou Yang suan jing* 夏侯陽算經.

[159] *Xiahou Yang suan jing* 夏侯陽算經.

[160] *Xiahou Yang suan jing,* one chapter 夏侯陽算經一卷.

[161] *Xiahou Yang suan jing,* one chapter 《夏侯陽算經》一卷.

[162] Han Yan, *Xiahou Yang suan jing,* one chapter 韓延夏侯陽算經一卷.

[163] Han Yan, *Xiahou Yang suan jing,* one chapter 韓延《夏侯陽算經》一卷.

[164] In ZKJD.

[165] *Sunzi suanfa* 孫子筭法. In the Ten Books, published in 1084. See note 64.

[166] Listed in *Sui zhi, Song zhi, Han zhi she bu.*

[167] *Sunzi suan jing,* one copy, one book, incomplete 孫子算經〔一部一冊闕〕.

[168] *Sunzi suan jing* (one) 孫子算經〔一〕.

[169] *Sunzi suanshu* (the character *shu* is the character *jing*) 孫子算術〔術字係經字〕.

[170] *Sunzi suan jing* 孫子算經.

[171] *Sunzi suan jing,* two chapters 孫子算經二卷. *Sui zhi, Song zhi, Han zhi she bu.*

[172] *Sunzi suan jing,* two chapters 《孫子算經》二卷.

[173] In ZKJD.

[174] *Hai dao suanfa* 海島筭法. Guangxu edition has *Hai dao suan jing* 海島筭經. In the Ten Books, published in 1084. See note 64.

[175] *Hai dao suan jing,* one copy, one book, incomplete 海島算經〔一部一冊闕〕.

[176] *Hai dao suan jing* (one) 海島筭經〔一〕.

[177] *Hai dao suan jing* 海島算經.

[178] *Hai dao suan jing,* one chapter (by Liu Hui); also one chapter (by Li Chunfeng) 《海島算經》一卷〔劉徽〕 又一卷〔李淳風〕.

[179] In ZKJD.

[180] *Yi gu yan duan,* one copy, three books, incomplete 益古衍段〔一部三冊闕〕.

[181] *Yi gu yan duan* (two) 益古衍段〔二〕.

[182] Li Ye, *Yi gu yan duan* 李冶益古衍段.

[183] *Yi gu yan duan* 益古衍段.

[184] In ZKJD.

[185] *Yi gu yan duan,* three books, complete (Yuan dynasty Zhiyuan period, methods of calculation by Li Ye) 益古衍段三冊全〔元至元閒李冶著算法也〕.

[186] Li Ye, *Ce yuan hai jing,* twelve chapters; also *Yi gu yan duan,* three chapters 李冶測圓海鏡十二卷又益古衍段三卷. Listed under the Yuan Dynasty.

[187] *Ce yuan hai jing,* twelve chapters (by Yuan dynasty Li Ye); *Yi gu yan yi,* three chapters (by Li Ye) 《測圓海鏡》十二卷〔元李冶〕 《益古衍義》三卷〔李冶〕.

[188] *Yi gu suanfa* 益古筭法. From after 1078. See note 56.
[189] At the Institute for the History of Natural Science Library, Nanjing Library, and Zhejiang Library.
[190] *Ma Jie gai zheng suanfa* (written by Wu Qiaoren of Hejian in Jiajing *wuxu* [1538]. Without multiplication and division, it only changes the methods of Wu Xinmin of Qiantang ...) 馬傑改正筭法〔河間吳橋人嘉靖戊戌作而無乘除只改錢塘吳信民法...〕.
[191] *Sui zhi.*
[192] *Poluomen suanfa,* three chapters; *Poluomen yin yang suan li,* one chapter; *Poluomen suan jing,* three chapters 《婆羅門算法》三卷《婆羅門陰陽算歷》一卷《婆羅門算經》三卷.
[193] *Sui zhi.*
[194] *Sui zhi.*
[195] *Yong zhang suanfa* (written by Zhu Yuanrong of Xin'an in Wanli *wuzi* [1588]) 庸章筭法〔萬曆戊子新朱元溶作〕.
[196] *Zhang Qiujian suanfa* 張丘建筭法. Cheng's bibliography states that this treatise is included in the Ten Books, published in 1084. See note 64. The Guangxu edition substitutes a *qiu* 邱 for *qiu* 丘—*Zhang Qiujian suanfa* 張邱建筭法.
[197] *Zhang Qiujian suan jing,* two chapters 張邱建算經二卷.
[198] *Zhang Qiujian suan jing,* two chapters (by Li Chunfeng); also three chapters (by Li Chunfeng) 《張邱建算經》二卷〔李淳風〕又三卷〔李淳風〕. The latter is not listed in this table as a separate title.
[199] *Zhang Qiujian suan jing* 張丘建算經.
[200] In ZKJD.
[201] *Xin Tang zhi, Song zhi.*
[202] *De yi suan jing,* seven chapters (by Chen Congyun) 《得一算經》七卷〔陳從運〕.
[203] *Jie yong suanfa* (one copy, one book, incomplete) 捷用算法〔一部一冊闕〕.
[204] *Jie yong suanfa* (one) 捷用算法〔一〕.
[205] *Jie yong suanfa* 捷用算法.
[206] *Qimeng faming suanfa* (written by Zheng Gaosheng of Fushan in Jiajing *bingxu* [1526]) 啟蒙發明筭法〔嘉靖丙戌福山鄭高昇作〕.
[207] *Wanjuan tang shumu* 萬卷堂書目: *Qimeng suanfa,* four books 啟蒙算法四冊. Listed under the heading "agriculture" (*nong jia* 農家).
[208] *Cao tang suanfa* 曹唐筭法. From after 1078. See note 56.
[209] *Zhong jia suan yin yang fa,* one chapter 眾家算陰陽法一卷.
[210] *Huang zhong suanfa,* thirty-eight chapters; *Suan lülü fa,* one chapter; *Zhong jia suan yin yang fa,* one chapter 《黃鐘算法》三十八卷 《算律呂法》一卷《眾家算陰陽法》一卷. This series of three entries is found verbatim in the *Tong zhi.* That these entries are grouped together may indicate that these texts were combined in printing; or it may indicate that the titles were merely copied, and from the same source.
[211] *Tong yuan suanfa* (two) 通原算法〔二〕.
[212] *Tong yuan suanfa* 通原算法.

[213] In *Yongle da dian suanxue can ben* 永樂大典算學殘本, 明刊本, 李儼所藏鈔本《諸家算法》中.

[214] *Suanshu bai ke zhu*, one chapter; *Suanfa tou lian cao*, one chapter; *Tong yuan suanfa*, two chapters 《算術百顆珠》一卷 《算法透簾草》一卷 《通原算法》二卷.

[215] *Tong yuan suanfa*, two chapters 通原算法二卷. Listed under the heading Yuan.

[216] *Tong wei ji* 通微集. From after 1078. See note 56.

[217] *Tong yuan suanfa*, one copy, two books, complete 通源算法〔一部二冊完全〕.

[218] *Tong ji ji* 通機集. From after 1078. See note 56.

[219] *Tong bian suan bao* (one copy, one books, incomplete) 通變算寶〔一部一冊闕〕.

[220] *Tong bian suan bao* (one) 通變算寶〔一〕.

[221] *Tong bian suan bao* 通變算寶.

[222] *Ce yuan hai jing* (written by the Academician Li Ye of Luancheng in Jiajing *gengxu* [1550], does not include multiplication and division) 測圓海鏡〔嘉靖庚戌學士欒城李冶公作無乘除〕. The Harvard edition states "Note: Li Ye is an Academician of the Yuan, here writing Jiajing *gengxu* is a mistake" 按李冶係元學士今作嘉靖庚戌誤. The *Suanfa tongzong jiaoshi* 算法統宗校釋, p. 1003 n. 55, states: "This sentence contains errors, since Li Ye did not live during the Ming dynasty Jiajing period. The treatise *Ce yuan hai jing* was already lost during the Jiajing period; this might be a work by a Ming writer explaining the *Ce yuan hai jing*" 『此句有誤, 因李冶不是明嘉靖人。《測圓海鏡》一書, 嘉靖年間已經失傳, 此書可能是明人解釋《測圓海鏡》之著作』. *Ce yuan hai jing*, as this table shows, was not lost at this time.

[223] *Ce yuan hai jing*, two volumes 測圓海鏡二本.

[224] *Ce yuan hai jing* 測圓海鏡.

[225] In ZKJD.

[226] Li Ye, *Ce yuan hai jing*, one copy, five books, incomplete 李冶測圓海鏡〔一部五冊闕〕.

[227] Li Ye, *Ce yuan hai jing* (five) 李冶測圓海鏡〔五〕.

[228] *Ce yuan hai jing*, four books, complete (Yuan dynasty Yuanzhi period Academician Li Ye of Luancheng wrote on the [Daoist] theories of *dongyuan jiurong*, namely that which today is called the mathematical arts" 測圓海鏡四冊全〔元至元間學士欒城李冶著洞淵九容之說即今之算法也〕.

[229] Li Ye, *Ce yuan hai jing* 李冶測圓海鏡.

[230] Li De, *Ce yuan hai jing* 李德測圓海鏡.

[231] In ZKJD.

[232] Li Ye, Preface to *Ce yuan hai jing fen lei shi shu* 測圓海鏡分類釋術序.

[233] Preface to *Ce yuan hai jing fen lei shi shu* 測圓海鏡分類釋術序.

[234] *Ce yuan suanshu*, one volume 測圓算術一本.

[235] *Ce yuan suanshu* 測圓算術.

[236] *Ce yuan suanshu* 測圓算術.

[237] At the Institute for the History of Natural Science Library; in ZKJD.

[238] *Zheng gu suanfa* 証古筭法. From after 1078. See note 56.
[239] *Qian jing* 鈐經. From after 1078. See note 56.
[240] *Qian shi* 鈐釋. From after 1078. See note 56.
[241] Institute for the History of Natural Science Library has a rare edition.
[242] *Huang di jiuzhang* 黃帝九章.
[243] *Huang zhong suanshu* 黃鍾算術.
[244] *Huang zhong suanfa*, thirty-eight chapters 黃鍾算法三十八卷.
[245] In *Gu jin suanxue congshu* 古今算學叢書.
[246] *Xin yi yi fa suan gui jiu li yao jue*, one chapter (written by Long Shouyi) 新易一法算軌九例要訣一卷〔龍受益撰〕.
[247] Yang Hui, *Jiuzhang* (one copy, one book, incomplete) 楊輝九章〔一部一冊闕〕.
[248] Yang Hui, *Jiuzhang* (one) 楊輝九章〔一〕.
[249] Yang Hui, *Jiuzhang* 楊輝九章.
[250] Yang Hui, *Jiuzhang*, one chapter 楊輝《九章》一卷.
[251] Yang Hui, *Jiuzhang*, one chapter 楊輝九章一卷. Incorrectly listed under the heading Yuan Dynasty.
[252] Yang Hui, *Jiuzhang*, one chapter 楊輝《九章》一卷. Listed under the category *xiaoxue* 小學.
[253] *Wanwu suan shu* 萬物筭數.
[254] *Xiang ming suanfa* 詳明筭法.
[255] *Xiang ming suanfa* (written by Yuan literati An Zhizhai and He Pingzi; includes multiplication and division, but nothing not in the *Nine Chapters*) 詳明筭法〔元儒安止齋何平子作有乘除而無九章不備〕. In the original, a popular variant is substituted for the character bei 備. On the authorship by An Zhizhai and He Pingzi, see *Xiang ming suanfa tiyao*, in ZKJD.
[256] In ZKJD.
[257] *Xiang ming suanfa* (one copy, one book, incomplete) 詳明算法〔一部一冊闕〕. Note: this treatise is listed as the last entry on p. 12a and as the first on p. 12b.
[258] *Xiang ming suanfa* (one) 詳明算法〔一〕.
[259] *Nei ban jing shu ji lue* 內板經書紀略: *Xiang ming suanfa* (one volume, one hundred and ten pages) 詳明算法〔一本百十葉〕. This is the only mathematical treatise listed in this bibliography (p. 615).
[260] *Xiang ming suanfa* 詳明算法.
[261] In ZKJD.
[262] *Xiang jie ri yang suanfa* 詳解日用筭法. From after 1208. Following these entries is the statement "The above all come from Yang Hui's *Zhai qi*." See note 152.
[263] *Xiang jie Huangdi jiuzhang* 詳解黃帝九章. Throughout the bibliography, the character 章 is incorrectly written. From after 1208. Following these entries is the statement "The above all come from Yang Hui's Zhai qi." See note 152.
[264] *Jia Xian jiuzhang* 賈憲九章. From after 1078. See note 56.
[265] *Suan lin ba fa* (written by the Taiyi Yang Pu of Wanling in Longqing *renshen* [1572]) 筭林拔法〔隆慶壬申宛陵太邑楊溥作〕.
[266] *Suanfa da quan*, four copies 算法大全四本.

[267] *Gu jin shu ke* 古今書刻: *Suanfa da quan* 筭法大全. Listed under the heading *Du cha yuan* 都察院; also under the heading *Hangzhou fu* 杭州府.

[268] *Suanfa bai ke zhu* (one) 筭法百課珠〔一〕.

[269] *Suanfa bai ke* 算法百課.

[270] *Zheng li ming jie* (written by Chen Bizhi of Jiangxi Ningdu in Jiajing *gengzi* [1540]) 〔正理明解〕嘉靖庚子江西寧都陳必智作. Again, substituting寧 for an idiosyncratic simplification.

[271] *Suan shu she yi* 筭術拾遺. In the Ten Books, published in 1084. See note 64.

[272] *Suanxue tong yan* (written by Liu Hong of Jingzhao in Chenghua *renchen* [1472]) 筭學通衍〔成化壬辰京兆劉洪作〕.

[273] Zheng prince [Zhu] Zaiyu, *Jia liang suan jing*, three chapters 鄭世子載堉嘉量算經三卷. The modern edition adds in a footnote that his surname is Zhu: 明史藝文志鄭世子作朱. Listed under the Ming.

[274] Zhu Zaiyu, *Jia liang suan jing*, three chapters 朱載堉《嘉量算經》三卷. Listed under the category *xiaoxue* 小學.

[275] *Jia liang suan jing*, three chapters, *Wen da*, one chapter, *Fan li*, one chapter 嘉量算經三卷問答一卷凡例一卷. In *Wan wei bie cang* 宛委別藏 and *Xuan yin Wan wei bie cang* 選印宛委別藏.

[276] *Zhai qi suanfa* (one copy, one book, incomplete) 摘奇算法〔一部一冊闕〕.

[277] *Zhai qi suanfa* 摘奇算法.

[278] *Zhai qi suanfa* 摘奇算法.

[279] *Suanfa jiao gu ji*, two chapters (Jia Xian) 算法斅古集二卷〔賈憲〕

[280] Apparently not extant.

[281] *Nanyong zhi jingji kao* 南廱志經籍考二卷 lists *Suanfa*, two chapters 算法二卷, with no annotations or other information, in the section "miscellaneous books" (*za shu lei* 雜書類). This is the only mathematical treatise listed in the bibliography. MDSMTBCK, p. 461.

[282] *Suanfa quanneng ji* (one copy, one book, incomplete) 算法全能集〔一部一冊闕〕.

[283] *Suanfa quanneng ji* (one) 算法全能集〔一〕.

[284] *Suanfa quanneng ji* 算法全能集.

[285] *Suanfa quanneng ji* 算法全能集.

[286] In ZKJD. Jia Hengji 賈亨輯, *Suanfa quanneng ji*, two chapters 算法全能集二卷 in *Xuanlan tang congshu san ji* 玄覽堂叢書三集.

[287] In ZKJD.

[288] *Suanfa qimeng*, one chapter 算法啟蒙一卷. Listed under the heading Yuan.

[289] In ZKJD. Several editions extant.

[290] *Suanfa xiu jue* (one) 算法袖訣〔一〕.

[291] *Suanfa tong zuan*, one volume 算法通纂一本.

[292] *Suanfa tou lian* (one copy, one book, incomplete) 算法透簾〔一部一冊闕〕.

[293] *Suanfa tou lian* (one) 筭法透簾〔一〕.

[294] *Suanfa tou lian* 算法透簾.

[295] *Suanfa tou lian cao*, one chapter 算法透簾草一卷. Listed under the heading Yuan.

[296] *Tou lian xi cao* 透簾細草 in ZKJD.

[297] *Suanfa buque* (one copy, one book, incomplete) 算法補缺〔一部一冊闕〕.
[298] *Suanfa buque* 算法補缺.
[299] In Beijing Library, modern reprint.
[300] Zhang Qujin, *Suan shu*, one chapter 張去斤算疏一卷.
[301] Zhang Qujin, *Suan shu*, one chapter 張去斤《算疏》一卷.
[302] *Suanshu*, twenty-six chapters (by Xu Shang) 《算術》二十六卷〔許商〕.
[303] *Suanshu*, sixteen chapters (by Du Zhong) 《算術》十六卷〔杜忠〕.
[304] *Suanshu bai ke zhu* (one copy, one book, incomplete) 算術百顆珠〔一部一冊闕〕.
[305] *Suanshu bai ke zhu*, one chapter 算術百顆珠一卷. Listed under the heading Yuan.
[306] Xie Chashu, *Suan jing*, three chapters 謝察術算經三卷.
[307] Xie Chawei, *Suan jing*, three chapters 《算經》三卷〔謝察微〕.
[308] In ZKJD.
[309] *Suan jing yi yi*, one chapter (written by Zhang Xu) 算經易義一卷〔張續撰〕.
[310] *Suan jing yi yi*, one chapter (by Zhang Xu) 《算經易義》一卷〔張續〕.
[311] Preface to *Suan jing biao*, one chapter 《算經表序》一卷.
[312] *Suan jing pin*, one chapter 算經品一卷. Listed under the heading Yuan.
[313] *Suan jing yao yong bai fa*, one chapter (written by Xu Yue) 算經要用百法一卷〔徐岳撰〕.
[314] *Suan jing yao yong bai fa*, one chapter (by Xu Yue) 《算經要用百法》一卷〔徐岳〕.
[315] *Suan jing tu shi*, nine chapters (written by Peng [Si]) 算經圖釋九卷〔彭錄〕 Listed under the heading *li jia* 歷家.
[316] Peng Si, *Suan jing tu shi*, nine chapters 彭絲算經圖釋九卷. Listed under the heading Yuan.
[317] In ZKJD.
[318] In *Yue lü quan shu* 樂律全書, SKQS, and others.
[319] *Suan xue yuan liu* (one copy, one book, incomplete) 算學源流〔一部一冊闕〕.
[320] *Suan xue yuan liu* (one) 算學源流〔一〕.
[321] *Suan xue yuan liu* 算學源流.
[322] In ZKJD.
[323] In ZKJD.
[324] *Zhui shu*, six chapters (written by Zu Chongzhi) 綴術六卷〔祖沖之撰〕. Later Han.
[325] *Zhui shu*, six chapters (by Zu Chongzhi) 《綴術》六卷〔祖沖之〕.
[326] In ZKJD.
[327] Yang Lian, *Zhui suan ju li*, one chapter; also *Shuxue tu jue faming*, one chapter 楊廉綴算舉例一卷又數學圖訣發明一卷. Listed under the Ming.
[328] Yang Lianzhui, *Zhui suan ju li*, one chapter; *Shuxue tu jue faming*, one chapter 楊廉《綴算舉例》一卷、《數學圖訣發明》一卷. Listed under the category *xiaoxue* 小學.

[329] Zhao Fei, *Suan jing*, one chapter 趙 𣣏算經一卷. The character I have composed here, 𣣏, is not in standard dictionaries; I have used the left-hand component, *fei* 匪, as the most likely pronunciation.
[330] Zhao Fei, *Suan jing*, one chapter 趙 𣣏算經一卷.
[331] *Shu shu* 數書.
[332] *Shu shu* 數書.
[333] *Shu shu jiu zhang* 數書九章.
[334] Qin Jiushao, *Shu shu jiu zhang* 秦九韶數書九章.
[335] In ZKJD.
[336] *Shushu ji yi*, annotated by Zhen Luan 甄鸞註數術紀遺.
[337] In ZKJD.
[338] *Shushu ji yi*, one chapter (by Xu Yue) 數術記遺一卷〔徐岳撰〕.
[339] *Shushu ji yi*, one chapter (by Xu Yue) 《數術記遺》一卷〔徐岳〕.
[340] Han dynasty Xu Yue, *Shushu ji yi* 漢徐岳數術紀遺. Also listed among the translations of Jesuit works is *Shu shu ji yi* 數術紀遺.
[341] Xu Yue, *Shu shu ji yi* 徐岳數術紀遺.
[342] *Shuxue jiuzhang*, one copy, three books, complete 數學九章〔一部三冊完全〕.
[343] *Shuxue jiuzhang* (three) 數學九章〔三〕.
[344] *Shuxue jiuzhang* 數學九章.
[345] *Shuxue jiuzhang*, nine chapters (by Qin Jiushao of the Song dynasty) 《數學九章》九卷〔宋秦九韶〕.
[346] *Shuxue jiuzhang*, three books, complete (written by Qin Jiushao of Lu prefecture during the Chunyou period of the Song dynasty, hand-copied edition) 數學九章三冊全〔鈔本宋淳祐閒魯郡秦九韶撰〕.
[347] Qin Jiushao (of Lu prefecture), *Shuxue jiuzhang*, nine chapters 秦九韶數學九章九卷〔魯郡人〕. This is the only treatise listed under the Song Dynasty.
[348] In ZKJD.
[349] *Pan zhu ji* 盤珠集. From after 1078. See note 56.
[350] In ZKJD.
[351] *Fan wei fen lei* 範圍分類.
[352] *Fan wei ge jue* 範圍歌訣.
[353] *Ji gu suanfa* 緝古筭法. In the Ten Books, published in 1084. See note 64.
[354] *Ji gu suanshu*, four chapters (Wang Xiaotong) 《緝古算術》四卷〔王孝通〕.
[355] Wang Xiaotong, *Ji gu suan jing* 王孝通緝古算經.
[356] In ZKJD.
[357] *Ying yang shu*, three chapters (by Xing Hepu) 《潁陽書》三卷〔邢和璞〕.
[358] Ni Zhi, *Ying yong suanfa* (written in the Yuan dynasty) 膩觶應用筭法〔元制〕.
[359] *Bian gu suanfa* 辨古筭法. From after 1078. See note 56.
[360] *Qian shi* (one copy, one book, incomplete) 錢式〔一部一冊闕〕.
[361] *Qian shi* (one) 錢式〔一〕.
[362] *Qian shi* 錢式.
[363] *Qian zhi* 錢誌〔一〕.

[364] *Qian zhi* 錢誌.
[365] *Qian pu* (one copy, one book, incomplete) 錢譜〔一部一冊闕〕.
[366] *Qian pu* (one) 錢譜〔一〕.
[367] *Qian pu* 錢譜.
[368] *Ying yong suanfa* 應用筭法. From after 1078. See note 56.
[369] *Xie jing suanshu*, three chapters 《謝經算術》三卷.
[370] *Shuang zhu suanfa*, two chapters 雙珠算法二卷. Listed under the Yuan.
[371] *Yi gu gen yuan* 議古根源. From after 1078. See note 56.
[372] *Xu gu zhai qi suanfa* 續古摘奇筭法. From after 1078. See note 56. Following these entries is the statement "The above all come from Yang Hui's *Zhai qi*." See note 152.
[373] In ZKJD.

Appendix C
Outlines of Proofs

This appendix presents brief justifications for several lemmas used in chapter 6. Because Sylvester's Identity is rarely included in current texts on linear algebra, for convenience I have summarized from standard sources a brief outline of a simple proof for the specific case of Sylvester's Identity used in chapter 6, Lemma C.3, together with Lemma C.1 (Schur Complement) and Lemma C.2 (Chio's Pivotal Condensation), which I will use as lemmas.

Lemma C.1 (Schur Complement). *Let A be an $n \times n$ matrix with* $\det A \neq 0$. *For r such that $1 \le r < n$, partition A into blocks such that A_{11} is the leading principal submatrix of order r and A_{22} is the trailing principal submatrix of order $(n-r)$, so that*

$$A = \begin{pmatrix} A_{11} & A_{12} \\ A_{21} & A_{22} \end{pmatrix}.$$

If $\det A_{11} \neq 0$, *then*

$$\det A = \det A_{11} \det(A_{22} - A_{21} A_{11}^{-1} A_{12}), \tag{C.1}$$

and $A_{22} - A_{21}A_{11}^{-1}A_{12}$ is called the Schur complement of A_{11} in A.

Remark C.1. This result can be generalized to any partition of A by subsequences $\{i_1, i_2, \ldots, i_r\}$ $\{j_1, j_2, \ldots, j_r\}$ and their complements.

Proof. Since $\det A \neq 0$, we can write

$$\begin{pmatrix} I_r & 0 \\ A_{21}A_{11}^{-1} & I_{n-r} \end{pmatrix} \begin{pmatrix} A_{11} & A_{12} \\ 0 & A_{22} - A_{21}A_{11}^{-1}A_{12} \end{pmatrix} = \begin{pmatrix} A_{11} & A_{12} \\ A_{21} & A_{22} \end{pmatrix}$$

Taking the determinant of both sides and simplifying proves equation (C.1).

The second lemma, following, provides a method for condensing an $n \times n$ determinant into an $(n-1) \times (n-1)$ determinant.

Lemma C.2 (Chio's Pivotal Condensation). *If* $a_{11} \neq 0$,

$$\begin{vmatrix} a_{11} & a_{12} & \cdots & a_{1n} \\ a_{21} & a_{22} & \cdots & a_{2n} \\ \vdots & \vdots & \ddots & \vdots \\ a_{n1} & a_{n2} & \cdots & a_{nn} \end{vmatrix} = \frac{1}{a_{11}^{n-2}} \begin{vmatrix} \begin{vmatrix} a_{11} & a_{12} \\ a_{21} & a_{22} \end{vmatrix} & \begin{vmatrix} a_{11} & a_{13} \\ a_{21} & a_{23} \end{vmatrix} & \cdots & \begin{vmatrix} a_{11} & a_{1n} \\ a_{21} & a_{2n} \end{vmatrix} \\ \begin{vmatrix} a_{11} & a_{12} \\ a_{31} & a_{32} \end{vmatrix} & \begin{vmatrix} a_{11} & a_{13} \\ a_{31} & a_{33} \end{vmatrix} & \cdots & \begin{vmatrix} a_{11} & a_{1n} \\ a_{31} & a_{3n} \end{vmatrix} \\ \vdots & \vdots & \ddots & \vdots \\ \begin{vmatrix} a_{11} & a_{12} \\ a_{n1} & a_{n2} \end{vmatrix} & \begin{vmatrix} a_{11} & a_{13} \\ a_{n1} & a_{n3} \end{vmatrix} & \cdots & \begin{vmatrix} a_{11} & a_{1n} \\ a_{n1} & a_{nn} \end{vmatrix} \end{vmatrix}. \tag{C.2}$$

Proof. A simple calculation shows that the determinant in the right-hand side of equation (C.2) is just the determinant of the Schur complement of (a_{11}) in A,

$$\begin{pmatrix} a_{22} & \cdots & a_{2n} \\ \vdots & \ddots & \vdots \\ a_{n2} & \cdots & a_{nn} \end{pmatrix} - \begin{pmatrix} a_{21} \\ \vdots \\ a_{n1} \end{pmatrix} (a_{11})^{-1} (a_{12}, a_{13}, \dots a_{1n})$$

$$= \frac{1}{a_{11}} \begin{pmatrix} \begin{vmatrix} a_{11} & a_{12} \\ a_{21} & a_{22} \end{vmatrix} & \begin{vmatrix} a_{11} & a_{13} \\ a_{21} & a_{23} \end{vmatrix} & \cdots & \begin{vmatrix} a_{11} & a_{1n} \\ a_{21} & a_{2n} \end{vmatrix} \\ \begin{vmatrix} a_{11} & a_{12} \\ a_{31} & a_{32} \end{vmatrix} & \begin{vmatrix} a_{11} & a_{13} \\ a_{31} & a_{33} \end{vmatrix} & \cdots & \begin{vmatrix} a_{11} & a_{1n} \\ a_{31} & a_{3n} \end{vmatrix} \\ \vdots & \vdots & \ddots & \vdots \\ \begin{vmatrix} a_{11} & a_{12} \\ a_{n1} & a_{n2} \end{vmatrix} & \begin{vmatrix} a_{11} & a_{13} \\ a_{n1} & a_{n3} \end{vmatrix} & \cdots & \begin{vmatrix} a_{11} & a_{1n} \\ a_{n1} & a_{nn} \end{vmatrix} \end{pmatrix}, \qquad \text{(C.3)}$$

where, that is, we have set

$$A_{11} = (a_{11}),\ A_{12} = (a_{12}, a_{13}, \dots a_{1n}),\ A_{21} = \begin{pmatrix} a_{21} \\ \vdots \\ a_{n1} \end{pmatrix},\ A_{22} = \begin{pmatrix} a_{22} & \cdots & a_{2n} \\ \vdots & \ddots & \vdots \\ a_{n2} & \cdots & a_{nn} \end{pmatrix}.$$

Taking the determinant of both sides of equation (C.3), and noting that the order of the matrix of determinants on the right-hand side is $n-1$, gives equation (C.2).

Lemma C.3 (Special Case of Sylvester's Identity). *Let A be an $n \times n$ matrix* $(n \geq 3)$ *with* $\det A \neq 0$. *Then for any $r \leq n-2$ and $i, j > r+1$,*

$$A\binom{1,\dots,r,r+1}{1,\dots,r,r+1} A\binom{1,\dots,r,i}{1,\dots,r,j} - A\binom{1,\dots,r,i}{1,\dots,r,r+1} A\binom{1,\dots,r,r+1}{1,\dots,r,j} = A\binom{1,\dots,r}{1,\dots,r} A\binom{1,\dots,r,r+1,i}{1,\dots,r,r+1,j}. \qquad \text{(C.4)}$$

Remark C.2. Though the lemma is true in general, we will assume here that for all $1 \leq k \leq n$, $A\binom{1,\dots,k}{1,\dots,k} \neq 0$.

Proof. We will prove this by induction on r. Given A of any order r, for $r = 1$, equation (C.4) is simply Chio's Pivotal Condensation,

$$\begin{vmatrix} a_{11} & a_{12} & a_{1j} \\ a_{21} & a_{22} & a_{2j} \\ a_{i1} & a_{i2} & a_{ij} \end{vmatrix} = \frac{1}{a_{11}} \begin{vmatrix} \begin{vmatrix} a_{11} & a_{12} \\ a_{21} & a_{22} \end{vmatrix} & \begin{vmatrix} a_{11} & a_{1j} \\ a_{21} & a_{2j} \end{vmatrix} \\ \begin{vmatrix} a_{11} & a_{12} \\ a_{i1} & a_{i2} \end{vmatrix} & \begin{vmatrix} a_{11} & a_{1j} \\ a_{i1} & a_{ij} \end{vmatrix} \end{vmatrix}.$$

Now we will use the inductive step to prove equation (C.4): assuming that equation (C.4) is true for $r < n-2$, we will show that it is then true for $r+1$. We take $i, j > r+2$ and apply the Chio's Pivotal Condensation to both terms on the right-hand side of equation (C.4) for $r+1$, $A\binom{1,\dots,r+1}{1,\dots,r+1} A\binom{1,\dots,r+1,r+2,i}{1,\dots,r+1,r+2,j}$, with a_{11} as the pivot, giving

$$\frac{1}{a_{11}^{2r}}\begin{vmatrix} \begin{vmatrix} a_{11} & a_{12} \\ a_{21} & a_{22} \end{vmatrix} & \cdots & \begin{vmatrix} a_{11} & a_{1,r+1} \\ a_{21} & a_{2,r+1} \end{vmatrix} \\ \vdots & \ddots & \vdots \\ \begin{vmatrix} a_{11} & a_{12} \\ a_{r+1,1} & a_{r+1,2} \end{vmatrix} & \cdots & \begin{vmatrix} a_{11} & a_{1,r+1} \\ a_{r+1,1} & a_{r+1,r+1} \end{vmatrix} \end{vmatrix}$$

$$\cdot \begin{vmatrix} \begin{vmatrix} a_{11} & a_{12} \\ a_{21} & a_{22} \end{vmatrix} & \cdots & \begin{vmatrix} a_{11} & a_{1,r+1} \\ a_{21} & a_{2,r+1} \end{vmatrix} & \begin{vmatrix} a_{11} & a_{1,r+2} \\ a_{21} & a_{2,r+2} \end{vmatrix} & \begin{vmatrix} a_{11} & a_{1j} \\ a_{21} & a_{2j} \end{vmatrix} \\ \vdots & \ddots & \vdots & \vdots & \vdots \\ \begin{vmatrix} a_{11} & a_{12} \\ a_{r+1,1} & a_{r+1,2} \end{vmatrix} & \cdots & \begin{vmatrix} a_{11} & a_{1,r+1} \\ a_{r+1,1} & a_{r+1,r+1} \end{vmatrix} & \begin{vmatrix} a_{11} & a_{1,r+2} \\ a_{r+1,1} & a_{r+1,r+2} \end{vmatrix} & \begin{vmatrix} a_{11} & a_{1j} \\ a_{r+1,1} & a_{r+1,j} \end{vmatrix} \\ \begin{vmatrix} a_{11} & a_{12} \\ a_{r+2,1} & a_{r+2,2} \end{vmatrix} & \cdots & \begin{vmatrix} a_{11} & a_{1,r+1} \\ a_{r+2,1} & a_{r+2,r+1} \end{vmatrix} & \begin{vmatrix} a_{11} & a_{1,r+2} \\ a_{r+2,1} & a_{r+2,r+2} \end{vmatrix} & \begin{vmatrix} a_{11} & a_{1j} \\ a_{r+2,1} & a_{r+2,j} \end{vmatrix} \\ \begin{vmatrix} a_{11} & a_{12} \\ a_{i1} & a_{i2} \end{vmatrix} & \cdots & \begin{vmatrix} a_{11} & a_{1,r+1} \\ a_{i1} & a_{i,r+1} \end{vmatrix} & \begin{vmatrix} a_{11} & a_{1,r+2} \\ a_{i1} & a_{i,r+2} \end{vmatrix} & \begin{vmatrix} a_{11} & a_{1j} \\ a_{i1} & a_{ij} \end{vmatrix} \end{vmatrix},$$

where we have a_{11}^{-2r}, because the order of the first determinant is $r+1$ and that of the second determinant is $r+3$. These determinants are scalar entries $\begin{vmatrix} a_{11} & a_{1l} \\ a_{k1} & a_{kl} \end{vmatrix}$, and by the inductive hypothesis, we can apply equation (C.4) to rewrite this as follows:

$$\frac{1}{a_{11}^{2r}}\left(\begin{vmatrix} \begin{vmatrix} a_{11} & a_{12} \\ a_{21} & a_{22} \end{vmatrix} & \cdots & \begin{vmatrix} a_{11} & a_{1,r+1} \\ a_{21} & a_{2,r+1} \end{vmatrix} & \begin{vmatrix} a_{11} & a_{1,r+2} \\ a_{21} & a_{2,r+2} \end{vmatrix} \\ \vdots & \ddots & \vdots & \vdots \\ \begin{vmatrix} a_{11} & a_{12} \\ a_{r+1,1} & a_{r+1,2} \end{vmatrix} & \cdots & \begin{vmatrix} a_{11} & a_{1,r+1} \\ a_{r+1,1} & a_{r+1,r+1} \end{vmatrix} & \begin{vmatrix} a_{11} & a_{1,r+2} \\ a_{r+1,1} & a_{r+1,r+2} \end{vmatrix} \\ \begin{vmatrix} a_{11} & a_{12} \\ a_{r+2,1} & a_{r+2,2} \end{vmatrix} & \cdots & \begin{vmatrix} a_{11} & a_{1,r+1} \\ a_{r+2,1} & a_{r+2,r+1} \end{vmatrix} & \begin{vmatrix} a_{11} & a_{1,r+2} \\ a_{r+2,1} & a_{r+2,r+2} \end{vmatrix} \end{vmatrix} \right.$$

$$\cdot \begin{vmatrix} \begin{vmatrix} a_{11} & a_{12} \\ a_{21} & a_{22} \end{vmatrix} & \cdots & \begin{vmatrix} a_{11} & a_{1,r+1} \\ a_{21} & a_{2,r+1} \end{vmatrix} & \begin{vmatrix} a_{11} & a_{1j} \\ a_{21} & a_{2j} \end{vmatrix} \\ \vdots & \ddots & \vdots & \vdots \\ \begin{vmatrix} a_{11} & a_{12} \\ a_{r+1,1} & a_{r+1,2} \end{vmatrix} & \cdots & \begin{vmatrix} a_{11} & a_{1,r+1} \\ a_{r+1,1} & a_{r+1,r+1} \end{vmatrix} & \begin{vmatrix} a_{11} & a_{1j} \\ a_{r+1,1} & a_{r+1,j} \end{vmatrix} \\ \begin{vmatrix} a_{11} & a_{12} \\ a_{i1} & a_{i2} \end{vmatrix} & \cdots & \begin{vmatrix} a_{11} & a_{1,r+1} \\ a_{i1} & a_{i,r+1} \end{vmatrix} & \begin{vmatrix} a_{11} & a_{1j} \\ a_{i1} & a_{ij} \end{vmatrix} \end{vmatrix}$$

$$- \begin{vmatrix} \begin{vmatrix} a_{11} & a_{12} \\ a_{21} & a_{22} \end{vmatrix} & \cdots & \begin{vmatrix} a_{11} & a_{1,r+1} \\ a_{21} & a_{2,r+1} \end{vmatrix} & \begin{vmatrix} a_{11} & a_{1,r+2} \\ a_{21} & a_{2,r+2} \end{vmatrix} \\ \vdots & \ddots & \vdots & \vdots \\ \begin{vmatrix} a_{11} & a_{12} \\ a_{r+1,1} & a_{r+1,2} \end{vmatrix} & \cdots & \begin{vmatrix} a_{11} & a_{1,r+1} \\ a_{r+1,1} & a_{r+1,r+1} \end{vmatrix} & \begin{vmatrix} a_{11} & a_{1,r+2} \\ a_{r+1,1} & a_{r+1,r+2} \end{vmatrix} \\ \begin{vmatrix} a_{11} & a_{12} \\ a_{i1} & a_{i2} \end{vmatrix} & \cdots & \begin{vmatrix} a_{11} & a_{1,r+1} \\ a_{i1} & a_{i,r+1} \end{vmatrix} & \begin{vmatrix} a_{11} & a_{1,r+2} \\ a_{i1} & a_{i,r+2} \end{vmatrix} \end{vmatrix}$$

$$\left. \cdot \begin{vmatrix} \begin{vmatrix} a_{11} & a_{12} \\ a_{21} & a_{22} \end{vmatrix} & \cdots & \begin{vmatrix} a_{11} & a_{1,r+1} \\ a_{21} & a_{2,r+1} \end{vmatrix} & \begin{vmatrix} a_{11} & a_{1j} \\ a_{21} & a_{2j} \end{vmatrix} \\ \vdots & \ddots & \vdots & \vdots \\ \begin{vmatrix} a_{11} & a_{12} \\ a_{r+1,1} & a_{r+1,2} \end{vmatrix} & \cdots & \begin{vmatrix} a_{11} & a_{1,r+1} \\ a_{r+1,1} & a_{r+1,r+1} \end{vmatrix} & \begin{vmatrix} a_{11} & a_{1j} \\ a_{r+1,1} & a_{r+1,j} \end{vmatrix} \\ \begin{vmatrix} a_{11} & a_{12} \\ a_{r+2,1} & a_{r+2,2} \end{vmatrix} & \cdots & \begin{vmatrix} a_{11} & a_{1,r+1} \\ a_{r+2,1} & a_{r+2,r+1} \end{vmatrix} & \begin{vmatrix} a_{11} & a_{1j} \\ a_{r+2,1} & a_{r+2,j} \end{vmatrix} \end{vmatrix} \right).$$

Then, again using Chio's Pivotal Condensation (in reverse), we can expand each of these determinants, and since each has the order $r+1$, the term $(\frac{1}{a_{11}})^{2r}$ vanishes, giving

$$A\begin{pmatrix}1,\dots,r+1,r+2\\1,\dots,r+1,r+2\end{pmatrix} A\begin{pmatrix}1,\dots,r+1,i\\1,\dots,r+1,j\end{pmatrix} - A\begin{pmatrix}1,\dots,r+1,i\\1,\dots,r+1,r+2\end{pmatrix} A\begin{pmatrix}1,\dots,r+1,r+2\\1,\dots,r+1,j\end{pmatrix}.$$

This then proves equation (C.4) for $r+1$, and completes the proof.

Proposition C.1. *Let A be an $n \times n$ matrix with* $\det A \neq 0$. *Let row reductions be performed following the* fangcheng *procedure (*fangcheng shu 方程術*) in the* Nine Chapters, *and let $a_{ij}^{(r)}$ denote entry in the ith row and j^{th} after r row reductions ($1 < r < n$). Further assume, without loss of generality, that the rows of A are arranged such that for all $1 \leq r < n$, $a_{rr}^{(r-1)} \neq 0$. Then for $r < i, j \leq n$,*

$$a_{ij}^{(r)} = \left(\prod_{l=1}^{r-1}\left(\frac{A\begin{pmatrix}1,\dots,l\\1,\dots,l\end{pmatrix}}{\prod_{m=1}^{l} k_{l+1,m}}\right)^{2^{r-l-1}}\right)\left(\frac{A\begin{pmatrix}1,\dots,r,i\\1,\dots,r,j\end{pmatrix}}{k_{i1}k_{i2}\dots k_{ir}}\right), \tag{C.5}$$

where for $1 \leq m < l \leq n$,

$$k_{lm} = \begin{cases} 1 \textit{ if } (a_{mm}^{(m-1)}, a_{lm}^{(m-1)}) = 1, \\ a_{mm}^{(m-1)} \textit{ if } a_{lm}^{(m-1)} = 0, \\ a_{mm}^{(m-1)} \textit{ if } a_{lm}^{(m-1)} = \pm a_{mm}^{(m-1)}, \\ d_{lm} \textit{ if } (a_{mm}^{(m-1)}, a_{lm}^{(m-1)}) \neq 1, \textit{for some } d_{lm} \in \mathbb{N} \textit{ with } d_{lm} | (a_{mm}^{(m-1)}, a_{lm}^{(m-1)}). \end{cases}$$

Remark C.3. If we write this out by expanding the product above in equation (C.5), we have

$$a_{ij}^{(r)} = \left(\frac{A\binom{1}{1}}{k_{21}}\right)^{2^{r-2}} \left(\frac{A\binom{1,2}{1,2}}{k_{31}k_{32}}\right)^{2^{r-3}} \cdots \left(\frac{A\begin{pmatrix}1,\dots,r-1\\1,\dots,r-1\end{pmatrix}}{k_{r1}k_{r2}\dots k_{r,r-1}}\right)\left(\frac{A\begin{pmatrix}1,\dots,r,i\\1,\dots,r,j\end{pmatrix}}{k_{i1}k_{i2}\dots k_{ir}}\right).$$

Remark C.4. Again, because the *fangcheng* procedure does not specify what to do, if $(a_{mm}, a_{lm}) \neq 1$, then k_{lm} might be any integer d_{lm} such that $d_{lm} | (a_{mm}, a_{lm})$: that is, although we cannot assume that the least common divisor is found or used, this does account for any possibility that some common divisor of a_{mm} and a_{lm} is used.

Proof. We will show this by induction. Assume we are given a matrix A of order n. If $r = 2$, equation (C.5) holds, as was shown previously (see the augmented matrix 6.26 on page 104). Then assume the proposition is true for a given r, $2 \leq r < n$. We want to show that this holds for the next row reduction, $r+1$. Let i, j be such that $r+1 < i \leq n$ and $r+1 < j \leq n$. The result of the next row reduction is then given by

$$a_{ij}^{(r+1)} = \frac{1}{k_{i,r+1}}\left(a_{r+1,r+1}^{(r)} a_{ij}^{(r)} - a_{i,r+1}^{(r)} a_{r+1,j}^{(r)}\right).$$

Gathering terms together, we have

$$\frac{1}{k_{i,r+1}}\left(\left(\frac{a_{11}}{k_{21}}\right)^{2^{r-2}}\left(\frac{A\binom{1,2}{1,2}}{k_{31}k_{32}}\right)^{2^{r-3}}\cdots\cdots\left(\frac{A\binom{1,\dots,r-1}{1,\dots,r-1}}{k_{r1}k_{r2}\dots k_{r,r-1}}\right)\right)^{2}$$
$$\cdot\left(\left(\frac{A\binom{1,\dots,r,r+1}{1,\dots,r,r+1}}{k_{r+1,1}k_{r+1,2}\dots k_{r+1,r}}\right)\left(\frac{A\binom{1,\dots,r,i}{1,\dots,r,j}}{k_{i1}k_{i2}\dots k_{ir}}\right)-\right.$$
$$\left.\left(\frac{A\binom{1,\dots,r,i}{1,\dots,r,r+1}}{k_{i1}k_{i2}\dots k_{ir}}\right)\left(\frac{A\binom{1,\dots,r,r+1}{1,\dots,r,j}}{k_{r+1,1}k_{r+1,2}\dots k_{r+1,r}}\right)\right).$$

Then, applying the determinant identity equation (C.4) shown above, we have

$$\left(A\binom{1,\dots,r,r+1}{1,\dots,r,r+1}A\binom{1,\dots,r,i}{1,\dots,r,j}\right)-\left(A\binom{1,\dots,r,i}{1,\dots,r,r+1}A\binom{1,\dots,r,r+1}{1,\dots,r,j}\right)=A\binom{1,\dots,r}{1,\dots,r}A\binom{1,\dots,r,r+1,i}{1,\dots,r,r+1,j},$$

and further simplifying, we get

$$a_{ij}^{(r+1)}=\left(\frac{A\binom{1}{1}}{k_{21}}\right)^{2^{(r+1)-2}}\left(\frac{A\binom{1,2}{1,2}}{k_{31}k_{32}}\right)^{2^{(r+1)-3}}\left(\frac{A\binom{1,2,3}{1,2,3}}{k_{41}k_{42}k_{43}}\right)^{2^{(r+1)-4}}\cdots$$
$$\cdot\left(\frac{A\binom{1,\dots,r-1}{1,\dots,r-1}}{k_{r1}k_{r2}\dots k_{r,r-1}}\right)^{2}\left(\frac{A\binom{1,\dots,r}{1,\dots,r}}{k_{r+1,1}k_{r+1,2}\dots k_{r+1,r}}\right)\left(\frac{A\binom{1,\dots,r+1,i}{1,\dots,r+1,j}}{k_{i1}k_{i2}\dots k_{i,r+1}}\right).$$

This then proves equation (C.5).

Proposition C.2. *For $n \geq 3$, we can write the final pivot $a_{nn}^{(n-1)}$ in terms of the remaining pivots $a_{rr}^{(r-1)}$ as follows:*

$$a_{nn}^{(n-1)}=\left(\prod_{r=1}^{n-2}\left(\frac{\left(a_{rr}^{(r-1)}\right)^{n-r-1}}{\prod_{i=r+1}^{n}k_{ir}}\right)\right)\frac{\det A}{k_{n,n-1}}. \tag{C.6}$$

Proof. We will show that

$$\left(\prod_{i=1}^{n-2}\left(\frac{A\binom{1,\dots,i}{1,\dots,i}}{\prod_{j=1}^{i}k_{i+1,j}}\right)^{2^{n-i-2}}\right)\frac{A\binom{1,\dots,n}{1,\dots,n}}{\prod_{j=1}^{n-1}k_{nj}}=\left(\prod_{r=1}^{n-2}\left(\frac{\left(a_{rr}^{(r-1)}\right)^{n-r-1}}{\prod_{i=r+1}^{n}k_{ir}}\right)\right)\frac{\det A}{k_{n,n-1}}, \tag{C.7}$$

where the left-hand side of equation (C.7) is just the preceding formula, equation (C.5), for $a_{nn}^{(n-1)}$, that is, $r=n-1$ and $i=j=n$.

Using equation (C.5), we can calculate the pivots $a_{rr}^{(r-1)}$,

$$a_{rr}^{(r-1)} = \begin{cases} A\binom{1}{1} \text{ if } r = 1, \\ \dfrac{A\binom{1,2}{1,2}}{k_{21}} \text{ if } r = 2, \\ \left(\prod_{i=1}^{r-2}\left(\dfrac{A\binom{1,\dots,i}{1,\dots,i}}{\prod_{j=1}^{i} k_{i+1,j}}\right)^{2^{r-i-2}}\right)\dfrac{A\binom{1,\dots,r}{1,\dots,r}}{\prod_{l=1}^{r-1} k_{rl}} \text{ if } r \geq 3. \end{cases} \tag{C.8}$$

We begin with the formula on the right-hand side of equation (C.7), and substitute for the pivots $a_{rr}^{(r-1)}$, using equation (C.8), yielding

$$\left(\left(\frac{\left(A\binom{1}{1}\right)^{n-2}}{\prod_{p=2}^{n} k_{p1}}\right)\left(\frac{\left(\frac{A\binom{1,2}{1,2}}{k_{21}}\right)^{n-3}}{\prod_{p=3}^{n} k_{p2}}\right)\right.$$
$$\left.\cdot\prod_{r=3}^{n-2}\left(\frac{\left(\left(\prod_{i=1}^{r-2}\left(\frac{A\binom{1,\dots,i}{1,\dots,i}}{\prod_{j=1}^{i} k_{i+1,j}}\right)^{2^{r-i-2}}\right)\frac{A\binom{1,\dots,r}{1,\dots,r}}{\prod_{l=1}^{r-1} k_{rl}}\right)^{n-r-1}}{\prod_{p=r+1}^{n} k_{pr}}\right)\right)\frac{A\binom{1,\dots,n}{1,\dots,n}}{k_{n,n-1}}.$$

Now we will show that this equals the left-hand side of equation (C.7) by direct computation.

First we will solve for the numerator, that is, the terms $A\binom{1,\dots,r}{1,\dots,r}$. Gathering terms, we have

$$\left(A\binom{1}{1}\right)^{n-2}\left(A\binom{1,2}{1,2}\right)^{n-3}\prod_{r=3}^{n-2}\left(\left(\prod_{i=1}^{r-2}\left(A\binom{1,\dots,i}{1,\dots,i}\right)^{2^{r-i-2}}\right)A\binom{1,\dots,r}{1,\dots,r}\right)^{(n-r-1)} A\binom{1,\dots,n}{1,\dots,n}.$$

Multiplying out the products, then rearranging and combining terms, gives

$$\left(\prod_{i=1}^{n-2}\left(A\binom{1,\dots,i}{1,\dots,i}\right)^{n-i-1}\right)\left(\prod_{r=3}^{n-2}\prod_{i=1}^{r-2}\left(A\binom{1,\dots,i}{1,\dots,i}\right)^{(2^{r-i-2})(n-r-1)}\right)A\binom{1,\dots,n}{1,\dots,n}.$$

Interchanging the order of the products, we have

$$\prod_{r=3}^{n-2}\prod_{i=1}^{r-2}\left(A\binom{1,\dots,i}{1,\dots,i}\right)^{(2^{r-i-2})(n-r-1)} = \prod_{i=1}^{n-4}\prod_{r=i+2}^{n-2}\left(A\binom{1,\dots,i}{1,\dots,i}\right)^{(2^{r-i-2})(n-r-1)}$$

and using the formula

$$\sum_{r=i+2}^{n-2} 2^{r-i-2}(n-r-1) = 2^{n-i-2} - (n-i-1) \tag{C.9}$$

to calculate the product,

$$\prod_{r=i+2}^{n-2}\left(A\binom{1,\dots,i}{1,\dots,i}\right)^{2^{r-i-2}(n-r-1)}=\left(A\binom{1,\dots,i}{1,\dots,i}\right)^{2^{n-i-2}-(n-i-1)},$$

and substituting into the previous equation yields

$$\left(\prod_{i=1}^{n-2}\left(A\binom{1,\dots,i}{1,\dots,i}\right)^{n-i-1}\right)\left(\prod_{i=1}^{n-4}\left(A\binom{1,\dots,i}{1,\dots,i}\right)^{2^{n-i-2}-(n-i-1)}\right)A\binom{1,\dots,n}{1,\dots,n}.$$

Since $2^{n-i-2}-(n-i-1)=0$ for $i=n-3$ and $n-2$, we can change the upper index on the product on the right from $n-4$ to $n-2$. Simplifying then gives the desired result,

$$\left(\prod_{i=1}^{n-2}\left(A\binom{1,\dots,i}{1,\dots,i}\right)^{2^{n-i-2}}\right)A\binom{1,\dots,n}{1,\dots,n},$$

that is, the numerator of the left-hand side of equation (C.7).

Now we must calculate the denominator. Gathering terms from the denominator on the right-hand side of equation (C.7), we have

$$k_{21}^{n-3}k_{n,n-1}\left(\prod_{r=2}^{n}k_{r1}\right)\left(\prod_{r=3}^{n}k_{r2}\right)\cdot\prod_{r=3}^{n-2}\left(\left(\prod_{i=r+1}^{n}k_{ir}\right)\left(\left(\prod_{i=1}^{r-2}\left(\prod_{j=1}^{i}k_{i+1,j}\right)^{2^{r-i-2}}\right)\prod_{j=1}^{r-1}k_{rj}\right)^{n-r-1}\right).$$

Distributing the products and rearranging terms yields

$$k_{21}^{n-3}k_{n,n-1}\left(\prod_{r=2}^{n}k_{r1}\right)\left(\prod_{r=3}^{n}k_{r2}\right)\left(\prod_{r=3}^{n-2}\prod_{i=r+1}^{n}k_{ir}\right)\cdot\left(\prod_{r=3}^{n-2}\prod_{i=1}^{r-2}\prod_{j=1}^{i}k_{i+1,j}^{2^{r-i-2}(n-r-1)}\right)\left(\prod_{r=3}^{n-2}\prod_{j=1}^{r-1}k_{rj}^{n-r-1}\right).$$

After some work—changing the order of the products, again using the sum in equation (C.9) to simplify the product, changing indices so that the terms in all the products are of the form $k_{i+1,j}$, and combining terms—we have the desired result,

$$\left(\prod_{i=1}^{n-2}\left(\prod_{j=1}^{i}k_{i+1,j}\right)^{2^{n-i-2}}\right)\prod_{j=1}^{n-1}k_{nj},$$

that is, the denominator of the right-hand side of equation (C.7), completing the proof.

Bibliography of Primary and Secondary Sources

Unless otherwise stated, all citations of Chinese primary sources are to editions reprinted in SKQS (see below). Texts that have been cited only as examples of classical Chinese usage have not been listed in the bibliography. Several important dictionaries are listed below in a separate section, "Standard References—Chinese and Japanese." Although many of the examples of classical Chinese usage come from these dictionaries, I have, following standard sinological conventions, cited not these dictionaries themselves but instead their original sources.

Collections—Chinese

[CSJC] *Congshu ji cheng chu bian* 叢書集成初編 [Complete collection of collectanea, first edition]. [1935] 1985–1991. Edited by Wang Yunwu 王雲五 (1888–1979). 4000 vols. Reprint of 3467 vols. published by Shanghai Shangwu yinshuguan 商務印書館 in 1935, together with 533 previously unpublished vols. Beijing: Zhonghua shuju 中華書局.

[GJTS] *Gu jin tu shu ji cheng* 古今圖書集成 [Collection of maps and books, ancient and contemporary]. [1725] 1934. Imperially sponsored. Compiled by Chen Menglei 陳夢雷, Jiang Tingxi 蔣廷錫, and others. Reprint, photolithographic reproduction of Palace edition. Beijing: Zhonghua shuju 中華書局.

[LSQS] Mei Wending 梅文鼎 (1633–1721). [c. 1720] *Li suan quan shu* 歷算全書 [Complete book of astronomy and mathematics]. Reprint, SKQS.

[SJSS] *Suan jing shi shu* 算經十書 [Mathematical classics, ten books]. [1774] 1930. Edited by Dai Zhen 戴震 (1724–1777). Reprint, *Wan you wen ku* 萬有文庫, Shanghai: Shangwu yinshuguan 商務印書館.

[SKQS] *Yingyin Wenyuan ge Siku quan shu* 景印文淵閣四庫全書 [Complete collection of the four treasuries, photolithographic reproduction of the edition preserved at the Pavilion of Literary Erudition]. [1773–1782] 1983–1986. Edited by Ji Yun 紀昀 and Lu Xixiong 陸錫熊. 1500 vols. Reprinted from the collection of the Guoli gugong bowuyuan 國立故宮博物院. Taipei: Taiwan shangwu yinshuguan 臺灣商務印書館. Digitized full text, searchable edition. Hong Kong: Chinese University Press.

[SKSJ] *Song ke suan jing liu zhong: fu yi zhong* 宋刻算经六种：附一种 [Six mathematical classics, with one appended, engraved during the Song Dynasty]. [1213] 1981. Beijing: Wenwu chubanshe 文物出版社.

[XXSK] *Xu xiu Siku quan shu* 續修四庫全書 [Continuation of the Complete collection of the four treasuries]. 1995–2002. Edited by *Xu xiu Siku quan shu* bian zuan weiyuanhui 《續修四庫全書》編纂委員會編. 1800 vols. Reprint. Shanghai: Shanghai guji chubanshe 上海古籍出版社.

[YLDD] *Yongle da dian* 永樂大典 [Great encyclopedia of the Yongle reign]. [1403–07] 1986. Compiled by Xie Jin 解縉 (1369–1415) et al. Reprint of extant fragments. 10 vols. Beijing: Zhonghua shuju 中華書局 .

[ZKJD] *Zhongguo kexue jishu dianji tonghui: shuxue juan* 中國科學技術典籍通彙：數學卷 [Comprehensive collection of the classics of Chinese science and technology: Mathematics volumes]. 1993. Edited by Guo Shuchun 郭書春. 5 vols. Zhengzhou: Henan jiaoyu chubanshe 河南教育出版社.

Primary Sources—Chinese

[CYFL] Gu Yingxiang 顧應祥 (1483–1565). *Ce yuan hai jing fen lei shi shu* 測圓海鏡分類釋術 [Sea mirror of circle measurement, arranged by categories, with explanations of the methods]. Reprint, SKQS; ZKJD.

[CYHJ] Li Ye 李冶 (1192–1279). *Ce yuan hai jing* 測圓海鏡 [Sea mirror of circle measurement]. Reprint, SKQS; ZKJD.

[FCL] Mei Wending 梅文鼎 (1633–1721). [c. 1674] 1995. *Fangcheng lun* 方程論 [On *fangcheng*]. Photolithographic reprint from the Mei Juecheng Chengxuetang 梅瑴成承學堂 printing of the *Mei shi congshu ji yao* 梅氏叢書輯要. Reprint, SKQS; ZKJD.

[JGG] *Jiguge ying Song chaoben Jiuzhang suanshu, wu juan* 汲古閣影宋抄本九章算經五卷 [Nine chapters on the mathematical arts, five chapters, Song Dynasty hand-copied manuscript, reprinted by the Drawing-from-the-Ancients Pavilion]. [1684] 1932. Reprint, *Tianlulinlang congshu, di yi ji* 天祿琳琅叢書，第一集 [13–14]. Beijing: Gugong bowuyuan 故宮博物院.

[JZSS] *Jiuzhang suanshu* 九章算術 [Nine chapters on the mathematical arts]. Editions include: Wuyingdian juzhen ban congshu 武英殿聚珍版叢書 edition (1774), reprint, ZKJD; Kong Jihan 孔繼涵 edition (1777), reprint, Shanghai: Shanghai guji chubanshe 上海古籍出版社, 1990; SKQS edition, first published in 1784. Recent critical editions include (see Secondary Sources) Chemla and Guo 2004; Guo 1998; Guo 1990; Qian 1984.

[JZXC] Li Huang (?-1812). [1812] 1995. *Jiuzhang suanshu xi cao tu shuo jiu juan* 九章算術細草圖說九卷 [Nine chapters on the mathematical arts, with detailed explanations and explanatory diagrams]. Reprint, ZKJD.

[SDY] Fang Zhongtong 方中通 (1634–1698). 1661. *Shu du yan* 數度衍 [Numbers and measurement, an amplification]. Reprint, SKQS.

[SFTZ] Cheng Dawei 程大位 (1533–1606). 1592. *Suanfa tong zong* 算法统宗 [Comprehensive source of mathematical arts]. Reprint, *Suanfa tong zong jiao shi* 算法统宗校释 [Comprehensive source of mathematical arts, collated, with explanatory notes], with prefaces by Yan Dunjie 嚴敦傑, Mei Rongzhao 梅榮照, and Li Zhaohua 李兆華. Hefei: Anhui jiaoyu chubanshe 安徽教育出版社, 1990.

[SSJZ] Qin Jiushao 秦九韶 (13th c.). *Shu shu jiu zhang* 數書九章 [Mathematical book in nine chapters]. Reprint, SKQS; ZKJD.

[SKCJ] *Siku cai jin shumu* 四庫採進書目. 1960. Edited by Wu Weizu 吳慰祖. Reprint.

[SKTY] *Siku quan shu zong mu tiyao ji Siku wei shou shumu, jinhui shumu* 四庫全書總目提要及四庫未收書目・禁燬書目. 1971. Edited by Wang Yunwu 王雲五 (1888–1979), Yong Rong 永瑢 (1744–1790), Ji Yun 紀昀 (1724–1805), et al. Reprint. Taibei: Taiwan shangwu yinshuguan 臺灣商務印書館.

[SLJY] *Yuzhi shu li jing yun* 御製數理精蘊 [The Emperor's collected essential principles of mathematics]. 1723. Attributed to the Kangxi 康熙 Emperor (r. 1662–1722). Reprint, ZKJD.

[SSS] *Suan shu shu* 算數書 [Book of computation]. [c. 186 B.C.E.] Transcribed into modern Chinese in Peng 2001.

[SZSJ] *Sunzi suan jing* 孫子算經 [Sunzi's mathematical classic]. Reprint, SKQS.

[TWSZ] Matteo Ricci (1552–1610) and Li Zhizao 李之藻 (1565–1630). *Tong wen suan zhi* 同文算指 [Guide to computation in the unified language]. Reprint, SKQS; ZKJD.

[WJJZ] Wu Jing 吳敬. [1450] 1995? *Jiuzhang xiang zhu bilei suanfa da quan 10 juan; Cheng chu kaifang qi li 1 juan* 九章詳註比類算法大全，10卷；乘除開方起例，1卷 [Complete compendium of mathematical arts of the Nine chapters, with detailed commentary, arranged by category, ten chapters; Examples of multiplication, division, and square roots, one chapter]. Photolithographic reprint of the edition engraved by Wang Jun 王均 in 1450, revised and supplemented by Wu Na 吳訥 in 1488, and preserved in the Shanghai Library 上海圖書館藏明景泰元年王均刻弘治元年吳訥補修本影印. Reprint, XXSK, Zi bu, Tianwen suanfa lei 續修四庫全書，子部，天文算法類, vol. 1043.

[YHJZ] Yang Hui 楊輝. [1261] 1993. *Xiang jie jiuzhang suanfa* 詳解九章算法 [Nine chapters on the mathematical arts, with detailed explanations]. Extant 5 juan, Yijiatang congshu 宜稼堂叢書 edition. Reprint, ZKJD.

[ZQJS] *Zhang Quijian suan jing* 張邱建算經 [Zhang Qiujian's mathematical classic]. Reprint, SKQS; ZKJD.

Primary Sources—Japanese

Masanobu Saka 坂正永 (fl. 1781–1789) and Ihe Murai 村井伊兵衛 (1755–1817). 1782. *Sanpo gakkai* 算法學海 [Sea of learning on mathematics]. 2 vols. Japanese Rare Book Collection, Library of Congress.

Seki Takakazu 關孝和 (ca. 1642–1708). [1680] 1974. *Seki Takakazu zenshu* 關孝和全集 [The complete works of Seki Takakazu]. Reprint. Edited by Hirayama

Akira 平山諦, Shimodaira Kazuo 下平和夫, and Hirose Hideo 広瀬秀雄. Tokyo: Ōsaka Kyōiku Tosho 大阪战育図書.

Yamamoto Tadayasu 山本格安 (d. ca. 1761). 1746. *Sanzui* 算髓 [The pith of mathematics]. Kyoto: Metogiya Kanbe han 蓍屋勘兵衛版.

Yasuaki Aida 會田安明 (1747–1817). 1785. *Kaisei sanpo* 改精算法. Edo: Shinshodo Suharaya Ichibe shi 申椒堂須原屋市兵衛梓. Japanese Rare Book Collection, Library of Congress.

Yoshida Mitsuyoshi 吉田光由 (1598–1672). 1977. *Jinkoki* 塵劫記. Edited by Oya Shin'ichi 大矢真一. Tokyo: Iwanami Shoten 岩波書店.

Standard References—Chinese and Japanese

DKWJ Morohashi Tetsuji 諸橋轍次. 1984. *Dai Kan-Wa jiten* 大漢和辞典 [Great Chinese-Japanese dictionary]. Tokyo: Taishūkan shoten 大修館書店.

KXZD Zhang Yushu 张玉書 et al. [1716] 1985. *Kang Xi zi dian* 康熙字典 [Kangxi dictionary]. Reprint, Shanghai: Shanghai shudian 上海书店.

SWJZ Xu Shen 許慎 (c. 55–c. 149). 121 C.E. *Shuo wen jie zi* 說文解字 [Explanation of simple graphs and analysis of compound characters]. Reprint, SKQS.

SWJZZ Xu Shen and Duan Yucai 段玉裁 (1735–1815). [1815] 1981. *Shuo wen jie zi zhu* 說文解字注 [Explanation of simple graphs and analysis of compound characters, with annotations]. Reprint, Shanghai: Shanghai guji chubanshe 上海古籍出版社.

Primary Sources—Western Languages

Cayley, Arthur. 1841. "On a Theorem in the Geometry of Position." *Cambridge Mathematical Journal* 2: 267–71. Reprint, *Collected Mathematical Papers of Arthur Cayley* 1: 1–4.

———. 1843. "On the Theory of Determinants." *Transactions of the Cambridge Philosophical Society* 8: 1–16. Reprint, *Collected Mathematical Papers of Arthur Cayley* 1: 63–79.

———. 1845. "On the Theory of Linear Transformations." *Cambridge Mathematical Journal* 4: 193–209. Reprint, *Collected Mathematical Papers of Arthur Cayley* 1: 80–112.

———. 1889. *The Collected Mathematical Papers of Arthur Cayley.* 13 vols. Cambridge: Cambridge University Press.

Clavius, Christoph. 1583. *Christophori Clauii Bambergensis e Societate Iesu Epitome arithmeticae practicae.* Rome.

Cramer, Gabriel. 1750. *Introduction à l'analyse des lignes courbes algébriques.* Genève: Freres Cramer & Cl. Philbert.

Diophantus, of Alexandria [fl. c. 250]. [c. 250] 1893–1895. *Diophanti Alexandrini Opera omnia, cum graecis commentariis.* Edited by Paul Tannery, *Bibliotheca scriptorum Graecorum et Romanorum Teubneriana.* Lipsiae.

Gauss, Carl Friedrich (1777–1855)). [1801] 1870. *Disquisitiones Arithmeticae.* Lipsiae. Reprint, *Carl Friedrich Gauss Werke,* vol. 1.

———. [1809] 1870. *Theoria motus corporum coelestium in sectionibus conicis solem ambientium.* Hamburgi: F. Perthes et I. H. Besser. Reprint, *Carl Friedrich Gauss Werke* 7: 1–288.

———. [1811] 1870. *Disquisitio de elementis ellipticis palladis: ex oppositionibus annorum 1803, 1804, 1805, 1807, 1808, 1809, Commentationes societatis regiae scientiarum Gottingensis, Tomus I, Classis mathematicae.* Göttingen. Reprint, *Carl Friedrich Gauss Werke* 6: 1–24.

———. [1821–1826] 1870. *Theoria combinationes observationum erroribus minimis obnoxia. Pars I, II, cum suppl.* Gottingae. Reprint, *Carl Friedrich Gauss Werke* 4: 55–104.

———. 1870. *Carl Friedrich Gauss Werke.* Edited by Ernst Christian Julius Schering, Martin Brendel, and Königlichen Gesellschaft der Wissenschaften zu Göttingen. Göttingen: Gedruckt in der Dieterichschen Universitätsdruckerei, W. F. Kaestner.

Jordan, Camille. 1870. *Traité des substitutions et des équations algébriques.* Paris: Gauthier-Villars. Reprint, electronic reproduction, Paris: BNF, Gallica, 2000.

Lagrange, J. L. 1759. "Recherches sur la méthode de maximis et minimis." *Miscellanea Taurinensia* 1: 1–20. Reprint, *Œuvres de Lagrange,* 1: 1–20.

———. 1867–1892. *Œuvres de Lagrange.* Edited by J. A. Serret, Gaston Darboux, J. B. J. Delambre, et al. 14 vols. Paris: Gauthier-Villars.

Leibniz, Gottfried Wilhelm, Freiherr von (1646–1716). 1923–. *Sämtliche Schriften und Briefe.* Darmstadt, O. Reichl.

———. [1849] 1962. *Leibnizens Mathematische Schriften.* Edited by K. Gerhardt. 7 vols. Berlin: A. Asher. Reprint, Hildesheim: G. Olms.

MacLaurin, Colin. [1748] 1756. *A Treatise of Algebra in Three Parts.* London: Printed for A. Millar and J. Nourse.

Secondary Sources—Western Languages

Abeles, Francine. 1986. "Determinants and Linear Systems: Charles L. Dodgson's View." *British Journal for the History of Science* 19: 331–35.

Aitken, A. C. 1946. *Determinants and Matrices.* Rev. ed. Edinburgh: Oliver and Boyel. Reprint, Westport, CT: Greenwood Press, 1983.

Althoen, Steven C., and Renate McLaughlin. 1987. "Gauss-Jordan Reduction: A Brief History." *American Mathematical Monthly* 94 (2): 130–42.

Andersen, Kirsti. 1994. "Precalculus, 1635–1665." In Grattan-Guinness 1994, 292–307.

Ang, Tian-Se [Hong Tianci 洪天賜]. 1972. "Chinese Interest in Indeterminate Analysis and Indeterminate Equations." *Bantai Xuebao (Journal of the Chinese Language Society, University of Malaya)* 5: 105–12.

Bapat, R. B., and T. E. S. Raghavan. 1997. *Nonnegative Matrices and Applications.* Encyclopedia of Mathematics and Its Applications 64. New York: Cambridge University Press.

Bareiss, Erwin H. 1968. "Sylvester's Identity and Multistep Integer-Preserving Gaussian Elimination." *Mathematics of Computation* 22 (103): 565–78.

Bellman, Richard Ernest. 1960. *Introduction to Matrix Analysis.* McGraw-Hill Series in Matrix Theory. New York: McGraw-Hill.

Benson, Donald C. 2003. *A Smoother Pebble: Mathematical Explorations.* New York: Oxford University Press.

Bhatia, Rajendra. 1997. *Matrix Analysis.* Graduate Texts in Mathematics 169. New York: Springer.

Bourbaki, Nicolas. 1994. *Elements of the History of Mathematics.* Translated by John Meldrum. New York: Springer. Originally published as *Éléments d'histoire des mathématiques* (Paris: Hermann, 1960).

Boyer, Carl B. [1949] 1959. *The History of the Calculus and Its Conceptual Development.* Reprint, New York: Dover.

Boyer, Carl B., and Uta C. Merzbach. 1991. *A History of Mathematics.* 2nd ed. New York: John Wiley and Sons.

Cardano, Girolamo. [1545] 1968. *The Great Art: Or, the Rules of Algebra.* Translated by T. Richard Witmer. Cambridge, MA: MIT Press. Originally published as *Artis magnae, sive de regulis algebraicis* in 1545; translation also includes material from the 1570 and 1663 editions.

Chabert, Jean-Luc, and E. Barbin, eds. 1999. *A History of Algorithms: From the Pebble to the Microchip.* New York: Springer. Originally published as *Histoire d'algorithmes: du caillou à la puce* (Paris: Belin, 1994).

Chemla, Karine. 1988. "La pertinence du concept de classification pour l'analyse de textes mathématiques chinois." *Extrême-orient, Extrême-occident* 10: 61–87.

———. 1991. "Theoretical Aspects of the Chinese Algorithmic Tradition (1st to 3rd Century)." *Historia Scientiarum: International Journal of the History of Science Society of Japan* 42: 75–98.

———. 1994. "Different Concepts of Equations in 'The Nine Chapters on Mathematical Procedures' 九章算術 and the Commentary on It by Liu Hui (Third Century)." *Historia Scientiarum: International Journal of the History of Science Society of Japan* 4 (2): 113–37.

———. 1996. "Relations between Procedure and Demonstration: Measuring the Circle in the 'Nine Chapters on Mathematical Procedures' and Their Commentary by Liu Hiu (3rd Century)." In *History of Mathematics and Education: Ideas and Experiences.* Edited by Hans Niels Jahnke, Norbert Knoche, Michael Otte, et al. Göttingen: Vandenhoeck and Ruprecht.

———. 1997a. "Qu'est-ce qu'un problème dans la tradition mathématique de la Chine ancienne? Quelques indices glanés dans les commentaires rédigés entre le 3ième et le 7ième siècles au classique Han *Les neuf chapitres sur les procédures mathématiques.*" *Extrême-Orient, Extrême Occident* 19: 91–126.

———. 1997b. "What Is at Stake in Mathematical Proofs from Third-Century China?" *Science in Context* 10 (2): 227–51.

———, ed. 2004. *History of Science, History of Text.* Boston Studies in Philosophy of Science 238. Boston: Kluwer Academic.

Chemla, Karine, and Guo Shuchun. 2004. *Les neuf chapitres: Le Classique mathématique de la Chine ancienne et ses commentaires.* Paris: Dunod.

Christianidis, Jean. 1994. "On the History of Indeterminate Problems of the First Degree in Greek Mathematics." In *Trends in the Historiography of Science.* Edited by Kostas Gavroglu. Dordrecht: Kluwer Academic.

Clagett, Marshall. 1999. *Ancient Egyptian Science,* Volume 3, *Ancient Egyptian Mathematics: A Source Book.* Memoirs of the American Philosophical Society 232. Philadelphia: American Philosophical Society.

Cooke, Roger. 2005. *The History of Mathematics: A Brief Course.* 2nd ed. Hoboken, NJ: Wiley-Interscience.

Crowe, Michael J. 1967. *A History of Vector Analysis: The Evolution of the Idea of a Vectorial System.* Notre Dame, IN: University of Notre Dame Press.

Cullen, Charles G. 1990. *Matrices and Linear Transformations.* 2nd ed. New York: Dover.

Cullen, Christopher. 1993. "Chiu chang suan shu." In *Early Chinese Texts: A Bibliographical Guide.* Edited by Michael Loewe. Early China Special Monograph Series 2. Berkeley: Society for the Study of Early China, Institute of East Asian Studies, University of California, Berkeley.

———. 1996. *Astronomy and Mathematics in Ancient China: The "Zhou bi suan jing."* New York: Cambridge University Press.

———. 2004. *The Suan shu shu* 筭數書 *"Writings on Reckoning": A Translation of a Chinese Mathematical Collection of the Second Century BC, with Explanatory Commentary.* Needham Research Institute Working Papers 1. Cambridge: Needham Research Institute.

Cullis, Cuthbert Edmund. 1913. *Matrices and Determinoids.* Cambridge: Cambridge University Press.

Dauben, Joseph W. 1985. *The History of Mathematics from Antiquity to the Present: A Selective Bibliography.* New York: Garland.

———. 1992. "The 'Pythagorean Theorem' and Chinese Mathematics: Liu Hui's Commentary on the [Gou-Gu] Theorem in Chapter Nine of the 'Jiu Zhang Suan Shu.'" In *Amphora: Festschrift für Hans Wussing zu seinem 65.* Edited by Sergei S. Demidov et al. Boston: Birkhäuser Verlag.

———. 2008. "Suan Shu Shu. A Book on Numbers and Computations: English Translation with Commentary." *Archive for History of Exact Sciences* 62 (2): 91–178.

Dauben, Joseph W., and Albert C. Lewis, eds. 2000. *The History of Mathematics from Antiquity to the Present: A Selective Annotated Bibliography.* Providence, RI: American Mathematical Society.

Dickson, Leonard E. [1919] 1966. *History of the Theory of Numbers.* Washington: Carnegie Institution of Washington. Reprint, New York: Chelsea.

Diophantus, of Alexandria (fl. c. 250 C.E.). [c. 250] 1890. *Die Arithmetik und die Schrift über Polygonalzahlen des Diophantus von Alexandria.* Translated by Gustav Wertheim. Leipzig: B. G. Teubner.

———. [c. 250] 1959. *Les six livres arithmétiques et Le livre des nombres polygones: oeuvres traduites pour la premiére fois du grec en français, avec une introduction et des notes.* Translated by Paul Ver Eecke. Paris: Albert Blanchard.

———. [c. 250] 1984. *Diophante. Les arithmétiques.* Translated by Rushdi Rashid. Collection des universités de France. Paris: Belles Lettres.

———. [c. 250] 2002. *Les arithmétiques. Collection des universités de France, Série grecque.* Paris: Les belles lettres.

Duff, Iain S., A. M. Erisman, and John Ker Reid. 1986. *Direct Methods for Sparse Matrices.* Monographs on Numerical Analysis. Oxford: Clarendon Press.

Edwards, C. H. 1979. *The Historical Development of the Calculus.* New York: Springer Verlag.

Elman, Benjamin A. 2000. *A Cultural History of Civil Examinations in Late Imperial China.* Berkeley: University of California Press.

———. 2005. *On Their Own Terms: Science in China, 1550–1900.* Cambridge, MA: Harvard University Press.

Eves, Howard Whitley. [1966] 1980. *Elementary Matrix Theory.* Boston: Allyn and Bacon. Reprint, New York: Dover.

Fowler, David. [1999] 2003. *The Mathematics of Plato's Academy: A New Reconstruction.* 2nd ed. Oxford: Clarendon. Reprint, Oxford University Press.

Gantmacher, F. R. 1959. *The Theory of Matrices.* Translated by K. A. Hirsch. 2 vols. New York: Chelsea.

Gauss, Carl Friedrich. 1855. *Méthode des moindres carrés.* Translated by Joseph Louis François Bertrand. Paris: Mallet-Bachelier.

———. 1887. *Abhandlungen zur Methode der kleinsten Quadrate.* Translated by A. Bhörsch and P. Simon. Berlin: P. Stankiewicz. Reprint, Würzburg: Physica-Verlag, 1964.

———. [1809] 1963. *Theory of the Motion of the Heavenly Bodies Moving About the Sun in Conic Sections: A Translation of Gauss's "Theoria Motus," with an Appendix.* Translated by Charles Henry Davis. Boston: Little, Brown, 1857. Reprint, New York: Dover. Originally published as *Theoria motus corporum coelestium in sectionibus conicis solem ambientium,* 1809.

———. [1801] 1966. *Disquisitiones Arithmeticae.* Translated by Arthur A. Clarke. New Haven: Yale University Press. Originally published in 1801.

———. [1821–1826] 1995. *Theory of the Combination of Observations Least Subject to Error: Part One, Part Two, Supplement.* Translated by G. W. Stewart. Classics in Applied Mathematics 11. Philadelphia: Society for Industrial and Applied Mathematics. Originally published as *Theoria combinationis observationum erroribus minimus obnoxiae: pars prior, pars posterior, supplementum.*

Golan, Jonathan S. 2004. *The Linear Algebra a Beginning Graduate Student Ought to Know.* Kluwer Texts in the Mathematical Sciences 27. Boston: Kluwer.

Goldstine, Herman Heine. 1977. *A History of Numerical Analysis from the 16th through the 19th Century.* Studies in the History of Mathematics and Physical Sciences 2. New York: Springer-Verlag.

Grattan-Guinness, Ivor. 1994. *Companion Encyclopedia of the History and Philosophy of the Mathematical Sciences.* 2 vols. Routledge Reference. New York: Routledge.

———, and W. Ledermann. 1994. "Matrix Theory." In Grattan-Guinness 1994, 775–86.

Greub, Werner Hildbert. 1975. *Linear Algebra.* 4th ed. Vol. 23, Graduate Texts in Mathematics. New York: Springer-Verlag.

Guicciardini, Niccolò. 1989. *The Development of Newtonian Calculus in Britain, 1700–1800.* New York: Cambridge University Press.

———. 1994. "Three Traditions in the Calculus: Newton, Leibniz, and Lagrange." In Grattan-Guinness 1994, 308–17.

Halmos, Paul R. 1974. *Finite-Dimensional Vector Spaces.* New York: Springer-Verlag.

Hart, Roger. 1999. "Beyond Science and Civilization: A Post-Needham Critique." *East Asian Science, Technology, and Medicine* 16: 88–114. Earlier version published as "On the Problem of Chinese Science," in *The Science Studies Reader,* edited by Mario Biagioli (New York: Routledge, 1999), 189–201.

———. 2000. "The Great Explanandum," essay review of *The Measure of Reality: Quantification and Western Society, 1250–1600,* by Alfred W. Crosby. *American Historical Review* 105 (2): 486–93.

———. 2000. "Translating the Untranslatable: From Copula to Incommensurable Worlds." In *Tokens of Exchange: The Problem of Translation in Global Circulations,* edited by Lydia H. Liu. Durham, NC: Duke University Press, 45–73. Earlier version published as "Translating Worlds: Incommensurability and Problems of Existence in Seventeenth-Century China," *Positions: East Asia Cultures Critique* 7 (1): 95–128.

———. 2006. "Universals of Yesteryear: Hegel's Modernity in an Age of Globalization." In *Global History: Interactions between the Universal and the Local.* Edited by A. G. Hopkins. London: Palgrave MacMillan, 66–97.

Hawkins, Thomas. 1975. "Cauchy and the Spectral Theory of Matrices." *Historia Mathematica* 2: 1–29.

———. 1977a. "Weierstrass and the Theory of Matrices." *Archive for History of Exact Sciences* 17: 119–163.

———. 1977b. "Another Look at Cayley and the Theory of Matrices." *Archives Internationale d'Histoire des Sciences* 27: 82–112.

Heath, Thomas Little, Sir. 1885. *Diophantus of Alexandria: A Study in the History of Greek Algebra.* Cambridge: Cambridge University Press.

Ho, Peng Yoke [He Bingyu 何丙郁]. 1970–. "Liu Hui" and "Yang Hui." In *Dictionary of Scientific Biography.* Edited by Charles Coulston Gillispie. New York: Scribner.

Hoe, John [Jock]. 1977. *Les systèmes d'équations polynômes dans le Siyuan yujian (1303).* Mémoires de l'Institut des Hautes Études Chinoises 6. Paris: Collège de France, Institut des Hautes Études Chinoises.

Hoffman, Kenneth, and Ray Alden Kunze. [1961] 1971. *Linear Algebra.* 2nd ed. Englewood Cliffs, NJ: Prentice-Hall.

Hohn, Franz Edward. 1973. *Elementary Matrix Algebra.* 3rd ed. New York: Macmillan.

Horiuchi, Annick. 1989. "Sur un point de rupture entre les traditions Chinoise et Japonaise des mathématiques." *Revue d'histoire des sciences et de leurs applications* 42: 375–90.

———. 1994a. *Les mathématiques Japonaises à l'époque d'Edo (1600–1868): Une étude des travaux de Seki Takakazu (?–1708) et de Takebe Katahiro (1664–1739).* Paris: J. Vrin.

———. 1994b. "The 'Tetsujutsu Sankei' (1722), an 18th Century Treatise on the Methods of Investigation in Mathematics." In *Intersection of History and Mathematics.* Edited by Sasaki Chikara et al. Basel: Birkhäuser.

Horn, Roger A., and Charles R. Johnson. 1985. *Matrix Analysis.* New York: Cambridge University Press.

———. 1991. *Topics in Matrix Analysis.* New York: Cambridge University Press.

Hummel, Arthur William. 1943. *Eminent Chinese of the Ch'ing Period (1644–1912).* 2 vols. Washington: U.S. Government Printing Office.

Hungerford, Thomas W. 1974. *Algebra.* New York: Holt, Rinehart, and Winston.

Høyrup, Jens. 2002. *Lengths, Widths, Surfaces: A Portrait of Old Babylonian Algebra and Its Kin.* Sources and Studies in the History of Mathematics and Physical Sciences. New York: Springer.

Imhausen, Annette. 2003. *Ägyptische Algorithmen: eine Untersuchung zu den mittelägyptischen mathematischen Aufgabentexten.* Ägyptologische Abhandlungen 65. Wiesbaden: Harrassowitz.

Jahnke, Hans Niels. 2003. *A History of Analysis.* Providence, RI: American Mathematical Society.

Katz, Victor J. 1993. *A History of Mathematics: An Introduction.* New York: HarperCollins.

Klein, Jacob (1899–1978). [1968] 1992. *Greek Mathematical Thought and the Origin of Algebra.* New York: Dover Publications. Originally published as *Griechische Logistik und die Entstehung der Algebra.*

Kleiner, Israel. 2007. *A History of Abstract Algebra.* Boston: Birkhauser.

Kloyda, Mary Thomas à Kempis, Sister. 1937. *Linear and Quadratic Equations, 1550–1660.* Osiris: Studies on the History and Philosophy of Science, and on the History of Learning and Culture 3.1. Bruges: Saint Catherine Press.

Knobloch, Eberhard. 1980. *Der Beginn der Determinantentheorie: Leibnizens nachgelassene Studien zum Determinantenkalkül.* Hildesheim: Gerstenberg.

———. 1990. "Erste europäische Determinantentheorie." In *Gottfried Wilhelm Leibniz: das Wirken des grossen Philosophen und Universalgelehrten als Mathematiker, Physiker, Techniker.* Edited by Erwin Stern and Albert Heinekamp. Hannover: Gottfried Wilhelm Leibniz Gesellschaft.

———. 1994a. "Determinants." In Grattan-Guinness 1994, 775–86.

———. 1994b. "From Gauss to Weierstrass: Determinant Theory and Its Historical Evaluations." In *The Intersection of History and Mathematics*. Edited by Sasaki Chikara, Sugiura Mitsuo, and Joseph W. Dauben. Science Networks Historical Studies 15. Boston: Birkhäuser-Verlag.

Kowalewski, Gerhard. 1909. *Einführung in die Determinantentheorie einschließlich der unendlichen und der Fredholmschen Determinanten*. Leipzig: Veit.

Lam, Lay-Yong [Lan Lirong 藍麗蓉]. 1977. *A Critical Study of the Yang Hui Suan Fa, a Thirteenth-Century Mathematical Treatise*. [Singapore]: Singapore University Press.

———. 1986. "The Conceptual Origins of Our Numeral System and the Symbolic Form of Algebra." *Archive for History of Exact Sciences* 36: 183–95.

———. 1987a. "Linkages: Exploring the Similarities Between the Chinese Rod Numeral System and Our Numeral System." *Archive for History of Exact Sciences* 37: 365–92.

———. 1987b. "The Earliest Negative Numbers: How They Emerged from a Solution of Simultaneous Linear Equations." *Archives internationale d'histoire des sciences* 37: 222–62.

———. 1988. "A Chinese Genesis: Rewriting the History of Our Numeral System." *Archive for History of Exact Sciences* 38: 101–08.

———. 1994. "Jiuzhang Suanshu: An Overview." *Archive for History of Exact Sciences* 47: 1–51.

———. 1997. "Zhang Qiujian Suanjing (The Mathematical Classic of Zhang Qiujian): An Overview." *Archive for History of Exact Sciences* 51: 201–40.

Lam, Lay-Yong, and Shen Kangshen. 1989. "Methods of Solving Linear Equations in Traditional China." *Historia Mathematica* 16: 107–22.

Lam, Lay-Yong, and Tian Se Ang. 2004. *Fleeting Footsteps: Tracing the Conception of Arithmetic and Algebra in Ancient China*. Rev. ed. River Edge, NJ: World Scientific.

Lang, Serge. [1966] 2004. *Linear algebra*. 3rd ed. New York: Springer-Verlag.

Lee, Hong R., and B. David Saunders. 1995. "Fraction Free Gaussian Elimination for Sparse Matrices." *Journal of Symbolic Computation* 19 (5): 393–402.

Leibniz, Gottfried Wilhelm. 1918, 1920. *The Early Mathematical Manuscripts of Leibniz*. Translated by Carl Immanuel Gerhardt. Chicago: Open Court.

———. 1977. *Discourse on the Natural Theology of the Chinese*. Translated by Henry Rosemont, Jr. and Daniel J. Cook. Honolulu: University Press of Hawaii.

———. 1994. *Writings on China*. Translated by Daniel J. Cook and Henry Rosemont. Chicago: Open Court.

Li Yan and Du Shiran. 1987. *Chinese Mathematics: A Concise History*. Translated by John N. Crossley and Anthony W.-C. Lun. Oxford: Clarendon Press.

Libbrecht, Ulrich. 1973. *Chinese Mathematics in the Thirteenth Century: The Shu-Shu Chiu-Chang of Ch'in Chiu-Shao*. MIT East Asian Series 1. Cambridge, MA: MIT Press.

———. 1982. "Mathematical Manuscripts from the Dunhuang Caves." In *Explorations in the History of Science and Technology in China.* Edited by Li Guohao. Shanghai: Chinese Classics Publishing House.

Ma, Li. 1994. *Studies of the Chinese Rectangular Array Algorithm in Nine Chapters.* Ph.D. dissertation. Göteborg, Sweden: Department of Mathematics, Chalmers University of Technology, 1994.

MacLane, Saunders, and Garrett Birkhoff. 1979. *Algebra.* 2nd ed. New York: Macmillan.

Marcus, Marvin, and Henryk Minc. 1964. *A Survey of Matrix Theory and Matrix Inequalities.* Allyn and Bacon Series in Advanced Mathematics. Boston: Allyn and Bacon.

Martzloff, Jean-Claude. 1981. *Recherches sur l'oeuvre mathématique de Mei Wending, 1633–1721.* Mémoires de l'Institut des Hautes Études Chinoises 16. Paris: Collège de France, Institut des Hautes Études Chinoises.

———. 1994. "Chinese Mathematics." In Grattan-Guinness 1994, 93–103.

———. [1987] 2006. *A History of Chinese Mathematics.* Translated by S. S. Wilson. New York: Springer, 1997. Corrected edition published in 2006. Originally published as *Histoire des mathématiques chinoises* (Paris: Masson, 1987).

May, Kenneth A. 1973. *Bibliography and Research Manual of the History of Mathematics.* Toronto: University of Toronto Press.

Meyer, Carl Dean. 2000. *Matrix Analysis and Applied Linear Algebra.* Philadelphia: Society for Industrial and Applied Mathematics.

Mikami, Yoshio. 1913. *The Development of Mathematics in China and Japan.* Abhandlungen zur Geschichte der mathematischen Wissenschaften mit Einschluss ihrer Anwendungen 30. Reprint, 2nd ed., New York: Chelsea, 1974.

———. 1914. On the Japanese Theory of Determinants. *Isis* 2: 9–36.

Morse, JoAnn S. 1981. "The Reception of Diophantus's Arithmetic in the Renaissance." Ph.D. dissertation, Princeton University.

Muir, Thomas. 1906–1923. *The Theory of Determinants in the Historical Order of Development.* 4 vols. London: Macmillan and Co. Reprint, 4 vols. bound as 2, New York: Dover, 1960.

———. 1933. *A Treatise on the Theory of Determinants.* Revised and enlarged by William H. Metzler. New York: Longmans, Green, and Co. Unabridged and corrected republication of 1933 edition, New York: Dover, 1960.

Neugebauer, O. 1951. *The Exact Sciences in Antiquity.* Princeton, NJ: Princeton University Press.

Neugebauer, O., Abraham Joseph Sachs, and Albrecht Götze, eds. 1945. *Mathematical Cuneiform Texts.* American Oriental Series 29. New Haven: American Oriental Society and the American Schools of Oriental Research.

Østerby, Ole, and Zahari Zlatev. 1983. *Direct Methods for Sparse Matrices.* Lecture Notes in Computer Science 157. New York: Springer-Verlag.

Peterson, Willard J. 1979. *Bitter Gourd: Fang I-Chih and the Impetus for Intellectual Change.* New Haven: Yale University Press.

Rashid, Rushdi, ed. 1984. *Les arithmétiques, Collection des universités de France.* Paris: Belles Lettres. Translation of Diophantus' *Arithmetica.*

Schafer, Edward H. 1977. *Pacing the Void: T'ang Approaches to the Stars.* Berkeley: University of California Press.

Séroul, Raymond. 2000. *Programming for Mathematicians.* Translated by Donal O'Shea. New York: Springer. Originally published as *Math-info: informatique pour mathématiciens* (Paris: InterEditions, 1995).

Serre, Denis. 2002. *Matrices: Theory and Applications.* New York: Springer.

Sesiano, Jacques. 1977. "Le traitement des équations indéterminées dans le 'Badi' fi'l-hisab' d'AbBakr al-Karaji." *Archive for History of Exact Sciences* 17: 297–379.

———. 1982. *Books IV to VII of Diophantus's Arithmetica in the Arabic Translation Attributed to Qusta Ibn Luqa.* Sources in the History of Mathematics and Physical Sciences 3. New York: Springer-Verlag.

Shen Kangshen, Anthony W. C. Lun, and John N. Crossley. 1999. *The Nine Chapters on the Mathematical Art: Companion and Commentary.* New York: Oxford University Press.

Shukla, Kripa Shankar, and K. V. Sarma, eds. 1976. *Aryabhatiya of Aryabhata.* New Delhi: Indian National Science Academy.

Smith, David Eugene. 1923. *History of Mathematics.* Boston.

———. 1936. *Algebra of Four Thousand Years Ago.* New York: Scripta Mathematica.

Smith, David Eugene, Augustus De Morgan, and George A. Plimpton. [1847–1939] 1970. *Rara Arithmetica: A Catalogue of the Arithmetics Written before the Year MDCI, with a Description of Those in the Library of George Arthur Plimpton of New York.* 4th ed. Compilation of Smith's *Rara Arithmetica* (Boston, 1908), brief addenda to the first edition (1910), De Morgan's *Arithmetical Books* (London, 1847), and *Addenda to Rara Arithmetica* (1939), respectively, pp. 1–494, 495–498, 499–548, and 551–703. New York: Chelsea.

Stewart, G. W. 1973. *Introduction to Matrix Computations.* New York: Academic Press.

———. 1998. *Matrix Algorithms.* Philadelphia: Society for Industrial and Applied Mathematics.

Stigler, Stephen M. 1986. *The History of Statistics: The Measurement of Uncertainty before 1900.* Cambridge, MA: Belknap Press of Harvard University Press.

Strang, Gilbert. [1976] 1988. *Linear Algebra and Its Applications.* 3rd ed. San Diego: Harcourt Brace Jovanovich.

Swetz, Frank. 1992. *The Sea Island Mathematical Manual: Surveying and Mathematics in Ancient China.* University Park, PA: Pennsylvania State University Press.

———. 2001. Review of *The Nine Chapters on the Mathematical Art: Companion and Commentary. American Mathematical Monthly* 108 (7): 673–75.

Swetz, Frank J., and Ang Tian Se. 1984. A Brief Chronological and Bibliographic Guide to the History of Chinese Mathematics. *Historia Mathematica* 11: 39–56.

Tropfke, Johannes, Kurt Vogel, Karin Reich, and Helmuth Gericke. [1902–1903] 1980. *Geschichte der Elementarmathematik, Bd. 1, Arithmetik und Algebra.* 3 vols. Berlin: Walter de Gruyter. First edition published Leipzig: Veit.

Turnbull, H. W. 1960. *The Theory of Determinants, Matrices, and Invariants.* 3rd ed. New York: Dover.

Van Egmond, Warren. 1980. *Practical Mathematics in the Italian Renaissance: A Catalog of Italian Abbacus Manuscripts and Printed Books to 1600.* Monografia, Istituto e museo di storia della scienza 4. Istituto e museo di storia della scienza.

Vogel, Kurt. 1968. *Neun Bücher arithmetischer Technik: Ein chinesisches Rechenbuch für den praktischen Gebrauch aus der frühen Hanzeit.* Braunschweig: Vieweg.

Waerden, B. L. van der. [1950] 1954. *Science Awakening.* Translated by Arnold Dresden. Groningen: P. Noordhoff. Revised translation of *Ontwakende wetenschap,* Groningen: P. Noordhoff.

———. 1983. *Geometry and Algebra in Ancient Civilizations.* Berlin: Springer-Verlag.

Wagner, Donald B. 1978. "Doubts Concerning the Attribution of Liu Hui's Commentary on the Chiu-Chang Suan-Shu." *Acta Orientalia* 39: 199–212.

———. 1985. "A Proof of the Pythagorean Theorem by Liu Hui (3rd Century A.D.)." *Historia Mathematica* 12: 71–73.

Watkins, David S. 1991. *Fundamentals of Matrix Computations.* New York: Wiley.

Wertheim, Gustav, ed. 1890. *Die Arithmetik und die Schrift über Polygonalzahlen des Diophantus von Alexandria.* Leipzig: B. G. Teubner.

Zhang, Fuzhen. 1999. *Matrix Theory: Basic Results and Techniques.* Universitext. New York: Springer.

Secondary Sources—Chinese and Japanese

Bai Shangshu 白尚恕. 1983. *Jiuzhang suanshu zhu shi* 《九章算术》注释 [Nine chapters on the mathematical arts, with annotations and explanatory notes]. Beijing: Kexue chubanshe 科学出版社.

———. 1990. *Jiuzhang suanshu jin yi* 九章算术今译 [Nine chapters on the mathematical arts, translated into modern Chinese]. Jinan: Shandong jiaoyu chubanshe 山东教育出版社.

Ding Fubao 丁福保 and Zhou Yunqing 周雲靑, editors. 1957. *Sibu zong lu suanfa bian* 四部總錄算法編 [Complete record of the four treasuries, mathematics volume]. Shanghai: Shang wu yin shu guan 商務印書館.

Guo Shuchun 郭书春. 1985. "*Jiuzhang suanshu* fangcheng zhang Liu Hui zhu xin tan" 《九章算術》方程章劉徽注新探 [New investigation of Liu Hui's commentaries on the *fangcheng* chapter of the Nine chapters on the mathematical arts]. *Ziran kexue shi yanjiu* 自然科學史研究 4 (1): 1–5.

———. 1988. "Jia Xian *Huangdi jiuzhang suan jing xi cao* chu tan" 賈憲《黃帝九章算經細草》初探 [A preliminary investigation of Jia Xian's The Yellow Emperor's nine chapters of mathematics classic, with detailed notes]. *Ziran kexue shi yanjiu* 自然科學史研究 7 (4): 328–34.

———. 1990. *Jiuzhang suanshu huijiaoben* 《九章算術》匯校本 [Nine chapters on the mathematical arts, critical edition]. Shenyang: Liaoning jiaoyu chuban she 辽宁教育出版社.

———. 1992. *Gudai shijie shuxue taidou Liu Hui* 古代世界数学泰斗刘徽 [Liu Hui, an outstanding mathematician of the ancient world]. Jinan: Shandong kexue jishu chubanshe 山东科学技术出版社.

———. 1994. *Zhongguo gudai shuxue* 中國古代數學 [Ancient Chinese mathematics]. Zhongguo wenhuashi zhishi congshu 中國文化史知識叢書 77. Taibei: Taiwan shangwu yinshuguan 臺灣商務印書館.

———, translator (into modern Chinese) and commentator. 1998. *Jiuzhang suanshu* 九章算术 [Nine chapters on the mathematical arts]. Zhongguo gudai keji mingzhu yi cong 中国古代科技名著译丛. Shenyang: Liaoning jiaoyu chuban she 辽宁教育出版社.

Jochi Shigeru 城地茂. 1988. "Chugoku Kohoku-sho Koryo-ken Chokasan iseki shutsudo Sansu-sho ni tsuite" 中国湖北省江陵県張家山遺跡出土『算数書』について (Suanshu shu unearthed in the Zhangjia shan ruins, Jiangling county, Hubei province, China). Sugaku shi kenkyū 数学史研究 117: 21–5.

———. 1991. "Nitchu no hoteiron saiko" 日中の方程論再考 [A reconsideration of higher degree equations in China and Japan]. Sugaku shi kenkyū 数学史研究 128: 26–35.

Kato Heizaemon 加藤平左エ門. 1954–1964. *Wasan no kenkyu* 和算ノ研究 [Research on Japanese mathematics]. 3 vols.: *Zatsuron* 雜論 [Miscellaneous topics]; *Hoteishikiron* 方定式論 [Theory of equations]; *Seisuron* 整数論 [Number theory]. Tokyo: Nihon gakujutsu shinkokai kan 日本学術振興会刊.

———. 1972. *Sansei Seki Takakazu no gyoseki: kaisetsu* 算聖関孝和の業績：解說 [An analysis of the accomplishments of the extraordinary mathematician Seki Takakazu]. Tokyo: Maki Shoten 槙書店.

Kawahara Hideki 川原秀城. 1980. "Kyu sho san jutsu kaisetu" 九章算術解說 [Comments on the Nine chapters on the mathematical arts]. In *Chugoku tenmongaku sugaku shu* 中国天文学・数学集. Edited by Yabuchi Kiyoshi 藪內清. Kagaku no meicho 科学の名著. Tokyo: Asahishuppansha 朝日出版社.

Li Di 李迪. 1984. *Zhongguo shuxue shi jianbian* 中國數學史簡編 [A concise history of Chinese mathematics]. Shenyang: Liaoning renmin chubanshe.

Li Di 李迪 and Guo Shirong 郭世荣. 1988. *Qing dai zhuming tianwen shuxue jia Mei Wending* 清代著名天文数学家梅文鼎 [Mei Wending, a famous astronomer and mathematician from the Qing Dynasty]. Shanghai: Shanghai kexue jishu wenxian chubanshe 上海科学技术文献出版社.

Li Jimin 李继闵. 1987. "Zhongguo gudai buding fenxi de chengjiu yu tese" 中国古代不定分析的成就与特色 [The achievements and special characteristics of indeterminate analysis in ancient China]. In Wu Wen-tsün [Wu Wenzun 吴文俊], *Qin Jiushao yu "Shu shu jiu zhang"* 秦九韶與《數書九章》 [Qin Jiushao and the Book of mathematics in nine chapters]. Beijing: Beijing shifan daxue chubanshe 北京大學出版社.

———. 1990. *"Jiuzhang suanshu" ji qi Liu Hui zhu yanjiu* 《九章筭術》及其劉徽注研究 [Research into the Nine chapters on the mathematical arts and the commentaries by Liu Hui]. Dongfang shuxue dianji 东方数学典籍. Xi'an: Shanxi renmin jiaoyu chubanshe 陕西人民教育出版社.

———. 1993. *Jiuzhang suanshu jiaozheng* 九章算术校证 [Nine chapters on the mathematical arts, with corrections and emendations]. Xi'an: Shanxi kexue jishu chubanshe 陕西科学技术出版社.

Li Wenlin 李文林 and Yuan Xiangdong 袁向東. 1982. "Zhongguo gudai buding fenxi ruogan wenti tantao" 中国古代不定分析若干問題探討 [A discussion of several problems of indeterminate analysis in ancient China]. In *Keji shi wenji* vol. 8. Shanghai: Shanghai kexue jishu chubanshe 上海科学技術出版社.

Li Yan 李俨. 1926. "Dunhuang shishi suanshu" 敦煌石室算書 [The mathematical treatises in the Dunhuang caves]. Reprint, Li Yan, *Zhong suan shi luncong* 中算史論叢. Beijing: 1954–55.

———. 1928. "*Yongle da dian* suanshu" 永樂大典算書 [The mathematical treatises in the Great encyclopedia of the Yongle reign]. Reprint, Li Yan, *Zhong suan shi luncong* 中算史論叢. Beijing: 1954–55.

———. 1929. "*Jiuzhang suanshu* bu zhu" 九章算術補註 [Supplementary commentary on the Nine chapters on the mathematical arts]. Reprint, Li Yan, *Zhong suan shi luncong* 中算史論叢. Beijing: 1954–55.

———. 1930. "Zhong suan jia zhi fangcheng lun" 中算家之方程論 [The theory of *fangcheng* of Chinese mathematicians]. Reprint, Li Yan, *Zhong suan shi luncong* 中算史論叢. Beijing: 1954–55.

———. 1935. "Dunhuang shi shi suanjing yi juan bing xu" 敦煌石室算經一卷并序 [One chapter and the preface from a mathematical treatise in the Dunhuang caves]. In *Guoli Beiping tushuguan guan kan* 國立北平圖書館館刊 9 (1): 39–46.

———. 1984. *Zhongguo suanxue shi* 中國算學史 [History of Chinese mathematics]. Beijing: Zhongguo kexueyuan 中國科學院, 1954–55. Reprint, Shanghai shu dian 上海書店.

———. 1990. *Zhong suan shi luncong* 中算史論叢. Shanghai: Shanghai shu dian 上海書店.

———, and Qian Baocong 钱宝琮. 1998. *Li Yan Qian Baocong kexueshi quanji* [Complete collection of the works of Li Yan and Qian Baocong on the history of science] 李俨钱宝琮科学史全集. Shenyang: Liaoning jiaoyu chubanshe 辽宁教育出版社.

Liu Hongtao 刘洪涛. 1997. *Shu suan da shi: Mei Wending yu tian wen li suan* 数算大师：梅文鼎与天文历算 [A great master of calculation: Mei Wending, and astronomy and calendrical calculations]. Shenyang: Liaoning renmin chubanshe 辽宁人民出版社.

Mei Rongzhao 梅榮照. 1984. "Liu Hui de fangcheng li lun" 劉徽的方程理論 [Liu Hui's theory on *fangcheng*]. *Kexueshi jikan* 科學史集刊 11: 77–95.

Ōya Shinichi 大矢眞一. 1975. *Kyū shō san jutsu* 九章算術 [Nine chapters on the mathematical arts]. In *Chugoku no kagaku* 中国の科学. Edited by Yabuuchi Kiyoshi 藪内清. *Sekai no meicho* 世界の名著. Tokyo: Chuo Koronsha 中央公論社 1975.

Peng Hao 彭浩. 2001. *Zhangjiashan Han jian* Suan shu shu *zhu shi* 張家山漢簡《算數書》註釋 [Zhangjia Mountain Han Dynasty bamboo strip Book of computation, with commentary and explanations]. Beijing: Kexue chubanshe 科学出版社.

Qian Baocong 钱宝琮. 1963. *Suan jing shi shu* 算經十書 [Mathematical classics, ten books]. Beijing: Zhonghua shuju 中華書局.

———. 1964. *Zhongguo shuxue shi* 中国数学史 [History of Chinese mathematics]. Beijing: Kexue chubanshe 科学出版社.

———. 1984. *Jiuzhang suan jing dianjiao* 九章算經點校 [Nine chapters on the mathematical arts, critical edition]. Taipei: Jiuzhang chubanshe 九章出版社.

Song Jie 宋杰. 1994. *Jiuzhang suanshu yu Han dai shehui jingji* 《九章算术》与汉代社会经济 [Nine chapters on the mathematical arts, and the society and economy of the Han Dynasty]. Dangdai Zhongguo xuezhe wenku 当代中国学者文库. Beijing: Shoudu shifan daxue chubanshe 首都师范大学出版社.

Takeda Kusuo 武田楠雄. 1954. *Dōbun sanshi* no seiritsu 同文算指の成立 [Creation of the Guide to calculation in the unified script]. *Kagakushi kenkyū* 科学史研究.

Wu Chengluo 吳承洛. 1937. *Zhongguo du liang heng shi* 中國度量衡史 [History of Chinese lengths, capacities, and weights]. Shanghai: Shangwu yinshuguan 商務印書館. Reprint, Shanghai: Shanghai shudian 上海书店, 1984.

Wu Wen-tsün [Wu Wenjun 吴文俊]. 1982. *Jiuzhang suanshu yu Liu Hui* 《九章算术》与刘徽 [Nine chapters on the mathematical arts and Liu Hui]. Zhongguo shuxue shi yanjiu congshu 中国数学史研究丛书. Beijing shifan daxue chubanshe 北京师范大学出版社.

———, ed. 1993. *Liu Hui yan jiu* 劉徽研究 [Research on Liu Hui]. Xi'an: Shanxi renmin jiaoyu chubanshe, 陕西人民教育出版社.

Xu Chunfang 許蒓舫. 1952. *Zhong suan jia de daishuxue yanjiu* 中算家的代數學研究 Beijing: Zhongguo qingnian chubanshe 中國青年出版社.

———, ed. 1998–. *Zhongguo shuxue shi da xi* 中国数学史大系 [Encyclopedic history of Chinese mathematics]. Beijing: Beijing shifan daxue chubanshe 北京师范大学出版社.

Yabuuchi Kiyoshi 藪內清, Jian Maoxiang 簡茂祥, Lin Guiying 林桂英, et al. 1981. *Zhongguo suanxue shi* 中國算學史 [History of Chinese mathematics]. Taibei xian Yonghe shi 台北縣永和市: Lianming wenhua youxian gongsi 聯鳴文化有限公司 .

Yang Kuan 楊寬. 1938. *Zhongguo lidai chidu kao* 中國歷代尺度考 [Investigation into measures throughout Chinese history]. Changsha: Shangwu yinshuguan 商務印書舘.

Index